P. J. Kuhn   A. P. J. Trinci   M. J. Jung
M. W. Goosey   L. G. Copping (Eds.)

Biochemistry of Cell Walls and Membranes in Fungi

With 88 Figures and 46 Tables

Springer-Verlag Berlin Heidelberg New York
London Paris Tokyo Hong Kong

Dr. PAUL J. KUHN
Shell Research Ltd.
Sittingbourne Research Centre
Sittingbourne, Kent ME9 8AG
United Kingdom

Dr. MICHAEL W. GOOSEY
Shell Research Ltd.
Sittingbourne Research Centre
Sittingbourne, Kent ME9 8AG
United Kingdom

Professor ANTHONY P. J. TRINCI
Microbiology Group
Department of Cell &
Structural Biology
School of Biological Sciences
University of Manchester
Manchester M13 9PL
United Kingdom

Dr. LEONARD G. COPPING
Dow Agricultural Products R&D
Letcombe Laboratory, Letcombe Regis
Wantage, Oxfordshire OX12 9JT
United Kingdom

Dr. MICHEL J. JUNG
Dow Agricultural Products
R & D, Letcombe Laboratory
Letcombe Regis, Wantage,
Oxfordshire OX12 9JT
United Kingdom

ISBN 3-540-50437-0 Springer-Verlag Berlin Heidelberg New York
ISBN 0-387-50437-0 Springer-Verlag New York Berlin Heidelberg

Library of Congress Cataloging-in-Publication Data
Biochemistry of cell walls and membranes in fungi / P. J. Kuhn ... [et al.], eds. p. cm. Includes
bibliographical references.
ISBN 0-387-50437-0 (alk. paper: U.S.)
1. Fungal cell walls. 2. Fungal membranes. 3. Fungi-Cytochemitry. I. Kuhn, P. J. (Paul John),
1954– . QK601.B54 1989 589.2′04875–dc20 89-21707

Typesetting: K+V Fotosatz GmbH, Beerfelden
Printing and binding: Brühlsche Universitätsdruckerei, Giessen
2131/3145-543210 – Printed on acid-free paper

Preface

Despite the many advances made during the last decade in various aspects of fungal biochemistry, there have been very few volumes devoted to the subject in recent years. This lack is all the more surprising in view of the increasing use of fungi in gene manipulation studies and in biotechnological applications, and of the current interest in the biorational discovery of novel agents for the control of fungal pathogens of plants and humans. We hope that this book goes some way to rectifying this situation by providing an up-to-date account of selected developments in two important areas, namely cell walls and membranes. Topics included in the book concern both yeasts and filamentous fungi.

Although the main emphasis is on biogenesis, functional aspects are also discussed, e.g. the role of glycoproteins in recognition of sterols in membranes and of calcium in regulation. Several contributions describe interference with the 'normal' biochemistry of cell walls and membranes with a view to increasing fundamental knowledge, but also highly relevant to the design of new fungicides and antimycotics. The steadily increasing impact of molecular biology on the study of fungal biochemistry is highlighted throughout.

We would like to thank the following sponsors for their generous support: Agricultural Genetics Company Ltd, BASF, Ciba-Geigy AG, Dow Chemical Co Ltd, Glaxo Group Research Ltd, ICI PLC, Monsanto, Pfizer Central Research, Schering AG and Shell Research Ltd. The help of the Biochemical Society and in particular advice from Dr. Peter Quinn are also gratefully acknowledged.

September 1989

P. J. KUHN
A. P. J. TRINCI
M. J. JUNG
M. W. GOOSEY
L. G. COPPING

Table of Contents

Index of Authors

The page numbers indicate the article to which each author has contributed.

Chapter 1

Cell Walls and Membranes in Fungi — An Introduction

P. J. KUHN[1] and A. P. J. TRINCI[2]

Situated between the plasma membrane and the external environment, the cell wall in fungi satisfies a number of vital functions (Chapter 2). It acts as a structural barrier maintaining cellular form, and preventing disruption of the protoplast by the uncontrolled entry of water from a normally hypotonic external milieu. The cell wall also serves as a site for a variety of enzymes that act in the provision of nutrients from the outside. In addition, macromolecules e.g. glycoproteins that are involved in recognition systems are also associated with the cell surface (Chapter 8).

The walls of filamentous fungi consist of an inner layer of crystalline microfibrils embedded in an amorphous matrix, together with one or more outer layers (Chapters 2 and 6; Trinci 1978). Rosenberger (1976) compared the two phase system of the inner wall with man-made composites such as glass-fibre-reinforced plastic. He pointed out that composites like this possess remarkable strength for their weight and that their mechanical properties vary with fibre orientation, fibre length and the degree of interaction between fibres and the matrix. Such factors may also influence the mechanical properties of fungal walls. Certainly, microfibrils form the main skeletal element of these walls.

Cell walls of the Basidiomycotina, Ascomycotina, Zygomycotina and some Mastigomycotina (the Chytridiomycetes and Hyphochytridiomycetes) contain chitin microfibrils whilst cell walls of other Mastigomycotina (the Oomycetes) contain cellulose microfibrils. The biosynthesis of chitin and of cellulose are discussed in Chapter 3 and 7, respectively. Certain vesicles, called chitosomes, have been isolated from chitin-containing fungi (Gooday and Trinci 1980). They are 40–70 nm in diameter, bounded by a membrane rich in sterols, and contain chitin synthase zymogen. When activated by proteolytic enzymes and incubated in the presence of substrate, chitosomes form chitin microfibrils in vitro which are indistinguishable from those formed in vivo. It is believed that active chitin synthase is located in the plasma membrane and accepts N-acetylglucosamine residues from uridine-diphospho-N-acetylglucosamine at the cytoplasmic face of the membrane and transfers them to the chitin chain spun out from the outer surface of the membrane.

A hypha increases in length by cell wall formation at the tip (the extension zone). Consequently, approx. 95% of the cell wall of a fungus is in a 'rigid' condition and

[1] Shell Research Ltd., Sittingbourne Research Centre, Sittingbourne, Kent ME9 8AG, UK
[2] Microbiology Group, Department of Cell and Structural Biology, Williamson Building, University of Manchester, Manchester M13 9PL, UK

only approx. 5% is in a 'plastic' condition. Thus, gross analysis of fungal cell walls are unlikely to provide useful information about the composition and properties of the extension zone wall. However, as pointed out many years ago by Castle (1942), the crucial events involved in hyphal growth and morphogenesis occur during extension growth at the tip, not in regions of rigidified wall below the tip. Fortunately, information about the extension zone wall can be obtained using cytological methods (Hunsley and Burnett 1970; Hunsley and Kay 1976) or from pulse chase experiments. Using the latter approach Wessels and colleagues (Chapter 6; Wessels 1986) were able to show that rigidification of the wall at the base of the extension zone results from the formation of an alkali insoluble chitin-β-glucan complex from the newly formed water-soluble/alkali-soluble $\beta(1\rightarrow3)$-glucan and chitin. Unfortunately, much less is known about wall rigidification than about wall synthesis.

Although fungal hyphae extend at rates up to 60 µm min^{-1}, their tip walls were thought to contain chitin microfibrils (Hunsley and Burnett 1970). Although the validity of this view is now less certain (Chapter 6), it is clear that microfibrils are formed rapidly and very near to the tip. Synthesis of microfibrils is supported by a large volume of distal cytoplasm which supplies the extension zone with vesicles (chitosomes?), wall precursor and enzymes required for cell wall biosynthesis (Trinci 1978). It is believed that rapid rates of hyphal extension can only be maintained because the tip wall is assembled from components produced in distal regions of the hypha and then transported to the tip, possibly by a system involving microfilaments or microtubules (McKerracher and Heath 1987). A novel and unifying model for fungal morphogenesis based on vesicles is described in Chapter 4.

Since chitin is not formed by plant or mammalian tissues, antifungal agents (agricultural fungicides or antimycotics) which affect either the biosynthesis of chitin microfibrils (Chapters 3 – 5) or the transport of vesicles to the tip are likely to exhibit selective toxicity. However, the observation that a mutant of *Aspergillus nidulans* lacking the secondary cell wall polymer, $\alpha(1\rightarrow3)$-glucan grows normally (Polacheck and Rosenberger 1977), suggests that compounds which affect synthesis of the secondary cell wall (Trinci 1978) are unlikely to be effective inhibitors of fungal growth. Knowledge about the biosynthesis of cell wall polymers, their assembly to form the tip wall, and wall rigidification at the base of the extension zone should facilitate the selection of compounds which can selectively inhibit this particular fungal target.

As with cell walls, the biochemistry of membranes in fungi can be broadly considered in terms of composition, biosynthesis and assembly, and function. Available evidence indicates that fungal membranes, like those of other organisms, are made up principally of lipid and protein. Unfortunately, although a wealth of information exists on the lipid composition of fungi (Weete 1974) much of the data derives from analysis of whole cell extracts rather than from purified membrane fractions. This state of affairs is not surprising since it is only during the last 10 – 20 years that reasonably well-defined membrane-enriched fractions have been obtained from fungi. Much of the work to date has centered on plasma membrane fractions, isolated principally from a number of yeasts and from *Neurospora crassa*. Recent reports of membrane preparations from filamentous representatives concern vacuolar membranes from *N. crassa* (Bowman et al. 1987) and plasma membranes from *Phytophthora megasperma* var. *sojae* (Baumer et al. 1987). Both of these studies are important, the former being one the few examples where comparisons have been made between the lipid compositions of different membrane fractions from the same fungus, and the latter being the first report of the isolation of a plasma membrane preparation from an Oomycete.

Information from the analysis of fungal membrane preparations has shown the principal lipids to be glycerophospholipids and sterols, along with generally smaller amounts of sphingolipids and glycolipids. The proportions of these components, and of associated acyl groups, can vary depending on among other factors the fungal species or strain, the type of membrane fraction, and the environmental conditions used in growth of the organism. Later contributions in this volume (Chapters 9, 10 and 16) describe the organisation and role of lipids in fungal membranes.

Knowledge of the biogenesis of membrane lipids in fungi is most highly developed for the sterol component. Indeed since studies using yeast have provided a significant input to fundamental knowledge on the biosynthesis of terpenes including sterols, it is appropriate that this area of metabolism is considered in some detail here (Chapters 11 – 15). The biosynthesis of phospholipids is also described (Chapters 16 and 17). Understanding of the biosynthesis and function of membrane lipids in fungi has benefitted greatly from studies using various inhibitors (Chapters 10 – 15 and 17), many originating from the agrochemical and pharmaceutical industries as agricultural fungicides and antimycotics, respectively. Inhibitors have been used to investigate enzyme mechanisms and to probe the role of lipids by altering their composition in the membrane.

Reported values for protein : lipid ratios in plasma membrane preparations from a number of fungi range from 0.96 – 2.1 (Weete et al. 1985). The total population of protein is heterogeneous and presumably includes both structural and enzymic components. Separation by electrophoresis of SDS-solubilized polypeptides from a plasma membrane fraction of *Taphrina deformans* revealed about 30 bands with a wide range of mol wts (Weete et al. 1985). Similarly, in preparations from *Saccharomyces cerevisiae*, 25 – 30 bands (mol wts 10000 – 300000) were detected following analysis by electrophoresis (Santos et al. 1978).

In the isolation of particular membrane fractions from cell-free preparations, one of the most widely applied criteria of purity is a progressive increase in the specific activity of a characteristic marker enzyme and the concomitant loss of extraneous activities associated with contaminating membranes. Ion-pumping ATPases are excellent examples of membrane-bound enzymes that can be used for this purpose. These enzymes, which are involved in many essential processes such as ATP synthesis and active transport, have been shown to be highly specific to different membrane fractions from *S. cerevisiae, Schizosaccharomyces pombe* and *N. crassa* (Bowman and Bowman 1986; Slayman 1987). As a result of intensive studies during recent years, ATPases are probably the best characterized proteins from fungal membranes (Chapter 19).

By analogy with other eukaryotes, it seems likely that some of the proteins in fungal plasma membranes function as receptors for external stimuli. In mammalian systems, and to a lesser extent those in plants, it has been shown that the transduction of signals into biochemical events within cells involves calcium as a second messenger, the turnover of inositol phospholipids, activation or inhibition of protein kinases or phosphatases, and altered phosphorylation states of intracellular proteins. Although comparable studies using fungal systems are at an early stage (Pitt 1984), some of the required elements e.g. calmodulin (Muthukumar et al. 1987) and a calcium- and phospholipid-dependent protein kinase (Favre and Turian 1987) have recently been described in fungi. This topical and important area is considered in Chapter 18.

In this introductory chapter we have tried to set the scene for the more detailed contributions which follow. Although each of these addresses a specific topic, it is important to stress that in fungi, as in other cellular and multicellular organisms, these various processes are interrelated and under strict regulatory control. For example, it is clear that during fungal growth and morphogenesis the rate at which the cell wall is synthesized and laid down must be closely linked to the biogenesis of the plasma membrane and other membrane elements. We are still far from having a complete picture of how this and other biochemical changes are co-ordinated, though some pieces of the jigsaw are undoubtedly to be found in this volume.

References

Baumer JS, Leonard RT, Erwin DC (1987) Isolation of plasma membrane of *Phytophthora megasperma f. sp. glycinea* and some properties of the associated ATPase. Exp Mycol 11:49–59

Bowman BJ, Bowman EJ (1986) H$^+$-ATPases from mitochondria, plasma membranes, and vacuoles of fungal cells. J Membr Biol 94:83–97

Bowman BJ, Borgeson CE, Bowman EJ (1987) Composition of *Neurospora crassa* vacuolar membranes and comparison to endoplasmic reticulum, plasma membranes, and mitochondrial membranes. Exp Mycol 11:197–205

Castle ES (1942) Spiral growth and reversal by spiralling in *Phycomyces*, and their bearing on primary wall structure. Am J Bot 29:664–672

Favre B, Turian G (1987) Identification of a calcium- und phospholipid-dependent protein kinase (protein kinase C) in *Neurospora crassa*. Plant Sci 49:15–21

Gooday GW, Trinci APJ (1980) Wall structure and biosynthesis in fungi. In: Gooday GW, Lloyd D, Trinci APJ (eds) 30th Symp Soc Gen Microbiol 1980. Cambridge Univ Press, Cambridge

Hunsley D, Burnett JH (1970) The ultrastructural architecture of the walls of some hyphal fungi. J Gen Microbiol 62:203–218

Hunsley D, Kay D (1976) Wall structure of the *Neurospora* hyphal apex: immunofluorescent localization of wall surface antigens. J Gen Microbiol 95:233–248

McKerracher LJ, Heath IB (1987) Cytoplasmic migration and intracellular organelle movements during tip growth of fungal hyphae. Exp Mycol 11:79–100

Muthukumar G, Nickerson AW, Nickerson KW (1987) Calmodulin levels in yeasts and filamentous fungi. FEMS Microbiol Lett 41:253–255

Pitt D (1984) Calcium in fungi. Plant Cell Environ 7:467–475

Polacheck Y, Rosenberger RF (1977) *Aspergillus nidulans* mutant lacking $\alpha(1-3)$-glucan, melanin and cleistothecia. J Bacteriol 132:650–656

Rosenberger RF (1976) The Cell Wall. In: Smith JE, Berry DR (eds) The filamentous fungi, vol 2. Arnold, London, pp 328–344

Santos E, Villanueva JR, Sentandreu R (1978) The plasma membrane of *Saccharomyces cerevisiae*. Biochim Biophys Acta 508:39–54

Slayman CL (1987) The plasma membrane ATPase of *Neurospora*: a proton-pumping electroenzyme. J Bioenerg Biomembr 19:1–20

Trinci APJ (1978) Wall and hyphal growth. Sci Prog (Oxford) 65:75–99

Weete JD (1974) Fungal lipid biochemistry. Plenum, New York London

Weete JD, Sancholle M, Touze-Soulet J-M, Bradley J, Dargent R (1985) Effects of triazoles on fungi. III. Composition of a plasma membrane-enriched fraction of *Taphrina deformans*. Biochim Biophys Acta 812:633–642

Wessels JGH (1986) Cell wall synthesis in apical hyphal growth. Inter Rev Cytol 104:37–79

Chapter 2

Fungal Cell Walls – A Review

J. F. PEBERDY [1]

1 Introduction

As the outermost part of the cell envelope the wall provides the interface between the organism and its environment. The cell wall has therefore several roles.

The wall confers shape to the cell it encloses and the diversity of cellular forms and structures found in the life cycles of many fungi is clearly a reflection of the versatility of the wall. The shape conferred by the wall is a function of its rigidity which in turn affects the osmotic integrity of the cell. Removal of the wall results in the liberation of protoplasts as discrete spherical entities which are highly sensitive to the osmotic environment. Therefore the wall indirectly affects the intracellular concentration of ions and solutes by restricting the water content of the cell.

The cell wall also acts as a filter controlling the secretion and uptake of molecules into the cell. Some enzymes are retained in the fabric of the wall or the periplasmic space and large molecules can be prevented from entering the cell. The wall therefore protects the fungus against hazards in the external environment as well as aiding its nutritional physiology, enabling the enzymatic conversion of nutrients into metabolisable forms prior to their entery into the protoplast. The wall may also be a store of carbon reserves.

For many fungi their environment is another organism, a plant or an animal, and in these cases the cell wall most probably has additional roles establishing what may be a crucial interaction for the survival of the fungus. Another type of cell-cell interaction involves random fusions of cells and fusions which represent the initial event in a reproductive process.

In the light of these diverse roles and functions it is clear that the fungal cell wall is a highly dynamic structure and is subject to change and modification at different stages in growth and development of an organism. During the past decade much interest has been focussed on these aspects. At the same time the more fundamental questions of wall structure and assembly are still largely unresolved.

In writing this review it is my intention to present a broad picture in the belief that it is timely to begin an assimilation of the information available. Other contributors to this volume will be dealing in depth with aspects of wall polymer synthesis and

[1] Department of Botany, Microbial Biochemistry & Genetics Group, University of Nottingham, Nottingham NG7 2RD, UK

morphogenesis (see Chapters 3, 4, 6, 7 and 8) and other reviews (Farkas 1979, 1985; Gooday and Trinci 1980; Aronson 1981; Gooday 1983; Sentendreu et al. 1984; Wessels 1984; Cabib 1987) have also dealt with some aspects of this subject. In consequence these topics will be considered, if only briefly, to ensure completeness. Furthermore an understanding of cell wall composition and organization are very pertinent to the understanding of its various functions.

2 The Chemistry and Architecture of the Fungal Cell Wall

2.1 Cell Wall Composition

In common with gram-positive bacteria and plants, the cell wall of fungi is composed primarily of polysaccharides; these may be both homo- and hetero-polymers. In some fungi proteins are also significant components of the cell and frequently are associated with one or more of the polysaccharide constituents. Lipids and melanins are minor wall components in many fungi, however, the quantitative contribution of the latter does not minimize their significance (see p. 21). A list of the wall components of fungi is given in Table 1.

Fungal wall polysaccharides can be divided into two groups on the basis of their presumed function and physical form. The skeletal polysaccharides are water insoluble and highly crystalline homopolymers and include chitin and β-linked glucans. By contrast the matrix polysaccharides are amorphous, or slightly crystalline, and are mostly water soluble. Some are homopolymers but there are also heteropolymers and often polysaccharide complexes.

Bartnicki-Garcia (1973) in drawing together all the then published data on fungal cell wall composition revealed a clear picture of the distribution of wall polymers and related this to the taxonomy of fungi. With few exceptions the many reports published subsequently have confirmed Bartnicki-Garcia's taxonomic conclusions. Thus, the

Table 1. Macromolecular components of fungal cell walls

---

1. Skeletal elements

    Chitin β-1-4-linked homopolymers of N-acetyl-D-glucosamine

    β-glucans β-1,3-glucan homopolymer comprised of D-glucose units with β-1,3- and
                β-1,6-glucosidic bonds (R-glucan)

2. Matrix components

    α-glucan α-1,3-homopolymer of glucose (S-glucan)
               α-1,3- and α-1,4-linked glucan (nigeran)

    Glycoproteins

3. Miscellaneous components

    Chitosan β-1,4-polymer of D-glucosamine

D-galactosamine polymers
Polyuronides
Melanins
Lipids

---

skeletal component of all the yeast fungi (Hemi-ascomycotina and Hemibasidiomyco-tina) consist of $(1-3)$-β-D-glucans, with $(1-6)$-β-linkages at intervals which form branch points. The degree of branching is significant in determining the crystallinity and therefore the solubility of the $(1-3)$-β-D-glucan. The walls of *Saccharomyces cerevisiae* contain two such β-glucans with differ which respect to the proportion of $(1-6)$-β-linkages. The predominant molecule is highly crystalline, having the larger molecular weight and has 3% $(1-6)$-β-linkages. The smaller glucan has 19% of these linkages and is soluble.

With the exception of the Oomycetes, all filamentous fungi have chitin together with $(1-3)$-β-glucan as the skeletal wall components. X-ray analyses of chitin from different biological sources have shown that this $(1-4)$-β-linked homopolymer of *N*-acetyl-D-glucosamine, with H-bonds linking adjacent chains, occurs in three forms designated α, β and γ. In fungal walls the polymer is of the first type which is distinguished by the antiparallel orientation of the chitin chains in the adjacent sheets. The Oomycetes are unique being the only fungi to have cellulosic walls. Like chitin, this is a linear homopolymer comprised of $(1-4)$-β-linked glucose residues. H-bonding between the chains creates a dense insoluble crystalline lattice.

In yeasts the matrix component is a mannan-protein complex. Drastic extraction procedures e.g. boiling in 20% sodium dodecyl sulphate (SDS) or exposure to 6 M urea, releases 60 different mannoproteins from isolated walls *S. cerevisiae* (E. Valentin et al. 1984). Some of the mannoproteins closely interact with the glucan component (Sanz et al. 1985; Pastor et al. 1984). Similar experiments using cell walls of *S. cerevisiae* and *Candida albicans* revealed some 40 different mannoproteins (Elorza et al. 1985). A similar mixture of mannoproteins has been isolated from other yeasts including *Zygosaccharomyces rouxii*, *Saccharomycopsis lipolytica*, *Hansenula wingei* and *Schizosaccharomyces pombe* (Herrero et al. 1987).

Treatment of walls of *S. cerevisiae* and *C. albicans* with Zymolase releases fewer mannoproteins, less than half in the case of *S. cerevisiae* and fewer in the case of *C. albicans*. Mannoproteins were also released from walls of *Z. rouxii*, *S. lipolytica* and *S. pombe*, however, the effect on *Rhodotorula glutinis* walls was limited (Herrero et al. 1987). Nevertheless, the release of mannoproteins by this enzyme, which has high $(1-3)$-β-glucanase activity suggests that some of these molecules are covalently linked to the β-glucan component of the wall. Conversely, the mannoproteins released by denaturing agents such as urea must be associated in some other manner, probably by H-bonds. Disulphide bridges may also have a role in forming associations between wall components; treatment with reducing agents appears to open up the wall making it more porous. Further support of the existence of specific interactions between specific mannoproteins and the $(1-3)$-β-glucan was presented by E. Valentin et al. (1986). Cells of *S. cerevisiae* treated with aculeacin A, which inhibits the synthesis of $(1-3)$-β-glucan, were found to have decreased incorporation of mannoproteins, particularly a high molecular molecule and a molecule of 33 KDa, into the cell wall. The supernatants from treated cells were found to contain the smaller mannoprotein.

The mannoproteins of yeast fall into two distinct families which differ in size and linkage characteristics. The largest molecules, which may have a molecular weight greater than 150 KDa have a branched polysaccharide moiety of up to 150 mannosyl units. The main chain or backbone of the molecule is composed of $(1-6)$-α-linkages from which short chains with mixed $(1-2)$-α- and $(1-3)$-α-linkages arise. Linkage to the protein component occurs through *N*-glucosidic bonds, involving diacetyl-chitobiose, from the inner core region of the polymer. The large mannoproteins from *Z. rouxii, S. lipolytica* and *S. pombe* would appear to have terminal sugar residues other than glucose, glucosamine or mannose because there is no binding with con-canavalin A (Herrero et al. 1987). The second group of mannoproteins are smaller molecules; some are composed of short $(1-2)$-α-/$(1-3)$-α-linked mannosides attached to a protein by *O*-glucosidic links. In *C. albicans* a small 30 KDa mannoprotein was released from Zymolase treated walls (Elorza et al. 1985). This has a single mannan chain with a *N*-glycosidic link to the protein. The small mannoproteins from *S. cerevisiae, C. albicans* and *H. wingei* show antigenic cross reactivity (Herrero et al. 1987), suggesting a degree of conservation of the protein component.

Changes in the mannoprotein pool of the wall of *S. cerevisiae* during the growth cycle of a population have been reported (E. Valentin et al. 1987). The most significant change concerned the larger molecules. In cells from the early exponential phase these molecules had a mean value for M_R of 200 000 (range 120 000 – 500 000). In older cultures, at late exponential phase, the mean size of the molecules was increased (M_R 300 000 – 350 000). The change in molecular size was due to quantitative changes in the *N*-glycosidic mannose residues; the peptide component and the *O*-glycosidic-linked components changed only slightly.

Wall mannoproteins may be important in the morphogenesis of yeast and mycelial forms of *C. albicans*. Walls of the yeast form have proportionately more mannan than the mycelial form (Elorza et al. 1983). SDS extracts of the two cell forms contained similar profiles of mannoproteins but quantitative differenecs were apparent (Elorza et al. 1985). Howveer, Zymolase digestion of the SDS extracted walls did give a different result; 4 mannoproteins were released from yeast cells compared to 2 from the mycelium.

Glycoproteins are also found in the walls of filamentous fungi, however, they are probably more heterogenous than those found in yeast, both with respect to the polysaccharide component and in the protein-carbohydrate linkage. The polysaccharide may be a homo- or heteropolymer, e.g. the glycoprotein from *Neurospora crassa* is composed of galactose and glucuronic acid residues and both *N*- and *O*-glycosidic linkages have been reported.

$(1-3)$-α-D-glucan is the major matrix polymer for most fungi. In some Aspergilli the α-glucan has alternating $(1-3)$-α- and $(1-4)$-α-linkages and is known specifically as nigeran. The α-glucan is normally found at the outer surface of the wall and occurs as thick irregular fibrils.

2.2 Cell Wall Organization

In any consideration of the role of the cell wall, the organization of polymers is as important as their chemical nature. Observations have revealed a distinct layering of materials in fungal walls. The layers may have several components. Normally, the skeletal component is located to the inner side of the wall and is frequently found embedded in amorphous matrix material. The inner surface of the wall is therefore characterised by its fibrillar appearance. The outer surface which is normally totally amorphous material is generally smooth or slightly reticulate.

Such a description is typical of filamentous species such as *Schizophyllum commune* (Seitsma and Wessels 1979), *Aspergillus nidulans* (Gibson 1973), and yeast (Cabib et al. 1982). In other fungi the wall is more complex; in *N. crassa* the inner layer of chitin fibrils is covered by protein and a glycoprotein reticulum with an outer layer of α- and β-glucans. The wall of *C. albicans* is also complex; several layers have been described, however, the fixation procedure used can affect wall morphology (Hilenski et al. 1986).

The introduction of new techniques for the rapid freezing of biological material has presented the opportunity to evaluate the surface of unfixed fungal cells. By employing these techniques, Tokunaga et al. (1986) demonstrated that the surface of *S. cerevisiae* and *C. albicans* cells was covered with a layer of "brush-like" fibrils some 150 nm in length arranged perpendicular to the cell surface. Somewhat longer fibrillar elements designated as fimbriae had been demonstrated on the surface of several fungi some years earlier (Poon and Day 1974, 1975).

3 Synthesis of Cell Wall Polymers

The publication by Glaser and Brown (1957) was the first description of chitin synthesis in fungi. Following this pioneering work our understanding of both chitin and glucan biosynthesis has been greatly extended particularly in the last decade through the work of Bartnicki-Garcia, Cabib and Gooday and their co-workers on chitin synthetase and Fevre on glucan synthetase (Farkas 1985).

The synthesis of chitin is catalyzed by chitin synthetase utilising uridine diphosphate *N*-acetylglucosamine (UDP-GlcNAc) as substrate which is linked to an oligosaccharide of the amino sugar. In all cases reported so far, the enzyme has been isolated from cell homogenates as a membrane or particulate preparation (Cabib 1987). Such extracts have a detectable level of activity when assayed which can be increased if the material is subjected to partial proteolytic treatment using either trypsin or an endogenous protease prepared from fungal homogenates. There are now several reports describing the presence of these active and zymogenic forms of chitin synthetase in the vegetative cells of many fungi.

First attempts to purify the particulate chitin synthetase involved solubilization with the detergent digitonin first by Gooday and de Rousset-Hall (1975) and subsequently by several other groups (Duran and Cabib 1978; Braun and Calderone 1978; Ruiz-Herrera et al. 1980; Hanseler et al. 1983; Vermeulen and Wessels 1983). Much purer preparations have now been described for *Coprinus cinereus* (Montgomery et al. 1984) and *S. cerevisiae* (Kang et al. 1984) which can be resolved by SDS-polyacrylamide gel electrophoresis (PAGE) to give a major band which corresponds to a molecular weight of 67000 and 63000 respectively.

Purified chitin synthetase preparations from *S. cerevisiae* (Kang et al. 1984) and *Schizophyllum commune* (Vermeulen and Wessels 1983) have a requirement for a phospholipid. The best additive for *S. cerevisiae* was phosphatidylserine (Duran and Cabib 1978), however, the phospholipid was not necessary to promote chitin synthetase activity run on a non-denaturing polyacrylamide gel (Kang et al. 1984).

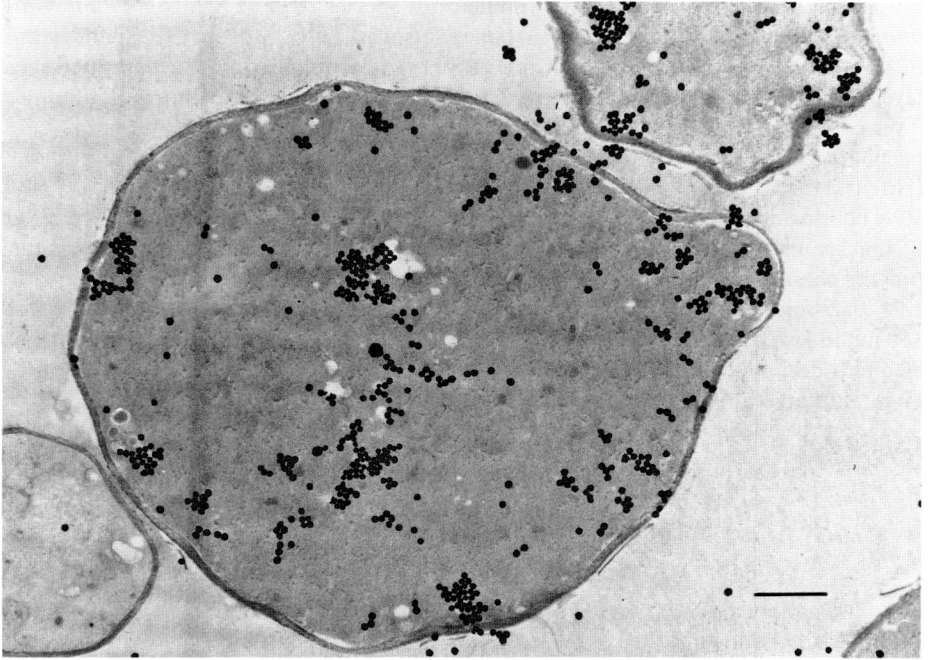

Fig. 1. Cell of *Mucor rouxii* permeabilized with toluene/ethanol and incubated with ^{14}C UDP-GlcNAc in the presence of trypsin. Silver grains are seen to accumulate in the cell. (From Sentendreu et al. 1984a)

The intracellular distribution and location of chitin synthetase is still a controversial matter. Evidence from studies on several fungi has indicated that the enzyme is located on the inner side of the plasma membrane. Two types of experiment have been performed; in the first purified preparations of plasma membranes which have enzyme activity were isolated by gradient centrifugation (Duran et al. 1975); secondly protoplast membranes labelled with concanavalin A were found to have activity (Braun and Calderone 1978; Vermeulen and Wessels 1983; Kang et al. 1985). The specific location of chitin synthetase to the inner or outer side of the membrane is unresolved, available experimental data indicating that it could be either (Duran et al. 1975; Cabib et al. 1983; Braun and Calderone 1978). On the contrary, Sentendreu et al. (1984a) have obtained results to indicate that chitin synthesis can occur inside the cell (Fig. 1); yeast cells of *Mucor rouxii* were permeabilised with toluene and ethanol and were found to incorporate UDP-GlcNAc. Clearly a definitive answer will not be possible until antibodies have been prepared using a purified chitin synthetase as antigen.

The controversy referred to above relates to the results obtained by Bartnicki-Garcia and his co-workers (Bartnicki-Garcia and Bracker 1984). Using gel filtration, sucrose density gradient centrifugation and isopycnic centrifugation they have isolated microvesicles or chitosomes, which are rich in zymogenic chitin synthetase (Ruiz-Herrera et al. 1975, 1977; Bracker et al. 1976; Bartnicki-Garcia et al. 1984). When activated and incubated with UDP-GlcNAc, chitosomes generate chitin fibrils (Bracker et al. 1976). However, Cabib (1987) has questioned whether the chitosomes are real structures or arise following the disruption of other organelles. It is possible that some of the microvesicles observed at the apex of a fungal hypha (Hoch and Howard 1980; Howard 1981) are chitosomes (Gooday 1983).

Synthases associated with the synthesis of $(1-3)$-β- and $(1-4)$-β-glucans have been reported in several different fungi. The enzyme is recovered as a particulate preparation (Szaniszlo et al. 1985) or as more purified plasma membrane material (Fevre and Rougier 1981; Girard and Fevre 1984). These preparations utilize uridine 5'-diphosphate glucose (UDP-Glc) as substrate, and unlike chitin synthesis a saccharide acceptor is not required. *Saprolegnia monoica* has both $(1-3)$-β and $(1-4)$-β activities; the former occurs in zymogenic form and is active with low UDP-Glc concentrations and in the presence of Mg^{2+} ions. The enzymes in non-soluble form are stimulated by ATP and GTP, the activity associated with both endoplasmic reticulum and plasma membrane fractions are so affected (Fevre 1984). $(1-3)$-β-glucan synthetase activity in *S. cerevisiae* is found associated with the plasma membrane, and there is no increase in activity of preparations following trypsin treatment. Maximal activity in vitro is dependent on the presence of glycerol, albumin and ATP.

Although the wall polymers are synthesized within the cytoplasm or at the plasma membrane a degree of processing takes place within the cell wall. These processes include the establishment of cross-linking between the polymers and possibly some assembly also. Data from *S. commune* (Seitsma et al. 1985) indicate that modification to the wall in sub-apical regions is important in conferring rigidity of hyphae. At the apex the newly formed wall contains only $(1-3)$-β-D-glucan, however, behind the apex a 'maturation' process occurs which involves the incorporation of a $(1-3)$-β-/$(1-6)$-β-D-glucan.

Elorza et al. (1983) demonstrated that the fluorochrome Calcofluor, which forms H bonds with cellulose and chitin, caused abnormal chitin deposition in *S. cerevisiae*. Regenerating protoplasts of *C. albicans* produced chitin as normal but the typical crystalline lattice was not formed. These workers concluded that the crystallization of chitin could involve the self assembly of nascent subunits within the fabric of the wall. In filamentous fungi this process could occur in sub-apical regions; thus chitin at the hyphal apex is sensitive to chitinase and to hot dilute mineral acid whereas that in a sub-apical position is not (Vermeulen and Wessels 1984).

4 Stability of the Cell Wall

Analyses of cell wall composition are typically made on material grown in culture for a pre-determined period and consequently only provide information about that sample. To what extent, however, is the cell wall subject to change? Information available at present is limited. Some reports e.g. Gomez-Miranda et al. (1984) indicate that there are no changes in cell wall composition during the different phases of growth of a batch culture. Although the chitin content of walls of some Aspergilli is affected by oxygen (Musilkova et al. 1982) and light (Feima 1983), in general it seems that the skeletal/fibrillar component shows no detectable turnover, despite the presence of chitinase and β-glucanase in the periplasmic space (Sentendreu et al. 1984b) and in *Aspergillus niger* and *A. aculeatus* the nigeran ($(1-3)$-α, $(1-4)$-α-glucan) content of the wall increases in cultures subjected to nitrogen deprivation (Gold et al. 1973; Bobbit and Nordin 1982).

Environmental factors such as the presence of certain metal ions can also influence the wall composition. Walls prepared from *Cunninghamella blakesleeana* grown in the presence of copper and cobalt ions showed clear differences from control material (Venkateswerlu and Stotzky 1986). In the presence of 30 μg ml^{-1} Cu mycelial biomass was reduced by 70% – 80% but the cell walls were 3-fold thicker than those formed in control cultures. The walls contained 40% less hexosamine than control cell walls with chitin and chitosan being present in equal amounts compared to a 40% : 60% ratio present in the control. Hyphal walls produced in the presence of Cu had 116% more protein than control walls and contained hydroxyproline, an amino acid not normally found in the wall. Differences, but not so extreme, were also found in the walls produced on the Co containing medium.

This interesting study clearly demonstrates the susceptibility of the wall synthetic system to environmental extremes and suggests that modification to the wall may be an important mechanism by which fungi respond to stress conditions.

5 Wall Formation During Protoplast Regeneration

The formation of a new cell wall is the crucial step in the regeneration of protoplasts prior to re-establishing the cellular form of the fungus. Several studies have shown that in the early stages at least, synthesis of the cell wall is unbalanced and therefore unlike the process in normal cells (de Vries and Wessels 1975; Peberdy 1979; Sonnenberg et al. 1982). Nevertheless, the process is an interesting morphogenetic model and does provide an experimental system for the synthesis of wall polymers. A recent example is the study reported by Douglas et al. (1984). Protoplasts were prepared from *A. niger* mycelium which was actively producing nigeran. The protoplasts rapidly developed a new cell wall, however, nigeran was not found in the wall until 12 – 24 h after the onset of regeneration and was correlated with the development of normal hyphae. It was concluded that nigeran was a secondary wall polymer and does not have a direct morphogenetic role.

Unbalanced growth is found in the most extreme case with protoplasts from budding yeasts which produce only glucan fibrils and secrete the mannoprotein into the liquid growth medium (Necas and Svoboda 1975). The formation of these naked fibrils has made it possible to make a detailed study of their assembly and physical characteristics (Kopecka and Kreger 1986).

Wall regeneration of protoplasts from *S. commune* and *A. nidulans* occurred in two stages, the first involved the deposition of a chitin network covering the surface of the protoplast, followed by a second in which chitin deposition continued in association with the biosynthesis of other wall polymers (de Vries and Wessels 1975; Peberdy 1979). In the former case it was found that the delayed synthesis of $(1-3)$-β-glucan

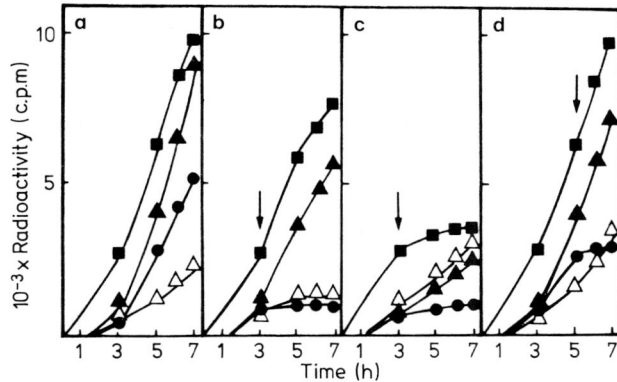

Fig. 2a–d. Distribution of radioactivity among the polysaccharide fractions during regeneration of protoplasts. Various suspensions (20 ml) of protoplasts (OD_{600} 0.22) were regenerated in osmotically stabilized Lee medium in the presence of 0.4 µCi[U-^{14}C]glucose ml^{-1}. **a** Control culture without antibiotics; **b** trichodermin or **c** nikkomycin added after 3 h of regeneration; **d** tunicamycin added after 5 h of regeneration. At the times indicated samples were taken, walls were isolated and the radioactivity in the polysaccharide fractions was determined: chitin, l-glucan, S-glucan, mannan. (From Elorza et al. 1987)

was a consequence of a need for de novo synthesis of the enzyme. A recent study has demonstrated a similar situation in *C. albicans* (Elorza et al. 1987). Protoplasts were treated with different inhibitors known to block processes involved in wall synthesis. Thus, protoplasts exposed to nikkomycin were impaired in chitin synthesis and the wall produced was enriched in alkali-soluble glucan. With papulocandin, which inhibits $(1-3)$-β-glucan synthetase, the protoplasts produced a much reduced wall with the glucan and mannoprotein content greatly reduced. The effect of tunicamycin depended on the timing of addition to the protoplast; if added at the start of regeneration, the level of all polymers produced was depressed but if added during regeneration then only mannoprotein and the alkali-insoluble glucan were affected (Fig. 2).

A further indication that walls produced during regeneration are abnormal was shown by Elorza et al. (1985) in an analysis of the mannoproteins produced by *C. albicans* protoplasts. The most significant difference was the absence of the high molecular weight molecules released by zymolase treatment. Instead these mannoproteins were found in the culture medium suggesting that covalent linking between the mannoprotein and glucan had not occurred.

6 Genetics and the Fungal Wall

The pioneering studies of Beadle and Tatum (1945) laid the foundation for the extensive application of genetic techniques to fungal biochemistry. This knowledge is proving even more important and useful as the techniques of molecular biology are applied to these organisms.

The first attempts to isolate potentially useful mutants were made with *N. crassa* and resulted in the isolation of some 100 strains with changed hyphal branching pattern and colonial form (Garnjobst and Tatum 1967). For several years Tatum and co-workers set about attempting to detect the primary biochemical lesions which were the basis of the morphological mutations, attempting to describe how defective enzymes might relate to cell wall formation (Scott 1976). Several mutants were found to have altered ratios of the wall polymers and investigations of various enzymes revealed that any detectable changes concerned four enzymes: glucose-6-phosphate dehydrogenase, phosphoglucomutase, 6-phosphogluconate dehydrogenase and phosphohexoisomerase. Alterations in these enzymes were found with respect to kinetic parameters, rates of heat inactivation, electrophoretic or electrofocussing behavior, specific activity and overall stability (Mishra 1977). In the dehydrogenase mutants, reduced levels of NADH and modified cAMP systems were also described (Scott et al. 1973). Thus in none of these mutants were the genetic defects correlated directly with cell wall synthesis observed, the morphological change being a consequence of a pleiotropic effect of the mutated gene.

By contrast mutants have been isolated in *A. nidulans* in which the biochemical lesion is more specifically wall associated. The strategy adopted by some groups has been to isolate temperature sensitive mutants; the value of these mutants is that the comparison of the normal condition, at the permissive temperature, and the defect, at the non-permissive temperature, can be made in one strain. One such mutant under non-permissive conditions has a reduced glucosamine and chitin content and for growth to occur it is necessary to provide some external osmotic 'support' suggesting possible enhanced autolysis in the wall (Katz and Rosenberger 1971). In another study (B. P. Valentine and Bainbridge 1975, 1978; Bainbridge et al. 1979) morphological mutants were isolated which showed various forms of hyphal swelling as the phenotype. In three types the defect could be relieved by different nutrients, namely mannose, glucosamine and choline leading to their designation *mnr, glc* and *cho*. In total, at least 15 genetic loci were found to be involved in the swollen hyphal phenotype (B. P. Valentine and Bainbridge 1978). Further evidence suggested that the walls of swollen hyphae were of normal thickness, implying change in the normal polarized growth process.

Other mutants of *A. nidulans* with grossly altered hyphal morphology were isolated by Thompson (1977). The characteristic feature of both, designated berry (*bry*) and coral (*crl*) (Fig. 3) was an apparent total loss of polarised growth. Further cytological and biochemical analysis of *crl* revealed that it had a much thicker cell wall than the wild type, and contained less chitin than the wild type. Parasexual genetic analysis indicated that the mutation arose as a single gene mutation.

The only reported mutant of a filamentous fungus with an obvious defective cell wall is the slime mutant of *N. crassa* (Emerson 1963). The slime phenotype results from the interaction of three mutations: (*fz* (fuzzy), *os* (osmotically sensitive) and *sg* (spontaneous germination of ascospores). Despite its unique property, it is only recently that information about the biochemical lesion has become available. Interestingly the mutant is not defective in the production of chitin synthetase (Selitrennikoff 1979) and although the level of active enzyme is reduced, the zymogenic form is present at higher levels than in the wild type (Bartnicki-Garcia et al. 1984; Leal-Morales and Ruiz-Herrera 1985). Leal-Morales and Ruiz-Herrera (1985) revealed what appears

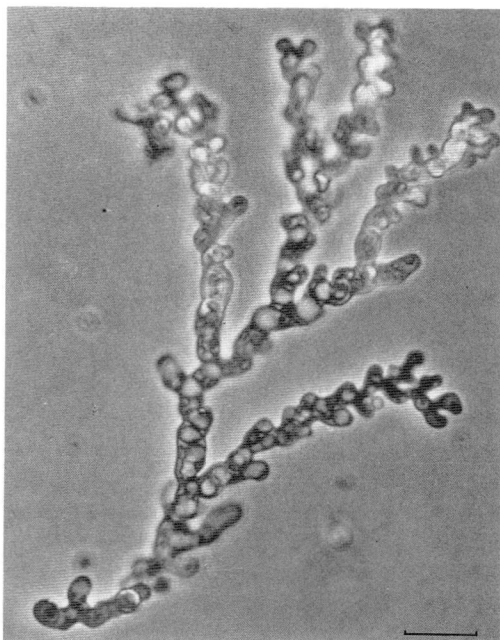

Fig. 3. Hyphal morphology of the coral mutant of *Aspergillus nidulans*. (From Thompson 1977)

to be significant defects, namely the production of β-glucan synthetase and activation of chitin synthetase. The slime mutant, therefore, lacks a cell wall because of the combined absence of the two skeletal polysaccharides; the remaining wall constituents being released into the medium, forming the slime. Of the three mutations which together confer the slime phenotype, it was shown that the alterations were not associated with the *os* mutation; a strain carrying this defect as a single mutation was normal for production of the two synthetases.

The application of recombinant DNA techniques in filamentous fungi present possibilities for raising the level of our understanding of the fungal cell wall still further. However, the limited experiences to date possibly should prepare us for the unexpected. A fragment of *S. cerevisiae* DNA which it was presumed included a structural gene for chitin synthetase (CHSI) was transformed into the yeast resulting in clones that produced 10-fold more of the zymogenic enzyme (Bulana et al. 1986). In a second experiment a disrupted form of CHSI was allowed to recombine with the normal gene to produce cells which lacked a functional CHSI locus. Although these cells lacked measurable enzyme, their chitin content and viability was not affected. Clearly these cells produced a chitin synthetase. Failure to detect activity under the previously established conditions implied that yeast produced two forms of the enzyme and the enzyme studied extensively over the years was not the one involved in chitin synthesis during septation (Cabib 1987; see also Chapter 3). This work raises tantalising questions regarding

(a) the occurrence of two forms of the enzyme in filamentous fungi with chitinous walls and septa, and

(b) why yeast makes a zymogenic enzyme that it apparently does not need.

7 The Cell Wall as a Target (see also Chapter 5)

The control of fungal infections of plants and animals depends on the availability of chemicals which are toxic to the fungus but cause minimal damage to the host organism.

The increasing incidence of fungal infections of man, particularly those associated with immunocompromised patients presents an important medical problem (Bodey and Fainstein 1985). The toxicity of most antifungal drugs currently in use underlines the need to develop new and safer drugs. In plant disease the problem is not one of more disease but resistance to existing control agents. The presence of chitin is clearly very relevant when identifying potential targets for growth inhibition in the fungal cell.

To date relatively few compounds which block chitin synthesis have been described. Two exceptions are the peptide nucleoside antibiotics, the polyoxins and the nikkomycins (Endo et al. 1970; Ohta et al. 1970). The polyoxins are a family of molecules (Isono et al. 1967) of which polyoxin D is the most effective. It is structurally similar to UDP-GlcNAc and is thus a strong competitive inhibitor of chitin synthetase (Hori et al. 1971). In vivo studies have demonstrated the effectiveness of this antibiotic against a range of fungi including *C. albicans* (Becker et al. 1983). Why the absence of chitin from the walls of this yeast should block cell division is not clear, but the results suggest that the polymer has a significant role in wall stability. The location of chitin in *C. albicans* cells was determined using wheat germ agglutinin complexed to colloidal gold as a probe. Dense labelling of the septa and bud scars was observed and labelling was also found in the innermost wall layers. Cells exposed to polyoxin D produced no detectable chitin as no label could be seen (Becker et al. 1983).

Differences in chitin levels is one of the distinguishing features seen in the yeast (Y) and mycelial (M) forms of *C. albicans*. In the Y form, the wall contains 1% chitin compared to 5% for the M form. A related factor is the level of activity of glucosamine-6-phosphate synthetase which is 4-fold higher in the M than in the Y form. As the M form is probably the most virulent (Borgers et al. 1983; Cassone 1984), prevention of the transition could be a useful means of chemotherapy to control infections. The increased chitin synthesis occurring in the M form is clearly a potential target (Chiew et al. 1980). The dipeptide antibiotic tetaine (bacilysin) is an effective growth inhibitor of both cell forms of *C. albicans* and of the yeast-mycelium transition (Milewski et al. 1983). Tetaine is made active by the action peptidases releasing the cleavage product anticapsin. More recent studies in vivo have revealed that the antibiotic inactivates glucosamine-6-phosphate synthetase (Milewski et al. 1986). Tetaine strongly inhibited the incorporation of [^{14}C] glucose into chitin particularly in the M form. Addition of N-acetylglucosamine reversed the inhibition. The overall effect of the antibiotic was therefore to block chitin synthesis.

The inhibition of glucosamine-6-phosphate synthetase by tetaine also affects the secretion of mannoproteins. This is demonstrated by the large (up to 70%) reduction in the release of enzymes such as invertase and alkaline phosphatase by isolated protoplasts. The reduced secretion is a consequence of impaired glycosylation which in turn is the consequence of reduced availability of N-acetylglucosamine.

Some reports have recorded the production of chitinase by plant tissues infected by fungal pathogens (Pegg and Young 1982; Boller et al. 1983). These observations led to the hypothesis that the production of the enzyme is a response to infection and

Table 2. Comparison of antifungal activity and enzymic activities between grain and bacterial chitinases

Activity assayed	Source of chitinase					
	Barley	Maize	Wheat	*P. stutzeri*	*Sa. marcescens*	*Sm. griseus*
Growth inhibition of P. blakesleeanus[a]	3	3	10	>50	>50	>50
Growth inhibition of T. reesei[a]	1	1	0.3	>50	>50	>50
Hydrolysis of chitin in agar[b]	10	10	10	0.1	1	1
Hydrolysis of bacterial cell walls in agar[b]	10	10	0.3	>50	>50	>50
Hydrolysis of radioactive chitin[c]	ND	ND	4.2	5.1	6.5	ND
Hydrolysis of p-nitrophenyl-β-D-N,N'-diacetylchitobiose[d]	<0.1	<0.1	<0.1	60	32	20

ND, not determined.

[a] Values given are minimum amounts of chitinase (μg protein per disc) required to produce detectable fungal growth inhibition.

[b] Values given are minimum amounts of chitinase (μg protein per disc) required to produce detectable polymer hydrolysis.

[c] Activities are expressed as μmol N-acetylglucosamine solubilized min^{-1} $(mg\ protein)^{-1}$.

[d] Activities are expressed as μmol p-nitrophenol released min^{-1} $(mg\ protein)^{-1}$.

(After Roberts and Selitrennikoff 1988).

provides a natural means of fungal control. However, conclusive supporting evidence is lacking despite some recent interesting reports. Roberts and Selitrennikoff (1986a, b) have isolated two proteins from barley grains which display anti-fungal activity. The protein with higher activity was found to have a chitinase function. A similar activity has subsequently been isolated from seeds of other cereals (Roberts and Selitrennikoff 1988). Comparison of these plant chitinases with enzymes of bacterial origin revealed that the former were more potent on fungal walls, a feature which was correlated with their endo-activity (Table 2).

In *Candida* chemotherapy, the azoles are proving to be the most useful antifungal drugs, despite some toxicity problems. Although the primary target for their action is ergosterol biosynthesis, several have a marked effect on cell morphology. In particular some reports suggested that various imidazoles inhibited the Y-M transition in fungi (Borgers 1980; Davies and Marriott 1981). This observation has not been confirmed, however, and the studies by Odds et al. (1985), which included ketoconazole, clotrimazole, itraconazole, thioconazole and miconazole, revealed that they had different effects on the Y and M forms. These included impaired branching by the hyphal form and suppression of hyphal emergence from yeast cells. Clearly these compounds have an indirect effect on the cell wall possibly at the membrane level.

8 The Cell Wall as an Interactive Surface

Cell-cell interactions are fundamental to the vegetative growth and reproductive development of many fungi. The level of these interactions is varied ranging from self fusions of cells or hyphae in the same population, fusions between cells or hyphae of compatible mating type and interactions between a fungal cell and a foreign cell, normally that of some host organism.

Self fusions involving hyphae within the same mycelium are important in several respects. They provide a means of establishing a hyphal network, in what is otherwise a radiating system, and so provide a means for a more extensive transmission of material throughout a colony. Hyphal fusions also occur to establish tissues which make up reproductive structures.

Self fusions are undoubtedly random in their occurrence without any pre-determined sites on the cell wall, although directed growth of a hyphal tip, sometimes seen to involve curvature through an arc of 90°, homing to a point on an adjacent hypha (Ainsworth and Rayner 1986) could be interpreted as a more specific interaction. Nevertheless, the physical event of a fusion involves changes in the wall at the site of fusion, namely a highly localized lysis (Aylemore and Todd 1986). Vesicles have been observed at the sites of fusion events in many fungi (Hawker and Beckett 1971; Harvey 1975; Rijkenberg and Truter 1975; van der Valk and Marchant 1978). What stimulus promotes the lysis reaction at the point of fusion remains unknown.

Matings between the mycelia of compatible strains which are part of a reproductive cycle also involve fusion events. We have no detailed information relating to this process. In *Aspergillus nidulans*, vegetative incompatibility, which affects hyphal fusion and viable heterokaryon formation, is controlled by several genes (the *het* genes). There is the suggestion that in a few of these the incompatibility barrier is located in the cell wall. Protoplast fusion crosses between the different incompatible strains did give rise to heterokaryons in a few cases (Dales and Croft 1977).

The most extensively investigated interaction between fungal cells is the mating process in *S. cerevisiae* (see also Chapter 8). Most strains of this yeast are heterothallic with two mating types designated **a** and α. Genetic control of this interaction is mediated by the *MAT* locus and the strains are genetically designated *MAT***a** and *MAT*α. These genes are primarily regulatory in function controlling several others which are involved in the mating process.

Mating between **a** and α cells is triggered by the secretion and reciprocal action of mating-type specific pheromones, the **a** factor and the α factor. When **a** cells are exposed to the α factor they exhibit a number of sequential responses (i) alterations to the cell surface which promotes adhesion of α cells, (ii) transient arrest of the cell cycle at the Gl phase and (iii) anisotropic wall and membrane synthesis, and other localised changes in the cell wall, which result in the formation of elongated cells. α cells respond similarly when exposed to the **a** factor.

Pheromone induced changes at the cell surface occur very rapidly, **a** cells develop agglutinability towards α cells within 15–20 minutes with maximal response at 50 minutes after pheromone addition. This new property of the cells is presumed to be associated with a surface glycoprotein. The wall of the expanding cell has a higher glucan to mannoprotein ratio than untreated cells (Lipke et al. 1976) and the mannoprotein itself is composed of shorter, unsubstituted $(1-6)$-α-linked oligosaccharide units (Tkacz and MacKay 1979). This modification to the mannoprotein is probably the reason for the observed immunological difference (Lipke and Ballou 1980). A further change in the pheromone "induced" cell is an increase in chitin synthetase activity and a more diffuse distribution of chitin (Schekman and Brawley 1979).

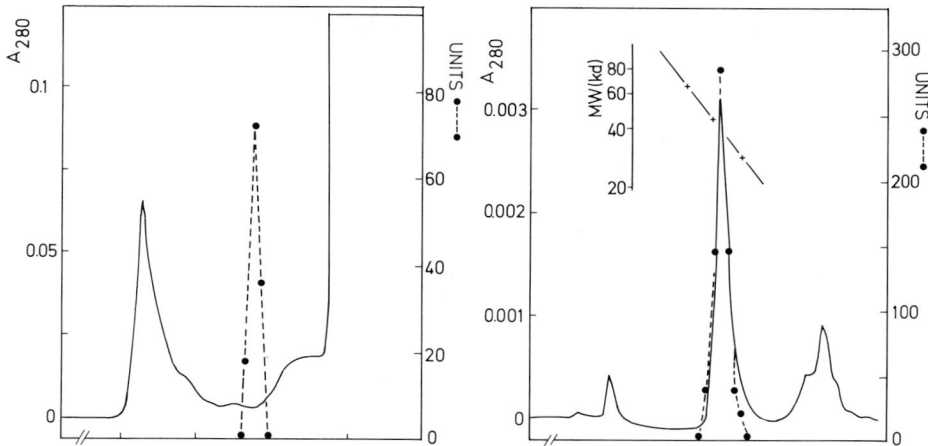

Fig. 4. Purification of mt-α agglutinin. *Left:* HP gel filtration of a concentrated crude DTT-extract from 5 g mt-α cells. A 600×7.5 mm Bio-Sil TSK-125 column was used with 100 ml mM NH₄Ac, pH 5.0 as eluent at a flow rate of 0.15 ml min⁻¹; fraction size: 0.45 ml. *Right:* the DTT-extract was fractionated by HP anion exchange chromatography; the unbound material was concentrated and applied to the Bio-Sil TSK-125 column. Bovine serum albumin (M_r 66000) ovalbumin (M_r 45000) and carbonic anhydrase (M_r 29000) were used as markers. (From Sijmons et al. 1987)

Recent studies (Sijmons et al. 1987; Terrance et al. 1987; see also Chapter 8) have confirmed the glycoprotein nature of the molecules involved in the surface binding of **a** and α cells of *S. cerevisiae*. The molecules, termed agglutinins, have been isolated from both cell types. The **a**-agglutinin was a glycoprotein with 3% mannose and has a molecular weight of 43 KDa (Fig. 4). **a**-agglutinin was released by treatment of the cells with diothiothreitol. However, the α-agglutinin could not be released by this reducing agent and was recovered by brief exposure of the cells to trypsin. It was a considerably larger molecule, 320 KDa and comprised 94% w/v mannose.

Electron micrographs of thin sections of α-factor treated **a** cells reveal a modified wall architecture in the elongating zone with a thinner inner layer and a more diffuse outer coating (Lipke et al. 1976). Although these physical alterations cannot be directly related to the biochemical changes described, it is clear that the mating reaction in yeast involves significant alterations at the cell surface.

Whether these *S. cerevisiae* agglutinins are synthesized de novo, or become exposed at the wall surface due to other changes, or become activated has still to be resolved. It appears that de novo RNA and protein synthesis are essential for agglutinability to develop in pheromone induced cells (Betz et al. 1978; Fehrenbacher et al. 1978), however, the fact that tunicamycin and other inhibitors of glycosylation do not affect the mating (Sprague et al. 1983) would at least indicate that glycoprotein assembly is not required for the induction of agglutinability.

For many pathogenic fungi the most important interaction is that with its host. Several fungi have been shown to have surface molecules which may be extensions of the wall such as fimbriae and fibrils (Fig. 5, Poon and Day 1974, 1975; Gardiner et al. 1981). As the fibrils were first described on the smut fungus (*Ustilago violacea*), a role for them in fungus-host interactions was an obvious hypothesis (Poon and Day

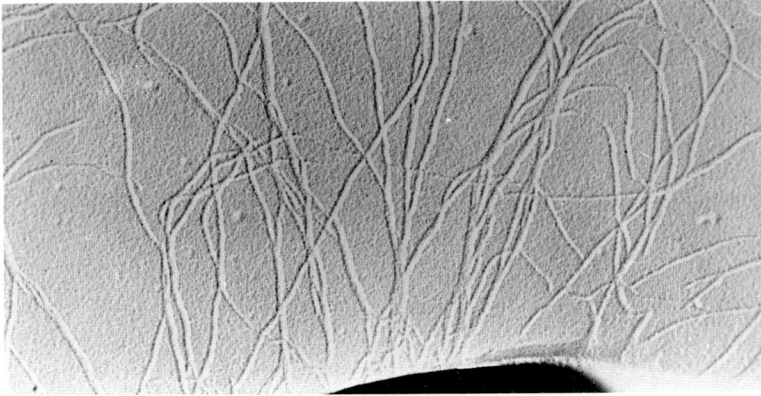

Fig. 5. Fimbriated cell of *Uromyces violacea* shadowed with tungsten oxide (×66800). (From Svircev et al. 1986a)

1974). Subsequently, fibrils have been found on the hyphae of other smuts and plant pathogens as well as ascomycete and basidiomycete yeasts (Gardiner et al. 1981, 1982; Gardiner and Day 1985; Day et al. 1986).

To determine a possible role for fimbriae in the fungus-plant interaction, Day et al. (1986) examined two disease situations using immunocytochemical techniques. Infected material was treated with protein A-gold labelled antigens produced against the protein fimbriae of *U. violacea* (Svircev et al. 1986a). Fimbriae were observed in both disease systems although the extent of the labelling and therefore of the presumed degree of fimbriation was different. It was greater for the fungus *Peronospora hyoscyami* f. sp. *tabacina* than for *U. heufleuri* and it was suggested by Day et al. (1986) that this might reflect the different nature of the disease caused by the two organisms. The former kills its host and destroys the plant tissue very rapidly, whereas the *Ustilago* is a systemic infection and does not kill its host. Further support for this correlation can be found with *Botrytis cinerea* which is heavily fimbriated and also kills its host (Svircev et al. 1986b).

In other plant pathogens it has been speculated that wall polysaccharides or glycoproteins have a role as determinants in the fungus-plant interaction. Two possible roles have been considered, as factors which determine compatibility or incompatibility between the fungus and the plant and as elicitors of phytoalexin synthesis by infected plants. Because spores are generally the infective agent the presence of such molecules in their walls might be expected. Some supportive evidence can be found in studies made on germinating uredospores of *Puccinia graminis* f. sp. *tritici*. Concanavalin A binding glycoproteins have been isolated from walls of the germlings (Kim et al. 1982) and the same class of molecule can be recovered from wheat leaves infected with the same rust (Holden and Rohringer 1985). Whether these molecules are the same is not known. A later study (Kim and Howes 1987) has shown that germling walls contain polypeptides which are very variable in their distribution with respect to several different races of the rust. This variation might be a factor in the interactive role of these wall components.

Another level of interaction which has a structural expression and so can be observed by microscopy involves the formation by the pathogen of specialised vegetative structures from which more invasive hyphae develop. One such structure is the apprressorium which arises as a swelling at the tip of a germ tube. What is significant is the precise positioning of the appressorium (the so-called thigmotropic response) over a suitable infection point e.g. a stomatal pore, through which a hypha can penetrate the leaf or stem tissue (Staples and Harvey 1987). The signal which initiates appressorium differentiation would appear to be a change in the surface topography over which the germ tube grows. The reaction can be mimicked in artificial systems by scratches or ridges on a surface (Tucker et al. 1987; Hoch et al. 1987). On a leaf the guard cells and pores of the stomata would undoubtedly provide such a signal. Proteins or glycoproteins in the wall are again implicated, as treatment with proteases blocks the response.

9 The Protective Role of the Cell Wall

The cell wall has an obvious role in protecting the protoplast simply by virtue of it being the outermost part of the cell. A fungus may be subject to stress from its physical and chemical environment and from other microorganisms. With regard to UV irradiation, γ irradiation and X-rays it is well known that cells and spores which have melanized cell walls are more resistant than hyaline cells. Evidence is also available to indicate that the degree of protection is a function of melanin concentration in the cell wall (Durell 1964; Lukiewicz and Abelwicz 1974; Murchink et al. 1968, 1972; Vasilevskaya et al. 1970; Zhadnova et al. 1973). Melanins apparently provide this protection by absorbing the radiation and dissipating the energy by undergoing irreversible increases in free radicals (Zhadnova et al. 1973).

Melanins in the cell wall are also important in conferring resistance to attack and lysis by enzymes and irradiation, the degree of resistance to attack by chitinases and β-glucanases is inversely related to the melanin content of the wall (Bloomfield and Alexander 1967; Bull 1970; Hurst and Wagner 1969; Kuo and Alexander 1967; Luther and Lipke 1980). Conversely a mutant of A. *nidulans* which lacks melanin is highly sensitive to lytic enzymes (Kuo and Alexander 1967). Melanin inhibition of lytic enzymes is non-competitive and results in the formation of an irreversible enzyme-melanin complex (Bull 1970). The α-glucan nigeran, by virtue of its lamellar-chain and folding characteristics, is also very resistant to enzymatic lysis and so also has a protective function (Marchessault et al. 1980).

The cell wall also has a protective property through its binding properties for heavy metals. Growth in the presence of heavy metals induces both chemical and structural changes in the cell wall as discussed previously. The binding of heavy metals is associated with these changes, for example in the case of C. *blakesleeana* it is thought that Cu binding is associated with the increased protein content of the wall (Venkateswerlu and Stotzky 1986). Other examples of Cu binding to cell walls have been reported in N. *crassa* (Subramanyan et al. 1983) and the binding of FeII by *Ceonoccum gramiforme* has also been described (Rodrigues et al. 1984).

10 Cell Wall Enzymes, Wall Porosity and Enzyme Secretion

Although there have been suggestions that many enzymes are associated with the cell wall, this claim is still controversial. A typical experiment has involved the analysis of enzyme distribution during protoplast release. Thus, many enzymes that were assumed to be associated with the wall are released during protoplast production, implying a periplasmic rather than a mural location. Confirmation of a truly wall-bound enzyme must involve a clear demonstration of activity in pure wall preparations and possibly loss during treatment of the wall with suitable, preferably pure, lytic enzymes.

Most reports of "periplasmic" enzymes have concerned S. cerevisiae and relate to the enzymes β-fructofuranosidase (Nurminen et al. 1970), asparaginase (Dunlop et al. 1978) and α-galactosidase (Lazo et al. 1977). A similar distribution for various carbohydrases has been found in Aspergillus species (M. Vainstein, L. Boddy and J. F. Peberdy, unpublished results).

The clearest examples of wall-bound enzymes are the autolysins, β-glucanases in both yeast and filamentous fungi and chitinase in filamentous fungi. S. cerevisiae produces 6 or 7 $(1-3)$-β-glucanases of which 4 have endo-activity and the remainder are exo-glucanases. Several enzymes have been detected in each of three locations namely a cell-free extract, the cell wall and the extracellular culture filtrate. Only one, endoglucanase IV, has been found in the cell wall (Hein and Fleet 1983a). This is a very minor component of the total glucanase pool occurring in log phase cells where it accounts for only 0.1% of the total activity. Presumably a larger amount of enzyme in the wall would be greatly destructive. The production of glucanase IV at this time corresponds with peak production of glucanases II, IIIA, and IIIB, which occur in both cell extracts and wall preparations, and it has been suggested that this is associated with budding (Hein and Fleet 1983b).

The role of yeast wall bound $(1-3)$-β-glucanases in the hydrolysis of wall bound glucans in vivo is based on observations of autolysis of walls suspended in buffer with the concomitant release of $(1-3)$-β-linked oligosaccharides (Hein and Fleet 1983a). However, the walls are never totally solubilized and the normal physical form remains. This suggests that the $(1-3)$-β-glucosidic linkage may be protected by its occurrence in complex forms and because of the association of glucans with the mannoproteins. The wall-bound enzyme may therefore have a function in weakening or plasticising the wall during processes such as budding and conjugation.

Wall-bound lytic enzymes have been described in several filamentous fungi including N. crassa (Mahadevan and Mahadkar 1970), A. nidulans (Polacheck and Rosenberger 1978) and S. monoica (Fevre 1976). In A. nidulans Polacheck and Rosenberger (1978) found that several enzymes became detached from isolated walls by digesting part of the wall fabric. The enzymes released were active on a range of substrates including chitin, $(1-3)$-β-glucan, glucosides, N-acetylglucosamine and proteins. As only cetyltrimethyl ammonium bromide was effective in releasing the lytic enzymes from isolated walls it was concluded that the association of wall and enzymes is mediated by a hydrophobic component, most probably a phospholipid (Rosenberger 1979).

Bartnicki-Garcia (1973) proposed a model employing a role of wall-bound lytic enzymes in the process of cell wall growth at the hyphal apex. It was postulated that a controlled lysis was required to allow for the insertion of new polymers. A model proposed by Rosenberger (1979) supports this view suggesting that the wall-bound lysins, which are found in older regions of hyphae, are enzymes enclosed in vesicles which did not reach the sites of wall synthesis and become associated with the wall in sub-apical regions. This model of balanced lysis and synthesis at the hyphal apex has been challenged by Wessels (1984; see also Chapter 6). He has suggested that wall-bound enzymes have a role in activating new growth points during branch formation, a view supported by studies on *N. crassa* (Mahadevan and Mahadkar 1970) and *S. monoica* (Fevre 1976).

Immunocytochemical methods were used by Sprey (1986) to demonstrate the localization of β-glucosidase in the outermost β-glucan layer of the *Trichoderma reesei* cell wall and at the plasma membrane. The enzyme was released from the wall by $(1-3)$-β-glucanase treatment, however, more detailed studies are necessary to confirm a truly mural location. The secretion of proteins into the external environment by fungi indicates that the wall, or regions of it must be porous to what in some cases are relatively large molecules. The implied involvement of vesicles and their apical distribution in filamentous fungi suggests that secretion may be localized to the nascent wall at the apex (Forster and Mendgen 1987). Conversely the porosity of the wall must be in some way controlled to restrict the entry of many substances some of which could be detrimental to the cell. An obvious possibility is that the amorphous material may act as a "filling" and control the size of channels through which penetrating molecules may enter. Some support for this idea may be seen in the experiments of Zlotnik et al. (1984) in which yeast cell walls treated with a β-glucanase, purified from the zymolyase complex, released mannoproteins. The same result was obtained with yeast cells but only after a pre-treatment with purified protease also derived from zymolyase. The protease hydrolysed the protein component and solubilised the mannoprotein. A consequence of this was an increase in porosity towards exogenous proteins. Thus horse radish peroxidase was found to penetrate the walls of cells treated in this way causing damage to the plasma membrane and cell lysis.

A more practical question concerning wall porosity concerns the uptake of DNA into intact yeast cells. Pre-treatment of cells with alkali metal cations (Ito et al. 1983a, 1984) or thiol compounds (Ito et al. 1983b, 1984) induces competency for DNA uptake and so makes the cells transformable. Brzobohaty and Kovac (1986) have found that pre-treatment of yeast cells with proteolytic enzymes including trypsin, proteinase K and pronase P also induces competency with a 3-fold increase in transformation frequency compared LiCl treatment. In view of the results of Zlotnik et al. (1984) it is possible that the proteases used also increased wall porosity by weakening the mannoprotein. In any event it is likely that the inducement of competency by the two treatments is the result of different mechanisms.

11 Concluding Statement

In general the chemistry of the cell wall correlates with the current taxonomy of fungi. The biochemistry of wall polymer synthesis, i.e. the functioning of the polysaccharide synthetases is well developed, however, the possibility of isozymes of these synthetases has to be considered in the light of recent work. Debate still ensues over the mechanism involved, particularly that concerning chitin synthesis. Genetical techniques still need further exploitation, some interesting mutants have been described but the eventual application of molecular biological techniques will undoubtedly bring significant advances in our knowledge of cell wall synthesis.

The protective role of certain wall components has been clearly defined and there are other indications that certain amorphous components may control wall porosity. Polymers that may play a role in interactions between fungal cells and between fungal cells and non-fungal cells have been identified. In the case of fungal-plant interactions such molecules may provide the key to an understanding of pathogen-host specificity.

Advances in biological understanding in the past decade have arisen from the development of a range of new biochemical, immunological and microscopical methods. All these are being applied to the study of the fungal cell wall and are the reason for the various advances already made. More exciting information will follow aiding our understanding of this important component of the fungal cell.

References

Ainsworth M, Rayner ADM (1986) Responses of living hyphae associated with self and non-self fusions in the basidiomycete *Phanerochaete velutina*. J Gen Microbiol 132:191–201

Aronson JM (1981) Cell wall chemistry, ultrastructure and metabolism. In: Cole GT, Kendrick B (eds) Biology of conidial fungi, vol. 2. Academic Press, London New York, p 459

Aylemore RC, Todd NK (1986a) Cytology of self fusions in hyphae of *Phanerochaete velutina*. J Gen Microbiol 132:571–579

Aylemore RC, Todd NK (1986b) Cytology of non-self fusions in hyphae of *Phanerochaete velutina*. J Gen Microbiol 132:581–591

Bainbridge BW, Valentine BP, Markham PA (1979) The use of temperature-sensitive mutants to study wall growth. In: Burnett JH, Trinci APJ (eds) Br Mycol Soc Symp, vol 2. Cambridge Univ Press, Cambridge, p 71

Bartnicki-Garcia S (1973) Fundamental aspects of hyphal morphogenesis. In: Ashworth JO, Smith JE (eds) Microbial differentiation. Symp Soc Gen Microbiol, vol 23. Cambridge Univ Press, Cambridge, p 245

Bartnicki-Garcia S, Bracker CE (1984) Unique properties of chitosomes. In: Nombela C (ed) Microbial cell wall synthesis and autolysis. Elsevier, Amsterdam, p 101

Bartnicki-Garcia S, Bracker CE, Lipmann E, Ruiz-Herrera J (1984) Chitosomes from the wall-less "slime" mutant of *Neurospora crassa*. Arch Microbiol 139:105–112

Beadle GW, Tatum EL (1945) Neurospora II. Methods of producing and detecting mutations concerned with nutritional requirements. Am J Bot 32:678–686

Becker JM, Covert NL, Shenbagamurthin P, Steinfeld AS, Naider F (1983) Polyoxin D inhibits growth of zoopathogenic fungi. Antimicrob Agents Chemother 23:926–929

Bell AA, Wheeler MH (1986) Biosynthesis and functions of fungal melanins. Annu Rev Phytopathol 24:411–451

Betz R, Duntze W, Manney JR (1978) Mating factor mediated sexual agglutination in *Saccharomyces cerevisiae*. FEMS Microbiol Lett 4:107–110

Bloomfield BJ, Alexander M (1967) Melanins and resistance of fungi to lysis. J Bacteriol 93:1276–1280

Bobbit TF, Nordin JH (1982) Production and composition of an exocellular nigeran-protein complex isolated from cultures of *Aspergillus awamori*. J Biochem 150:365–376

Bodey GP, Fainstein V (1985) Candidasis. Raven Press, New York

Boller T, Gehri A, Mauch F, Vogeli U (1983) Chitinase in bean leaves: induction by ethylene, purification, properties and possible function. Planta 157:22–31

Borgers M (1980) Mechanism of action of antifungal drugs, with special reference to the imidazole derivatives. Rev Infect Dis 2:520–534

Borgers M, Bossche H van den, Brabander M de (1983) The mechanism of action of the new antimycotic ketoconazole. Am J Med 74:2–8

Bracker CE, Ruiz-Herrera J, Bartnicki-Garcia S (1976) Structure and transformation of chitin synthase particles (chitosomes) during microfibril synthesis in vitro. Proc Natl Acad Sci USA 73:4570–4574

Braun PC, Calderone RTA (1978) Chitin synthesis in *Candida albicans*: comparison of yeast and hyphal forms. J Bacteriol 133:1472–1477

Brzobohaty B, Kovac LK (1986) Factors enhancing genetic transformation of intact yeast cells modify cell wall porosity. J Gen Microbiol 132:3089–3093

Bulana CE, Slater M, Cabib E, A-Young J, Burlati A, Adair WL, Robbins PW (1986) The *S. cerevisiae* structural gene of chitin synthase is not required for chitin synthesis in vivo. Cell 46:213–225

Bull AT (1970) Inhibition of polysaccharases by melanin: enzyme inhibition in relation to Mycolysis. Arch Biochem Biophys 137:345–356

Burnett JH (1979) Aspects of the structure and growth of hyphal walls. In: Burnett JH, Trinci APJ (eds) Fungal walls and hyphal growth. Br Mycol Soc Symp, vol 2. Cambridge Univ Press, Cambridge, p 1

Cabib E (1987) The synthesis and degradation of chitin. Adv Enzymol 59:59–101

Cabib E, Ulane RE, Bowers B (1973) Yeast chitin synthetase separation of the zymogen from its activating factor and recovery of the latter in the vacuole fraction. J Biol Chem 248:1451–1458

Cabib E, Roberts R, Bowers B (1982) Synthesis of the yeast cell wall and its regulation. Annu Rev Biochem 52:763–793

Cabib E, Bowers B, Roberts RL (1983) Vectorial synthesis of a polysaccharide by isolated plasma membranes. Proc Natl Acad Sci USA 80:3318–3321

Cassone A (1984) Cell wall of pathogenic yeasts and implications for antimycotic chemotherapy. Molecular aspects of chemotherapy. Post Symp IUPAC. Abstr 14th Int Symp Chem Nat Prod, Gdansk, p 34

Chiew YY, Shepherd MG, Sullivan PA (1980) Regulation of chitin synthesis during genera tube formation in *Candida albicans*. Arch Microbiol 125:97–104

Dales RP, Croft JH (1977) Protoplast fusion and the isolation of heterokaryons and diploids from vegetatively incompatible strains of *Aspergillus nidulans*. FEMS Microbiol Lett 1:201–204

Davies AR, Marriott MS (1981) Inhibitory effects of imidazole antifungals on the yeast-mycelial transformation in *Candida albicans*. Mykosen 25:481–486

Day AW, Gardiner RB, Smith R, Svircev AM, KcKeen WE (1986) Detection of fungal fimbriae by protein A-gold immunocytochemical labelling in host plants infected with *Ustilago heufleuri* or *Peronospora hyoscyami* f. sp. *tabacina*. Can J Microbiol 32:577–584

Douglas CM, Synan TR, Bobbitt TF, Nordin JH (1984) Nigeran synthesis by regenerating protoplasts of *Aspergillus awamori* correlates with formation of hyphae. Exp Mycol 8:146–160

Dunlop PC, Meyer GM, Ban D, Ron RJ (1978) Characterization of two forms of asparaginase in *Saccharomyces cerevisiae*. J Biol Chem 253:1297–1304

Duran A, Cabib E (1978) Solubilization and partial purification of yeast chitin synthetase. Confirmation of the zymogenic nature. J Biol Chem 253:4419–4425

Duran A, Bowers B, Cabib E (1975) Chitin synthetase zymogen is attached to the yeast plasma membrane. Proc Natl Acad Sci USA 72:3952–3955

Durrell LW (1964) The composition and structure of walls of dark fungus spores. Mycopathol Mycol Appl 23:339–345

Elorza MV, Rico H, Sentendreu R (1983) Calcofluor white alters the assembly of chitin fibrils in *Saccharomyces cerevisiae* and *Candida albicans* cells. J Gen Microbiol 131:2209–2216

Elorza V, Murgui A, Sentendreu R (1985) Dimorphism on *Candida albicans*: Contribution of mannoproteins to the architecture of yeast and mycelial cell walls. J Gen Microbiol 131:2209–2216

Elorza MV, Murgui A, Hortensia R, Miragall F, Sentendreu R (1987) Formation of a new cell wall by protoplasts of *Candida albicans*: Effect of papulacandin B, tunicamycin and nikkomycin. J Gen Microbiol 133:2315–3225

Emerson (1963) Slime — a plasmodioid variant of *Neurospora crassa*. Genetica 34:162–182

Endo A, Kakiki K, Misato T (1970) Mechanism of action of the antifungal agent polyoxin D. J Bacteriol 104:189–196

Farkas V (1979) Biosynthesis of cell walls of fungi. Microbiol Rev 43:117–144

Farkas V (1985) The fungal cell wall. In: Peberdy JF, Ferenczy L (eds) Fungal protoplasts: applications in genetics and biochemistry. Dekker, New York Basel, p 3

Fehrenbacher G, Perry K, Thorner J (1978) Cell-cell recognition in *Saccharomyces cerevisiae*: regulation of γ mating-specific adhesion. J Bacteriol 134:893–901

Feima J (1983) Some aspects of nitrogen metabolism in *Aspergillus giganteus* mut. *alba*. I Chitin content in the cell walls. Acta Physiol Plant 5:123–128

Fevre M (1976) Recherches sur le determinisme de la morphologenese hyphae. Thesis, Univ Claude Bernard, Lyon, France

Fevre M (1984) Action of nucleotides on membrane bound and solubilized β-glucan synthases from *Saprolegnia monoica*. In: Nombela C (ed) Microbial cell wall synthesis and autolysis. Elsevier, Amsterdam, p 131

Fevre M, Rougier M (1981) β-1–3- and β-1–4-glucan synthesis by membrane fractions from the fungus *Saprolegnia*. Planta 151:232–241

Forster H, Mendgen K (1987) Immunocytochemical localization of pectinases in hyphae of *Phytophthora infestans*. Can J Bot 65:2607–2613

Gardiner RB, Day AW (1985) Fungal fimbriae-IV Composition and properties of fimbriae from *Ustilago violacea*. Exp Mycol 9:334–350

Gardiner RB, Canton M, Day AW (1981) Fimbrial variation in smuts and heterobasidiomycete fungi. Bot Gaz 142:147–150

Gardiner RB, Podogorski C, Day AW (1982) Serological studies on the fimbriae of yeasts and yeastlike species. Bot Gaz 143:534–541

Garnjobst L, Tatum EL (1967) A survey of new morphological mutants in *Neurospora crassa*. Genetics 57:579–604

Gibson R (1973) Studies on protoplasts of *Aspergillus* and their regeneration. Thesis Univ Nottingham, Nottingham

Girard V, Fevre M (1984) β-1–4- and β-1–3-glucan syntheses are associated with the plasma membrane of the fungus *Saprolegnia*. Planta 160:400–406

Glaser L, Brown DH (1957) The synthesis of chitin in cell free extracts of *Neurospora crassa*. J Biol Chem 228:729–742

Gold MH, Mitzel DL, Segel IH (1973) Regulation of nigeran accumulation by *Aspergillus aculeatus*. J Bacteriol 113:856–862

Gomez-Miranda B, Guerrero C, Leal JA (1984) Effect of culture age on cell wall polysaccharides of *Penicillium allahabadense*. Exp Mycol 8:298–303

Gooday GW (1983) The hyphal tip. In: Smith JE (ed) Fungal differentiation. Dekker, New York Basel, p 315

Gooday GW, Rousset-Hall A de (1975) Properties of chitin synthetase from *Coprinus cinereus*. J Gen Microbiol 89:137–145

Gooday GW, Trinci APJ (1980) Wall structure and biosynthesis in fungi. In: Gooday GW, Lloyd D, Trinci APJ (eds) The eukaryotic microbial cell. Symp Soc Gen Microbiol, vol 30. Cambridge Univ Press, Cambridge, p 207

Hanseler E, Nyhlen LE, Rast DM (1983) Dissociation and reconstitution of chitosomes. Biochim Biophys Acta 745:121–133

Harvey IC (1975) Development and germination of chlamydospores in *Pleiochaete setosa*. Trans Br Mycol Soc 64:489–495

Hawker LE, Beckett A (1971) Fine structure and development of the zygospore of *Rhizopus sexualis* (Smith) Callen. Philos Trans R Soc London Ser B 263:71–100

Hein NH, Fleet GH (1983a) Separation and characterisation of six (1–3)-β-glucanases from *Saccharomyces cerevisiae*. J Bacteriol 156:1204–1213

Hein NH, Fleet GH (1983b) Variation of (1–3)-β-glucanases in *Saccharomyces cerevisiae* during vegetative growth, conjugation and sporulation. J Bacteriol 156:1214–1220

Herrero E, Sanz P, Sentendreu R (1987) Cell wall proteins liberated from zymolyase from several ascomycetous and imperfect yeasts. J Gen Microbiol 133:2895–2903

Hilenski LL, Naider F, Becker JM (1986) Polyoxin D inhibits colloidal gold-wheat germ agglutinin labelling of chitin in dimorphic forms of *Candida albicans*. J Gen Microbiol 132:1441–1451

Hoch HC, Howard RJ (1980) Ultrastructure of freeze-substituted hyphae of the basidiomycete *Laetisaria arvalis*. Protoplasma 103:281–297

Hoch HC, Staples RC, Whitehead B, Corneau J, Wolf E (1987) Signaling for growth orientation and cell differentiation by surface topography in *Uromyces*. Science 235:1659–1662

Holden DW, Rohringer R (1985) Proteins in intercellular washing fluids from noninoculated and rust-infected leaves of wheat and barley. Plant Physiol 78:715–723

Hori M, Kakaki K, Suzuki S, Misato T (1971) Studies on the mode of action of polyoxins. Part III. Relation of polyoxin structure to chitin synthetase. Agric Biol Chem 35:1280–1291

Howard RJ (1981) Ultrastructural analysis of hyphal tip cell growth in fungi: Spitzenkörper cytoskeleton and endomembranes after freeze-substitution. J Cell Sci 48:89–103

Hurst HM, Wagner GH (1969) Decomposition of ^{14}C labeled cell wall and cytoplasmic fractions from hyaline and melanic fungi. Soil Sci Am Proc 33:707–711

Isono K, Nagatsu J, Kobinata K, Sasaki K, Susuki S (1967) Studies on polyoxins, antifungal antibiotics. Part IV. Isolation and characterization of polyoxins C, D, E, F, G, H and I. Agric Biol Chem 31:190–199

Ito H, Fukuda Y, Murata K, Kimura A (1983a) Transformation of intact yeast cells treated with alkali cations. J Bacteriol 153:163–168

Ito H, Murata K, Kimura A (1983b) Transformation of yeast cells treated with 2 mercaptoethanol. Agric Biol Chem 47:1691–1692

Ito H, Murata K, Kimura A (1984) Transformation of intact yeast cells treated with alkali cations or thiol compounds. Agric Biol Chem 48:341–347

Kang MS, Elango N, Mattia E, Au-Young J, Robbins PW, Cabib E (1984) Isolation of chitin synthetase from *Saccharomyces cerevisiae*. Purification of the enzyme by entrapment in the reaction product. J Biol Chem 259:14966–14972

Kang MS, Au-Young J, Cabib E (1985) Modification of yeast plasma membrane density by concanavalin A attachment. J Biol Chem 260:12680–12684

Katz D, Rosenberger RF (1971) Lysis of an *Aspergillus nidulans* mutant blocked in chitin synthesis and its relation to wall assembly. Arch Mikrobiol 80:284–292

Kim WK, Howes NK (1987) Localization of glycopeptides and racevariable polypeptides in urediosporelings and urediosporeling walls of *Puccinia graminis tritici*; affinity to concanavalin A, soybean agglutinin, and *Lotus* lectin. Can J Bot 65:1785–1791

Kim WK, Rohringer R, Chong J (1982) Sugar and amino acid composition of macromolecular constituents released from walls of urediosporlings of *Puccinia graminis tritici*. Can J Plant Pathol 4:317–327

Kopecka M, Kreger DR (1986) Assembly of microfibrils in vivo and in vitro from $(1-3)$-β-D-glucan synthesised by protoplasts of *Saccharomyces cerevisiae*. Arch Microbiol 143:387–395

Kuo M-J, Alexander M (1967) Inhibition of the lysis of fungi by melanins. J Bacteriol 94:624–629

Lazo PS, Ochoa AG, Gascon S (1977) α-galactosidase from *Saccharomyces cerevisiae*. Cellular localization and purification of the external enzyme. Eur J Biochem 77:375–382

Leal-Morales CA, Ruiz-Herrera J (1985) Alterations in the biosynthesis of chitin and glucan in the slime mutant of *Neurospora crassa*. Exp Mycol 9:28–38

Lipke PN, Ballou CE (1980) Altered immunochemical reactivity of *Saccharomyces cerevisiae* a cells after α factor induced morphogenesis. J Bacteriol 141:1170–1177

Lipke PN, Taylor A, Ballou CE (1976) Morphogenetic effects of α-factor on *Saccharomyces cerevisiae* a cells. J Bacteriol 127:610–618

Lukiewicz S, Abelwicz Z (1974) EPR studies on the radioprotective role of melanins. Radiat Res 59:220–221

Luther JP, Lipke H (1980) Degradation of melanin by *Aspergillus fumigatus*. Appl Environ Microbiol 40:145–155

Mahadevan PR, Mahadkar UR (1970) Role of enzymes in growth and morphology of *Neurospora crassa*: cell bound enzymes and their possible role in branching. J Bacteriol 104:318–332

Marchessault RH, Revol JF, Bobbitt JF, Nordin JH (1980) Enzymic depolarization of lamellar single crystals of nigeran. Biopolymers 14:1069–1080

Milewski S, Chmara H, Borowski E (1983) Growth inhibitory effect of antibiotic tetaine on yeast and mycelial forms of *Candida albicans* 135:130–136

Milewski S, Chmara H, Borowski E (1986) Antibiotic tetaine a selective inhibitor of chitin and mannoprotein biosynthesis in *Candida albicans*. Arch Microbiol 145:234–240

Mishra NC (1977) Genetics and biochemistry of morphogenesis in *Neurospora*. Adv Genet
19:341−405

Montgomery GWG, Adams DJ, Gooday GW (1984) Studies on the purification of chitin synthase
from *Coprinus cinereus*. J Gen Microbiol 130:291−297

Murchink TG, Kashkine GB, Abaturov YuD (1968) Resistance of the dark-coloured fungi *Stemphylium botryosum* Waller and *Cladosporium cladosporoides* (Fries) de Fries to γ radiation.
Microbiol USSR 37:724−727

Murchink TG, Kashkina GB, Abaturov YuD (1972) The resistance of fungi with various pigments to
γ radiation. Microbiol USSR 41:67−69

Musilkova M, Ujcova E, Seichert L, Fencl Z (1982) Effect of changed cultivation conditions on the
morphology of *Aspergillus niger* and on acid biosynthesis in laboratory conditions. Fol Microbiol
27:328−332

Necas O, Svoboda A (1985) Cell wall regeneration and protoplast reversion. In: Peberdy JF, Ferenczy
L (eds) Fungal protoplasts: applications in biochemistry and genetics. Dekker, New York Basel, p
115

Nurminen T, Oura E, Sunalainen H (1970) The enzymatic composition of the isolated cell wall and
plasma membrane of baker's yeast. Biochem J 116:61−69

Odds FC, Cockayne A, Hayward J, Abbott AB (1985) Effects of imidazole- and triazole-derivative
antifungal compounds on the growth and morphological development of *Candida albicans* hyphae.
J Gen Microbiol 131:2581−2589

Ohta N, Kakiki K, Musato T (1970) Studies on the mode of action of polyoxin D. Part II. Effect of
polyoxin D on the synthesis of fungal cell wall chitin. Agric Biol Chem 34:1224−1234

Pastor FIJ, Herrero E, Sentendreu R (1984) Structure of the *Saccharomyces cerevisiae* cell wall: mannoproteins released by Zymolyase and their contribution to wall architecture. Biochim Biophys Acta
802:292−300

Peberdy JF (1979) Wall biogenesis by protoplasts. In: Burnett JH, Trinci APJ (eds) Fungal walls and
hyphal growth. Br Mycol Soc Symp, vol 2. Cambridge Univ Press, Cambridge, p 49

Pegg GF, Young DH (1982) Purification and characterization of chitinase enzymes from healthy and
Verticillium albo-atrum infected tomato plants, and from *V. albo-atrum*. Physiol Plant Pathol
21:389−409

Polacheck Y, Rosenberger RF (1975) Autolytic enzymes in hyphae of *Aspergillus nidulans*: their action on old and newly-formed walls. J Bacteriol 121:332−337

Polacheck Y, Rosenberger RF (1978) The distribution of autolysins in hyphae of *Aspergillus nidulans*:
existence of a lipid mediated attachment to hyphal walls. J Bacteriol 135:741−754

Poon NH, Day AW (1974) Fimbriae in the fungus *Ustilago violacea*. Nature (London) 250:648−649

Poon NH, Day AW (1975) Fungal fimbriae (I). Structure origin and synthesis. Can J Microbiol
21:537−546.3

Rijkenberg FHJ, Truter SJ (1975) Cell fusion in the alcuim of *Puccinia sorghi*. Protoplasma
83:233−246

Roberts WK, Selitrennikoff CP (1986a) Isolation and partial characterization of two antifungal proteins from barley. Biochim Biophys Acta 880:161−170

Roberts WK, Selitrennikoff CP (1986b) Isolation and partial characterization of two antifungals from
barley. J Cell Biol Suppl 10C:26

Roberts WK, Selitrennikoff P (1988) Plant and bacterial kitinases differ in antifungal activity. J Gen
Microbiol 134:169−176

Rodrigues RK, Kleman DS, Barton LL (1984) Iron metabolism by an ectomycorrhizal fungus,
Cenococcum graniforme. J Plant Nutr 7:459−468

Rosenberger RF (1979) Endogenous lytic enzymes and wall metabolism. In: Burnett JH, Trinci APJ
(eds) Fungal walls and hyphal growth. Cambridge Univ Press, Cambridge, p 265

Ruiz-Herrera J, Sing VO, Wonde WJ van der, Bartnicki-Garcia S (1975) Microfibril assembly by granules of chitin synthetase. Proc Natl Acad Sci USA 72:2706−2710

Ruiz-Herrera J, Lopez-Romero E, Bartnicki Garcia S (1977) Properties of chitin synthetase in isolated
chitosomes from yeast cells of *Mucor rouxii*. J Biol Chem 252:3338−3343

Ruiz-Herrera J, Bartnicki-Garcia S, Bracker CE (1980) Dissociation of chitosomes by digitonin into
165 subunits with chitin synthetase activity. Biochim Biophys Acta 745:121−133

Sanz P, Herrero E, Sentendreu R (1985) Autolytic release of mannoproteins from cell walls of *Saccharomyces cerevisiae*. J Gen Microbiol 131:2925−2932

Schekman R, Brawley VL (1979) Localised deposition of chitin on the yeast cell surface in response to mating pheromone. Proc Natl Acad Sci USA 76:645 – 649

Scott AW (1976) Biochemical genetics of morphogenesis in *Neurospora*. Annu Rev Microbiol 30:85 – 104

Scott AW, Mishra NC, Tatum EL (1973) Biochemical genetics of morphogenesis in *Neurospora*. Brookhaven Symp Biol 25:1 – 18

Seitsma JH, Wessels JGH (1979) Wall structure and growth in *Schizophyllum commune*. In: Burnett JH, Trinci APJ (eds) Fungal walls and hyphal growth. Br Mycol Soc Symp, vol 2. Cambridge Univ Press, Cambridge, p 27

Seitsma JH, Sonnenberg AMS, Wessels JGH (1985) Localization by autoradiography of synthesis of $(1-3)$-β and $(1-6)$-β linkages in a wall glucan during hyphal growth of *Schizophyllum commune*. J Gen Microbiol 131:1331 – 1337

Selitrennikoff CP (1979) Chitin synthase activity from the slime variant of *Neurospora crassa*. Biochim Biophys Acta 571:224 – 232

Sentendreu R, Martinez-Ramon A, Ruiz-Herrera J (1984a) Localization of chitin synthase in *Mucor rouxii* by an autoradiographic method. J Gen Microbiol 130:1193 – 1199

Sentendreu R, Herrero E, Elorza MV (1984b) The assembly of wall polymers in yeast. In: Nombela C (ed) Microbial cell wall synthesis and autolysis. Elsevier, Amsterdam, p 51

Sijmons PC, Nederbragt AJA, Klis FM, Ende H van den (1987) Isolation and composition of the constitutive agglutinins from haploid *Saccharomyces cerevisiae* cells. Arch Microbiol 148:208 – 212

Sonnenberg ASM, Seitsma JH, Wessels JGH (1982) Biosynthesis of an alkali-insoluble cell wall glucan in *Schizophyllum commune* protoplasts. J Gen Microbiol 128:2667 – 2674

Sprague GF, Blair LC, Thorner J (1983) Cell interactions and regulation of cell type in the yeast *Saccharomyces cerevisiae*. Annu Rev Microbiol 37:623 – 660

Sprey B (1986) Localisation of β-glucosidase in *Trichoderma reesei* cell walls with immunoelectron microscopy. FEMS Microbiol Lett 36:287 – 292

Staples RC, Harvey HC (1987) Infection structures — form and function. Exp Mycol 11:163 – 169

Subramanyan C, Venkateswerlu G, Rao SLN (1983) Cell wall composition of *Neurospora crassa* under conditions of copper toxicity. Appl Environ Microbiol 46:585 – 590

Svircev V, Smith R, Gardiner RB, Racki IM, Day AW (1986a) Fungal fimbriae. V. Protein A-gold immunocytochemical labelling of the fimbriae of *Ustilago violacea*. Exp Mycol 10:19 – 27

Svircev AM, Gardiner RB, McKeen WE, Day AW (1986b) *Botrytis cinerea* antigens in cytoplasm of infected *Vicia faba*. Phytopathology 76:1111 – 1112

Szaniszlo PJ, Kang MS, Cabib E (1985) Stimulation of $\beta(1-3)$ glucan synthetase of various fungi by nucleoside triphosphates: Generalised regulatory mechanism for cell wall biosynthesis. J Bacteriol 161:1188 – 1194

Terrance K, Heller P, Wu Y-S, Lipke PN (1987) Identification of glycoprotein components of α-agglutinin, a cell adhesion protein from *Saccharomyces cerevisiae*. J Bacteriol 169:475 – 482

Thompson K (1977) Biochemical genetics of morphogenesis in *Aspergillus nidulans*. Thesis, Univ Nottingham, Nottingham

Tkacz JS, MacKay VL (1979) Sexual conjugation in yeast cell surface changes in response to the action of mating hormones. J Cell Biol 80:326 – 333

Tokunaga M, Kusamichi M, Korke H (1986) Ultrastructure of outermost layer of cell wall in *Candida albicans* observed by rapid-freezing techniques. J Electron Microsc 35:237 – 246

Tucker BE, Hoch HC, Staples RC (1987) The involvement of F-actin in *Uromyces* cell differentiation: The effects of cytochalasin E and phalloidin. Protoplasma 135:88 – 101

Valentin E, Herrero E, Pastor FIJ, Sentendreu R (1984) Solubilization and analysis of mannoprotein molecules from the cell wall of *Saccharomyces cerevisiae*. J Gen Microbiol 130:1419 – 1428

Valentin E, Herrero E, Sentendreu R (1986) Incorporation of mannoproteins into the walls of aculeacin A. Treated yeast cells. Arch Microbiol 146:214 – 220

Valentin E, Herrero E, Rico H, Miragall F, Sentendreu R (1987) Cell wall mannoproteins during the population growth phases in *Saccharomyces cerevisiae*. Arch Microbiol 148:88 – 94

Valentine BP, Bainbridge BW (1975) Mutations affecting the incorporation of mannose into the cell wall of *Aspergillus nidulans*. Proc Soc Gen Microbiol 2:90

Valentine BP, Bainbridge BW (1978) The relevance of a study of a temperature-sensitive ballooning mutant of *Aspergillus nidulans* defective in mannose metabolism to our understanding of mannose as a wall component and carbon/energy source. J Gen Microbiol 109:155 – 168

Valk P van der, Marchant R (1978) Hyphal ultrastructure in the fruit body of the basidiomycetes *Schizophyllum commune* and *Coprinus cinereus*. Protoplasma 95:57−72

Vasilevskaya AI, Zhdanova NM, Pokhodenko VD (1970) Character of survival of some gamma irradiated species of dark coloured Hypomycetes. Mikrobiol Zh 33:438−441

Venkateswerlu G, Stotzky G (1986) Copper and cobalt alter the cell wall composition of *Cunninghamella blakesleeana*. Can J Microbiol 32:654−662

Vermeulen CA, Wessels JGH (1983) Evidence for a phospholipid requirement of chitin synthetase in *Schizophyllum commune*. Curr Microbiol 8:67−81

Vermeulen CA, Wessels JGH (1984) Ultrastructural differences between wall apices of growing and non-growing hyphae of *Schizophyllum commune*. Protoplasma 120:123−131

Vries OMH de, Wessels JGH (1975) Chemical analysis of cell wall regeneration and reversion of protoplasts of cell wall regeneration and reversion of protoplasts of *Schizophyllum commune*. Arch Microbiol 102:209−218

Wessels JGH (1984) Apical hyphal wall extension. Do lytic enzymes play a role? In: Nombela C (ed) Microbial cell wall synthesis and autolysis., FEMS Symp no 27. Elsevier, Amsterdam, p 31

Zhadnova NM, Gavryushina AI, Vasilevskaya AI (1973) Effect of γ- and UV-irradiation on survival of *Cladosporium* sp. and *Oidiodendron cerealis*. Mikrobiol Zh 35:449−452

Zlotnik H, Fernandez MP, Bowers B, Cabib C (1984) *Saccharomyces cerevisiae* mannoproteins from an external cell wall layer that determines wall porosity. J Bacteriol 159:1018−1026

Chapter 3

Chitin Synthesis in Yeast (*Saccharomyces cerevisiae*)

E. Cabib[1], S. J. Silverman[1], A. Sburlati[1], and M. L. Slater[2]

1 Introduction

Chitin is a major structural component of fungal cell walls (Bartnicki-Garcia 1968; Cabib and Shematek 1981). In filamentous fungi, it contributes to the structure of both lateral walls and septa. In the yeast, *Saccharomyces cerevisiae*, it is mainly found in the primary septum, although a small amount appears to be dispersed throughout the cell wall (Molano et al. 1980; Horisberger and Vonlanthen 1977; Mol and Wessels 1987). This localization suggested that the formation of chitin and of the chitinous septum would make a good model for the study of mechanisms of morphogenesis. We approached this problem by investigating the synthesis of chitin at the molecular level in the hope that the mechanisms by which chitin synthetase activity is controlled would also be operative in regulating in vivo the formation of the structure supported by the polysaccharide.

2 Zymogen Nature of Chitin Synthetase

Chitin is a linear polymer consisting of *N*-acetylglucosamine residues joined to each other by $\beta(1 \rightarrow 4)$ linkages. Chitin synthetase catalyzes the transfer of N-acetylglucosamine from UDP-*N*-acetylglucosamine to a growing chain of *N*-acetylaminosugar residues (Fig. 1; Kang et al. 1984). Chitin synthetase was detected in yeast as a membrane-bound enzyme about 20 years ago, and it soon became apparent that most of the enzyme was in a cryptic state from which activity could be elicited by treatment with different proteases (Fig. 1; Cabib and Farkas 1971). These results suggested that chitin synthetase exists in the yeast cell as a zymogen, a hypothesis that was confirmed by the finding that the enzyme did not lose its cryptic nature after solubilization from the membranes (Duran and Cabib 1978). Activation by proteases appears to be a general mechanism for regulation of fungal chitin synthetases (Cabib 1987) although

---

[1] Laboratory of Biochemistry and Metabolism, National Institute of Diabetes and Digestive and Kidney Diseases, Bethesda, Maryland 20892, USA
[2] Division of Research Grants, National Institutes of Health, Bethesda, Maryland 20892, USA

Chitin synthetase zymogen

|
Protease
↓

Active chitin synthetase

|
|
↓
n UDP-N-acetylglucosamine ⟶ n UDP + (β(1→4)
                                 N-acetylglucosamine)$_n$

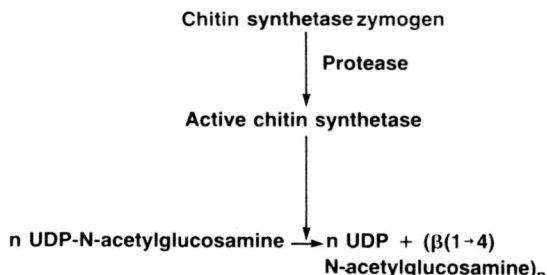

Fig. 1. Activation of chitin synthetase zymogen and stoichiometry of catalyzed reaction

the possibility of physiological activation by some other protein modification has not been discarded. In *S. cerevisiae*, it was thought that the vacuolar enzyme proteinase B might be the physiological activator (Ulane and Cabib 1976), but this hypothesis was disproved by the finding that in mutants lacking proteinase B, cell division and chitin content were normal (Zubenko et al. 1979; Wolf and Ehmann 1979).

3 Intracellular Distribution of Chitin Synthetase

The bulk of chitin synthetase activity in *S. cerevisiae* was found to be attached to the plasma membrane (Duran et al. 1975; Kang et al. 1985), although some enzyme was recovered in a lighter fraction that might correspond to the "chitosomes" isolated by Bartnicki-Garcia and his coworkers (Bartnicki-Garcia et al. 1978). Furthermore, the intact membranes that were isolated after stabilization with concanavalin A were able to carry out a vectorial synthesis of chitin: the polysaccharide was laid down on the external face of the membranes, as it is found in vivo (Cabib et al. 1983).

4 Cloning of a Chitin Synthetase Structural Gene

A mutant of *S. cerevisiae* apparently lacking chitin synthetase activity was isolated by M. Slater in our laboratory. Because the mutant grew and divided normally and had a normal chitin content, the nature of the mutation was not well understood at that time. This mutant was transformed with a high-copy plasmid carrying a DNA library from yeast. Among the transformants, one was isolated that overproduced chitin synthetase, as detected by an autoradiographic assay in situ (Bulawa et al. 1986). The gene carried in the transforming plasmid was isolated, and mapped with restriction endonucleases (Fig. 2; Bulawa et al. 1986).

Subsequent cloning of the gene in *Schizosaccharomyces pombe* resulted in expression of zymogenic chitin synthetase activity in this organism which normally lacks both the polysaccharide and the synthetase. It was concluded that the cloned gene is the structural gene for the enzyme.

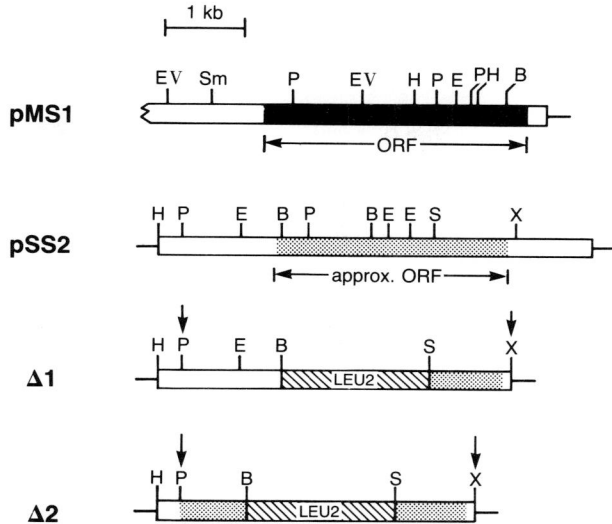

Fig. 2. Restriction maps of different DNA inserts described in the text. *pMSI*, *CHSI* gene in vector YEp 24; *pSS2*, *CHS2* gene in vector YEp 351; *Δ*1, disruption obtained by first deleting the region downstream of the Xba I site and that between the leftmost Bgl II and the Sal I site in pSS2, followed by insertion of the *LEU2* gene. In *Δ*2, both the region downstream of the Xba I site and that between the two Pst I sites were excised, followed by deletion of the sequence between the remaining Bgl II site and the Sal I site and ligation of the *LEU2* gene. In both cases, the DNA fragments used for transformation were excised from the plasmid with Pst I and Xba I, as indicated by the arrows. In all cases, the direction of transcription is from left to right. Abbreviations: ORF, open reading frame; B, Bgl II; E, Eco RI; EV, Eco RV; H, Hind III; P, Pst I; S, Sal I; Sm, Sma I; X, Xba I

The gene sequence predicts a protein of 1131 amino acids with a molecular weight of 130000. By comparison, a determination of the molecular weight of purified enzyme, after activation by trypsin and under nondenaturing conditions, yielded a value of 570000 (Kang et al. 1984). It is clear then, that under those conditions, i.e. in the presence of 0.1% digitonin, the enzyme forms fairly large aggregates. Considering that the protein had been partially trimmed by trypsin, the number of monomers in the aggregate must be larger than it appears at first sight. It is attractive to speculate that such subunits, if associated in the original membrane, could co-operate in the synthesis of a chitin fibril; at present no information is available, however, about the organization of the synthetase in the membrane.

A surprising result was obtained when the gene was disrupted and then introduced into diploids so as to displace one of the two normal chitin synthetase genes. After sporulation, each of the resulting tetrads showed 2 to 2 segregation with respect to chitin synthetase activity but all four haploid strains were viable and had normal chitin levels (Table 1). This result clearly showed that the chitin synthetase activity studied in yeast up to that point was not essential for cell division or for chitin synthesis in vivo and suggested that another chitin synthetase may be present in the cell. On that expectation, the cloned gene was named *CHSI* and the corresponding enzyme, chitin synthetase 1 (Chs1).

Table 1. *CHS1* gene disruption

Spore	Chitin synthetase activity	Chitin % wet weight × 10³
A	−	94
B	−	108
C	+	104
D	+	89

5 A New Chitin Synthetase

The search for a new chitin synthetase soon met with success. Such an activity, at a relatively low level, was found in two different laboratories in strains that harbored a disrupted *CHS1* gene and were therefore devoid of Chs1 activity (Sburlati and Cabib 1986; Orleans 1987). Because the preparations obtained in both laboratories are not identical in their properties, it cannot be concluded, as yet, whether they correspond to one or to two different enzymes. Both, however, clearly differ from Chs1 in some properties. The enzyme isolated in our laboratory, which we name chitin synthetase 2 (Chs2), shows a much higher stimulation by Co^{2+} than by Mg^{2+}, whereas Chs1 requires Mg^{2+} but is not affected by Co^{2+} (Table 2; Sburlati and Cabib 1986). Furthermore, the pH optimum of Chs2 is between 7 and 8 whereas that of Chs1 is between 6 and 6.5 (Fig. 3A). Chs2 is somewhat more resistant to inhibition by polyoxin D and by NaCl than Chs1, a property that was used to demonstrate the presence of Chs2 in normal strains containing Chs1 (Sburlati and Cabib 1986). On account of these differences in behavior, especially those in metal ion stimulation and pH optimum, it is

Table 2. Comparison of properties of Chs1 and Chs2

Strain	Enzyme measured	Chitin synthetase activity[a] (mU/mg protein)			
		Mg^{2+}		Co^{2+}	
		No trypsin	+ trypsin	No trypsin	+ trypsin
Wild type	Chs1 [b]	0.13	0.85	−	0.22
CHS1:: *ura3* [c]	Chs2	0.013	0.093	0.085	0.29
CHS1::*ura3* (pSS1) [d]	Chs2	0.54	7.6	1.0	10

[a] Membrane fractions were obtained from protoplasts and assayed as described (Sburlati and Cabib 1986). The concentration of Mg^{2+} was 4 mM and that of Co^{2+} 2.5 mM. One unit (U) is defined as the amount of enzyme that catalyzes the incorporation of 1 μmol of *N*-acetylglucosamine per min into chitin at 30°C.

[b] For comparison, both enzymes were measured under the same conditions at pH 7.5. At optimum pH (6.5) and in the presence of phosphatidylserine, the activity of Chs1 is two to three times that shown.

[c] Strain containing a disrupted *CHS1* gene.

[d] Strain containing a disrupted *CHS1* gene and transformed with a plasmid carrying the *CHS2* gene.

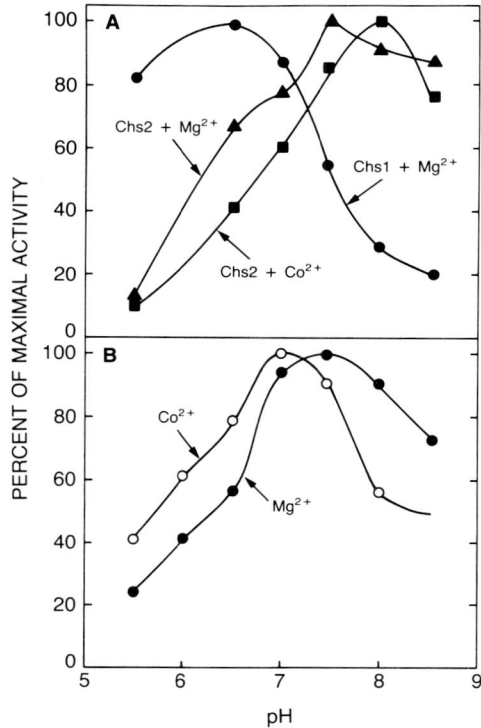

Fig. 3 A, B. Effect of pH on activity of Chs1 and Chs2. **A** Enzymes from wild type cells (Chs1) or cells containing a disrupted *CHS1* gene (Chs2). **B** Enzyme from cells containing a disrupted *CHS1* gene, after transformation with plasmid carrying *CHS2*

relatively easy to distinguish between the two activities. On the other hand, Chs1 and Chs2 share certain properties: both enzymes are attached to the plasma membrane and both are stimulated by proteolytic treatment, although the specificity of proteases is not the same for the two synthetases (Sburlati and Cabib 1986).

6 Cloning of the Chitin Synthetase 2 Structural Gene

To establish whether Chs2 was the enzyme required for septum formation in vivo, it was necessary to clone the corresponding structural gene. This was done by transforming a strain containing a *CHS1* disruption with a high-copy plasmid carrying a yeast DNA genomic library and by screening for transformants overproducing chitin synthetase activity (Silverman et al. 1988). One transformant was isolated that overproduced, by a factor of about 50 compared to wild type, a chitin synthetase activity with similar properties to Chs2 (Table 2). It can be observed that the activation of the cloned enzyme by trypsin was higher than in untransformed cells, i.e. the synthetase showed more markedly its zymogenic character.

The cloned gene was mapped with restriction endonucleases (Fig. 2). The restriction map clearly differed from that of the *CHS1* gene. Furthermore, in direct blots,

Table 3. Expression of Chs2 in *S. pombe*

Transforming plasmid	Chitin synthetase activity[a] (mU/mg protein)			
	Mg^{2+}		Co^{2+}	
	No trypsin	After trypsin	No trypsin	After trypsin
YEp 13[b]	0.007	0.007	0	0
pSS1[c]	0.008	0.04	0.017	0.21
pSS2X[d]	0.009	0.12	0.045	0.6
pSS2XP[e]	0.13	2.6	0.52	4.2

[a] Membranes were isolated from extracts obtained by disruption of cells with glass beads in a Vortex mixer. See Table 2 for definition of units and concentration of cations.
[b] Vector without DNA insert.
[c] DNA insert similar to that of pSS2 (Fig. 2), but the vector was YEp13 rather than YEp351.
[d] Plasmid obtained by deleting the insert of pSS2 downstream from the Xba I site.
[e] Plasmid obtained by deleting the sequence between the two Pst I sites in pSS2X.

the two genes did not hybridize; in genomic blots, each fragment hybridized to a single gene. At this level, therefore, there is no evidence for homology of the two genes, although some similarity may be revealed by the nucleotide sequence.

Partial deletions of the DNA insert were carried out and the resulting plasmids were used to transform yeast cells. By measuring the Chs2 activity of the transformants it was possible to ascertain what part of the sequence was necessary for synthesis of the enzyme. In this way, it was determined that the minimal fragment for at least partial Chs2 activity extended from the rightmost of the two Pst I sites (Fig. 2) to a point between the Sal I and the Xba I sites. By Northern gel analysis (Thomas 1980) of yeast Poly A^+ RNA, a 3.1 kilobase transcript corresponding to the insert was detected. The approximate position of the sequence hybridizing to the mRNA, as determined by ribonuclease protection analysis, is shown by the stippled area in Figure 2 that covers the region required for Chs2 expression and probably corresponds approximately to the open reading frame.

Because the strains in which the gene had been cloned harbored a normal chitin synthetase 2 gene, it was not possible at this point to decide whether the clone insert contained the structural gene for the enzyme or a regulatory gene that enhanced its expression. To distinguish between these possibilities, plasmids containing the original insert or partial deletions thereof were used to transform *S. pombe*, as was done previously with the *CHS1* gene. Again, expression of chitin synthetase activity resulted (Table 3). Surprisingly, the highest activity was obtained with one of the partial deletions, presumably because it was carried in the cells at a high copy number. The properties of the enzymes corresponded to those of Chs2, both with regard to metal stimulation (Table 3) and to pH optimum (not shown). The activity was highly dependent on previous trypsin treatment, thus confirming that the enzyme is in a zymogenic form. From all these data it may be concluded that the cloned gene is the structural gene for Chs2.

7 Requirement of Chitin Synthetase 2 for Cell Division

To determine whether *CHS2* was essential for cell growth, the gene was partially deleted and a LEU2 gene was ligated into the gap so created. Two such disruptions, $\Delta 1$ and $\Delta 2$, were carried out (Fig. 2). The corresponding DNA fragments were excised with appropriate restriction nucleases and used for integrative transformation of diploids carrying a *leu2* mutation in both chromosomal complements. The transformants that grew in the absence of leucine were sporulated and tetrads were analyzed. No tetrad gave rise to more than 2 colonies (Table 4), and all surviving progeny were leucine auxotrophs, i.e. they contained a normal *CHS2* gene. Therefore, the presence of a disrupted *CHS2* gene resulted in inability to form a colony. It did not, however, prevent spore germination. Observation of the plates under the microscope revealed that spores harboring the disrupted gene did germinate and gave rise to some aberrant growth that later stopped (Fig. 4). The resulting cells were much larger than normal and grossly deformed (Figs. 4 and 5). After staining for chitin with Calcofluor White, some fluorescence was observed at constrictions (Fig. 5), but clearly no septa were present. That no cell separation had occurred was confirmed by the fact that the aberrant microcolonies behaved as a unit when picked up and again laid down with a

Table 4. Tetrad analysis of sporulating diploids with a *CHS2* disruption[a]

Strain	Number of colonies per tetrad				
	4	3	2	1	0
Wild type	8	2	0	0	0
$\Delta 1$[b]	0	0	30	8	1
$\Delta 2$[b]	0	0	21	5	3

[a] All diploids in this experiment contained normal *CHS1* genes.
[b] For disruptions see Fig. 2.

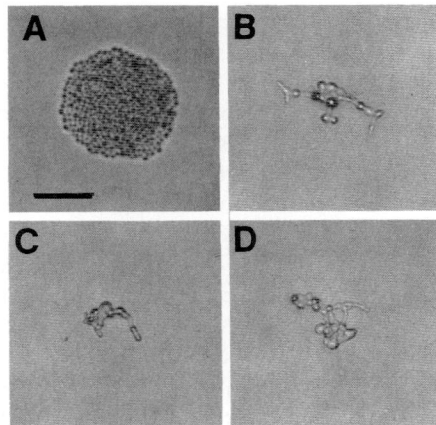

Fig. 4A–D. Aspect of colonies, 16 hours after tetrad dissection. **A** wild type. **B, C,** and **D** spores containing a *CHS2* disruption. Upon continued incubation, wild type colonies continued to grow, whereas those with the gene disruption were unchanged. *Bar,* 50 µm

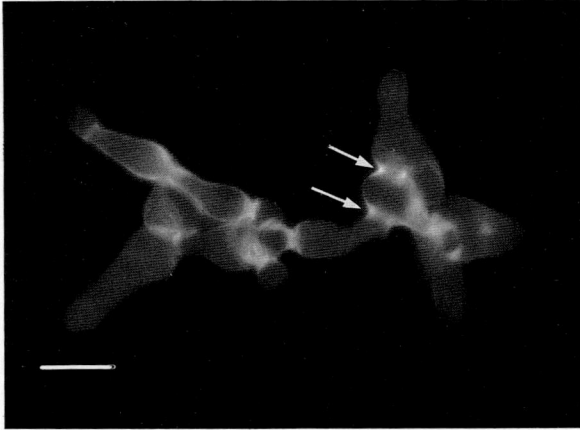

Fig. 5. Aspect of one of the aberrant colonies with the *CHS2* disruption at high magnification and after staining with Calcofluor White. Staining was performed by micromanipulating the microcolony to agar containing the dye and later to clean agar. *Arrows* point to sites where the absence of a septum is manifest. *Bar*, 10 nm

micromanipulator. It is concluded that *CHS2* is essential for cell division and septum formation. It is of interest that in the disruptions of *CHS2*, the same results were obtained whether the transformed diploids contained either normal or disrupted *CHS1* genes, that is, *CHS1* could not substitute for *CHS2* when the latter was deleted.

8 Chitin Synthetase 1 — A Repair Enzyme?

Although the experiments described above clearly indicated the role of Chs2 in cell division, that of Chs1 remained obscure. The results of the *CHS1* gene disruption establish that this is not an essential gene under normal conditions, but certain abnormalities in strains containing the disrupted gene have been noted. When these strains are allowed to grow in minimal rather than rich medium, many buds and single cells appear very bright when viewed under phase contrast (Fig. 6). These cells are permeable to trypan blue, i.e. they are dead (Bulawa et al. 1986). The difference observed between yeast cells grown in rich or minimal medium was found to be due simply to the poor buffering capacity of the latter that allowed the pH to drop to values of 3 and below during growth. Addition of a buffer, such as succinate or 2-(N-morpholino)ethanesulfonate at pH 5.7, to minimal medium sufficed to eliminate almost completely the occurrence of bright cells. Because the internal pH of yeast cells is maintained within narrow margins in the face of large external changes (Den Hollander et al. 1981), it seemed probable that the observed cell lysis might be due to the effect of pH on a system involved in the metabolism of chitin and exposed on the outside of the cell. We described a system that meets those requirements several years ago (Correa et al. 1982; Elango et al. 1982). It is an endochitinase, localized in the periplasmic space of the cell and endowed with a very acidic pH optimum (2.5–3), a range in which the lysis of buds was observed. To ascertain whether this enzyme was involved in the damage to cells, a chitinase inhibitor, allosamidin (Koga

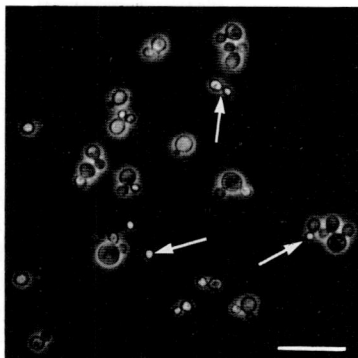

Fig. 6. Bright cells observed in culture of a strain harboring a *CHS1* disruption, after growth in minimal medium. Arrows point to some of the bright buds and single cells. *Bar,* 20 nm

et al. 1987; see also Chapter 5) was used. At a concentration of 0.15 mg/ml, which inhibits the yeast chitinase by about 80%, the occurrence of bright cells was reduced 50% to 75% (A. Sburlati, S. Silverman and E. Cabib, unpublished results). Although preliminary, these results suggest that the chitinase, which under normal circumstances probably helps in the separation of cells after division, may actually over-digest chitin when its activity is greatly increased at low pH values. No cell lysis is observed, however, unless chitin synthetase 1 is absent, which suggests that this enzyme is able to repair the damage brought about by the chitinase.

Another property of cells harboring a *CHS1* disruption is a greater sensitivity to killing by polyoxin D, an inhibitor of chitin synthetases (Bulawa et al. 1986; see also Chapter 5). In normal cells, the decrease in chitin formation caused by action of the inhibitor on Chs2 may elicit activation of Chs1, which, although partially inhibited, could add to total chitin synthesis. A higher concentration of polyoxin would then be needed to overcome the added activity. Again, Chs1 would serve as an emergency enzyme.

9 The Present Status of Chitin Synthesis in Yeast

The last few years have witnessed dramatic changes in our knowledge of chitin synthesis in *Saccharomyces*. The enzyme that was believed for a long time to be instrumental in the formation of the polysaccharide suddenly lost its status, soon to be substituted by another synthetase. The essential role of the new enzyme, chitin synthetase 2, in septum formation has now been unequivocally established by the genetic evidence. Despite these new and exciting developments, our general beliefs about the synthesis of chitin have not changed greatly. Chs2, as Chs1, is also a zymogen, activatable by proteases, and its intracellular distribution is also similar, i.e. the bulk of it is attached to the plasma membrane. Therefore, we see no need to modify our previous working hypothesis that chitin deposition in the cell is brought about by localized activation of a zymogen more or less uniformly distributed on the plasma membrane (Cabib et al. 1982).

As discussed in the preceding section, the function of Chs1 may not be fundamentally different from that of Chs2, except that it would act only under exceptional circumstances. Thus, although not essential for cell division, Chs1 would probably have survival value under certain conditions, such as in a low pH environment. It would be of interest to determine whether two (or more) chitin synthetases exist in other fungi, especially those of the filamentous type, and to compare their functions with those found in *S. cerevisiae*. The availability of genomic probes from *CHS1* and *CHS2* should facilitate the search for such isoenzymes.

In order to function, both Chs1 and Chs2 need to be activated, be it by partial proteolysis or by some other modification. We still know nothing about the physiological mechanism for the regulation of these activities. The clarification of this problem is a major challenge for future studies.

References

Bartnicki-Garcia S (1968) Cell wall chemistry, morphogenesis and taxonomy of fungi. Annu Rev Microbiol 22:87–108

Bartnicki-Garcia S, Bracker CE, Reyes E, Ruiz-Herrera J (1978) Isolation of chitosomes from taxonomically diverse fungi and synthesis of chitin microfibrils in vitro. Exp Mycol 2:173–192

Bulawa CE, Slater M, Cabib E, Au-Young J, Sburlati A, Adair WL, Robbins PW (1986) The S. cerevisiae structural gene for chitin synthase is not required for chitin synthesis in vivo. Cell 46:213–225

Cabib E (1987) The synthesis and degradation of chitin. Adv Enzymol 59:59–101

Cabib E, Farkas V (1971) The control of morphogenesis: an enzymatic mechanism for the initiation of septum formation in yeast. Proc Natl Acad Sci USA 68:2052–2056

Cabib E, Shematek EM (1981) Structural polysaccharides of plants and fungi: comparative and morphogenetic aspects. In: Ginsburg V, Robbins PW (eds) Biology of carbohydrates, vol I. Wiley, New York, p 51

Cabib E, Roberts R, Bowers B (1982) Synthesis of the yeast cell wall and its regulation. Annu Rev Biochem 51:763–793

Cabib E, Bowers B, Roberts RL (1983) Vectorial synthesis of a polysaccharide by isolated plasma membranes. Proc Natl Acad Sci USA 80:3318–3321

Correa JU, Elango N, Polacheck I, Cabib E (1982) Endochitinase, a mannan-associated enzyme from *Saccharomyces cerevisiae*. J Biol Chem 257:1392–1397

Den Hollander JA, Ugurbil K, Brown TR, Shulman RG (1978) Phosphorus-31 nuclear magnetic resonance studies of the effect of oxygen upon glycolysis in yeast. Biochemistry 20:5871–5880

Duran A, Cabib E (1978) Solubilization and partial purification of yeast chitin synthetase. Confirmation of the zymogenic nature of the enzyme. J Biol Chem 253:4419–4425

Duran A, Bowers B, Cabib E (1975) Chitin synthetase zymogen is attached to the yeast plasma membranes. Proc Natl Acad Sci USA 72:3952–3955

Elango N, Correa JU, Cabib E (1982) Secretory character of yeast chitinase. J Biol Chem 257:1398–1400

Horisberger M, Vonlanthen M (1977) Location of mannan and chitin on thin sections of budding yeasts with gold markers. Arch Microbiol 115:1–7

Kang MS, Elango N, Mattia E, Au-Young J, Robbins PW, Cabib E (1984) Isolation of chitin synthetase from *Saccharomyces cerevisiae*. Purification of an enzyme by entrapment in the reaction product. J Biol Chem 259:14966–14972

Kang MS, Au-Young J, Cabib E (1985) Modification of yeast plasma membrane density by concanavalin A attachment. Application to study of chitin synthetase distribution. J Biol Chem 260:12680–12684

Koga D, Isogai A, Sakuda S, Matsumoto S, Suzuki A, Kimura S, Ide A (1987) Specific inhibition of *Bombyx mori* chitinase by allosamidin. Agric Biol Chem 51:471−476

Mol PC, Wessels JGH (1987) Linkages between glucosaminoglycan and glucan determine alkali-insolubility of the glucan in walls of *Saccharomyces cerevisiae*. FEMS Microbiol Lett 41:95−99

Molano J, Bowers B, Cabib E (1980) Distribution of chitin in the yeast cell wall. An ultrastructural and chemical study. J Cell Biol 85:199−212

Orleans P (1987) Two chitin synthases in *Saccharomyces cerevisiae*. J Biol Chem 262:5732−5739

Sburlati A, Cabib E (1986) Chitin synthetase 2, a presumptive participant in septum formation in *Saccharomyces cerevisiae*. J Biol Chem 261:15147−15152

Silverman S, Sburlati A, Slater ML, Cabib E (1988) Chitin synthase 2 is essential for septum formation and cell division in *Saccharomyces cerevisiae*. Proc Natl Acad Sci USA 85:4735−4739

Thomas PS (1980) Hybridization of denatured RNA and small DNA fragments transferred to nitrocellulose. Proc Natl Acad Sci USA 77:5201−5205

Ulane RE, Cabib E (1976) The activating system of chitin synthetase from *Saccharomyces cerevisiae*. Purification and properties of the activating factor. J Biol Chem 251:3367−3374

Wolf DH, Ehmann C (1979) Studies on a proteinase B mutant of yeast. Eur J Biochem 98:375−384

Zubenko GS, Mitchell AP, Jones EW (1979) Septum formation, cell division and sporulation in mutants of yeast deficient in proteinase B. Proc Natl Acad Sci USA 76:2395−2399

Chapter 4

A Novel Computer Model for Generating Cell Shape: Application to Fungal Morphogenesis

S. BARTNICKI-GARCIA, F. HERGERT, and G. GIERZ[1]

1 Introduction

The question of how fungal cells attain their characteristic morphology has long been of great interest to experimental and theoretical biologists. Much of this attention has centered on hyphae, the tip-growing tubular cells typical of fungi (for a recent comprehensive review of fungal apical growth see Wessels 1986).

Nearly a century ago, Reinhardt (1892) published a classic analysis of apical growth of fungal hyphae. Since then, several models have been advanced to try to describe in mathematical terms the pattern of surface growth that leads to generation of a tubular shape from a tip-growing cell (Green 1969; da Riva Ricci and Kendrick 1972; Trinci and Saunders 1977; Prosser and Trinci 1979; Koch 1982). Parallel studies have been made with tip-growing cells of green plants (Green and King 1966; Picton and Steer 1982).

2 Current Concepts of Wall Growth in Fungi

2.1 Gradients of Wall Expansion

From early studies with surface markers (e.g. Castle 1958), it became clear that apical growth entailed a gradient of surface expansion that was maximum at the tip and declined towards the base of the extension zone (Green 1969). Quantitative autoradiographic studies, in which fungal hyphae were briefly exposed to a specific cell wall precursor ($[^3H]$-GlcNAc to label chitin/chitosan), confirmed the existence of a sharp descending gradient of wall synthesis centered at the hyphal apex (Bartnicki-Garcia and Lippman 1969; Gooday 1971).

[1] Department of Plant Pathology, and Department of Mathematics and Computer Science, University of California, Riverside, California 92521, USA

2.2 Previous Models

Green (1969) proposed that the gradient of wall expansion needed to generate a tubular cell by tip growth followed a cosine function of α, an angle that measures the displacement of a growing point along the apical dome. Such a relationship, derived for a tip of hemispherical shape, was revised by Trinci and Saunders (1977) who took into account that hyphal tips are not truly hemispherical but have a tapered "half-ellipsoidal" shape. They proposed instead a modified cotangent function as a more appropriate description of such shapes. In a different model, the physical processes believed to be involved in apical growth, namely the rates of wall "setting" (da Riva Ricci and Kendrick 1972) were also included. For an incisive review of mathematical models see Prosser (1979). Despite the fact that these models afforded equations that approximated the shape of fungal hyphae, they appear to have a serious shortcoming. They are basically geometric exercises that formulate equations from artificial co-ordinates and reference points (e.g. apical pole, base of the apical dome, angle α) for which there are no corresponding subcellular structures in a hypha. Note that any dimensions given to a hyphal apex are somewhat arbitrary since, except for the unlikely hemispherical dome, the base of the apical dome can not be precisely defined because of the lack of a sharp demarcation between the acutely tapered portion of the apex and the quasi-cylindrical portion of the hypha.

3 A Model of Wall Growth Based on Vesicles

To circumvent the limitation of above models, we decided to formulate a model that would take into account what is clearly the main subcellular structure involved in cell wall growth in fungi, namely secretory vesicles.

3.1 Vesicular Basis for Wall Growth

From pioneering ultrastructural studies of fungal hyphae (McClure et al. 1968; Girbardt 1969; Grove and Bracker 1970; Grove et al. 1970; Heath et al. 1971), it became abundantly clear that vesicles were primarily responsible for wall growth. Vesicles carry to the cell surface precursors and enzymes needed for plasma membrane and wall formation. This vesicular concept of cell wall growth is now universally accepted not only for fungi (Gooday and Trinci 1980) but for the walled cells of all eucaryotic kingdoms (Bartnicki-Garcia 1984).

3.2 An Earlier Qualitative Model

The vesicular theory of growth was incorporated in a previous *qualitative* model of cell wall growth and morphogenesis of fungi (Bartnicki-Garcia 1973). This model was designed to explain wall growth in terms of individual submicroscopic events or "units of wall growth." Each unit of growth was postulated to correspond to "the amount of growth obtained from the discharge of a single vesicle (or minimum combination of different vesicles) carrying components essential for wall growth." In this model, overall surface expansion was believed to result from the sum total of minute units of wall growth, but no effort was made to quantify this relationship. This incipient model predicted that morphogenesis depended on the pattern of spatial distribution of these wall units, and these in turn depended on the pattern of migration of vesicles to the cell surface. A random pattern would produce uniform expansion (= spherical cells), a polarized pattern would be responsible for tubular cells.

The model presented here represents a major refinement and extension of this qualitative model. Note that Prosser and Trinci (1979) also elaborated a mathematical model based on vesicle distribution to explain septation and branching of hyphae.

4 Computer Simulation of Morphogenesis

For a vivid representation of the principle under which the model was built, imagine the fungus as being a container under pressure continuously bombarded from *within* by a myriad of tiny vesicles. Upon impact, each vesicle 'missile' would insert itself in the periphery of the cell and increase its surface area by one unit.

4.1 Programming

The model was done in two dimensions. The cell shapes created represent median cross sections through a three-dimensional living cell. The simulation is done on a grid with squares representing a unit of area (Fig. 1). A square can be either *full* or *empty*. The union of full squares constitutes the fungal cell, the empty squares are the surrounding medium. Each filled square is derived from a "vesicle". The vesicles are

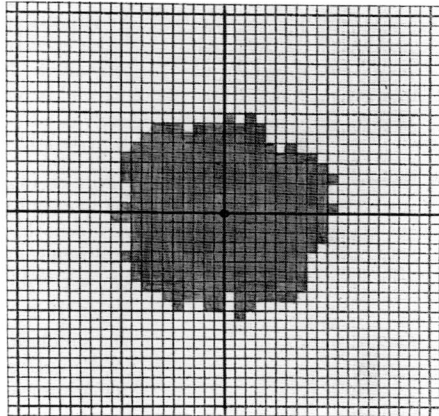

Fig. 1. A magnified representation of the programming grid. Individual growth events are shown as squares. Each event is triggered by a "vesicle" emanating from the center (VSC) and traveling in a randomly chosen direction

released in all directions from an idealized point source: the Vesicle Supply Center or VSC. The program chooses an angle at random. This angle defines a ray emanating from the VSC and represents the direction in which the vesicle will travel. The probability distribution is uniform, i.e, all angles have the same likelihood. The program searches for the first empty square along the ray and changes its status from empty to full. Cell area is thus increased by one unit.

4.2 Assumptions

For the purposes of simulating an ultra-simplified version of a growing fungal cell a number of assumptions were made (see Table 1).

4.3 Simulation of Spherical Growth: Stationary VSC

Figures 2a−h show the simplest case of growth simulation in which vesicles are released continuously and randomly in all directions from a stationary central VSC.

Table 1. Assumptions in the computer simulation of fungal cell development

Assumptions	Explanations
1. The cell is a two-dimensional object.	The conclusions can be readily extrapolated to the corresponding three-dimensional solid of revolution.
2. Vesicles are released from a point source called the Vesicle Supply Center or VSC.	This idealized point source could actually be the geometric center of a complex vesicle-generating apparatus present in fungal cells, and whose overall effect is to release vesicles in all directions. See text for other alternatives. The actual structure of the VSC is not critical for the model.
3. From the VSC, vesicles migrate randomly to the cell surface.	Although the direction is randomly chosen, the model assumes a straight path to the cell surface.
4. Upon reaching the cell periphery, vesicles insert their mass into the wall and expand the cell surface by one unit.	No distinction between wall and plasma membrane is made. The entire mass of the vesicle is assumed to be incorporated to the cell surface. The model does not take into account that much of the wall is made in situ. However, any mass produced in situ is likely to be proportional to the amount of enzyme delivered by the vesicles.
5. Each vesicle supplies all ingredients needed for surface growth.	Although there is evidence for more than one kind of vesicle being involved in wall growth (Bartnicki-Garcia 1987), the model considers that one type of vesicle delivers all necessary materials to produce a unit of growth, including ingredients needed to give the wall a localized measure of transient plasticity (Bartnicki-Garcia 1973).
6. Growth rate is linear.	The simulation would work equally well if the vesicle parameters (N, V) were programmed to change in exponential fashion to mimic the basic growth mode of fungi.

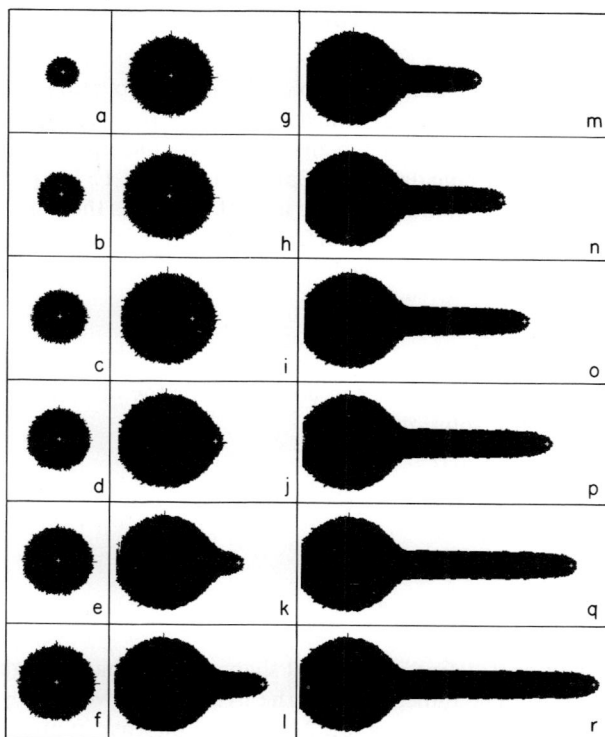

Fig. 2a–r. Computer simulation of spherical growth and hyphal growth. In each frame, 8000 "vesicles" were released from the VSC (+). The amount of growth in each frame is that shown in frame a. In frames a–h a gradually expanding circle results from the vesicles released by a fixed VSC. In frames i–r the VSC is made to advance continuously while releasing vesicles at the same frequency. As the VSC advances towards the surface (i–k) a protrusion emerges from the circle, which on continued growth a acquires the characteristic shape of a hyphal tube (l–r)

The resulting shape is a growing circle, i.e., a two-dimensional simulation of spherical growth.

4.4 Simulation of Cylindrical (Hyphal) Growth: Moving VSC

After a period of spherical growth generated by vesicles released from a stationary VSC, the VSC is programmed to move in a fixed direction while releasing vesicles at the same rate (Fig. 2i–r). The circle keeps enlarging but with increasing asymmetry. Although vesicles are continually released by the VSC, uniformly in all directions, the number of vesicles reaching different points on the cell surface becomes increasingly disproportionate; the frequency of vesicle impacts per unit of surface will be increasingly greater on the advancing side of the VSC than on the rest of the cell surface. As the VSC approaches the cell surface, the deformation becomes more and more apparent until, eventually, a conspicuous bud protrudes from the circular shape (Fig. 2k). The protrusion grows longer until a well defined tube is generated (Fig. 2l–r). As long as the simulation continues with the same parameters, namely the same frequency of vesicle release and same rate of displacement of the VSC, the tube will elongate and produce the two-dimensional equivalent of a long cylinder with a slight taper. This entire process is a realistic two-dimensional simulation of spore germination.

5 Mathematical Model for Hyphal Growth

With the same premises employed for the above computer simulation of hyphal tube growth, namely, a two-dimensional figure that grows from particles emitted randomly from a continuously advancing source, the geometric function generated by such system was elucidated (S. Bartnicki-Garcia, F. Hergert, and G. Gierz, to be published). By assuming that the axis of growth is the y-axis and letting $y = f(x, t)$ be the curve that describes the surface of the cell at times t, the following ordinary differential equation was derived:

$$\frac{N}{V} y - (x^2 + y^2) = \frac{N}{V} xy'$$

where N is the increase in area and V is the rate of linear displacement of the VSC, and whose solution yielded

$$y = x \cot \frac{xV}{N} \qquad (1)$$

5.1 Hyphal Shape

Equation 1 defines the overall shape of a hypha in longitudinal, median cross section (Fig. 3). The dimensions of the hypha are governed by two parameters which have *physiological* significance: N = rate of increase in area = number of vesicles released by the VSC per unit time; V = rate of linear displacement of the VSC. The ratio

$$\frac{N}{V} = d \qquad (2)$$

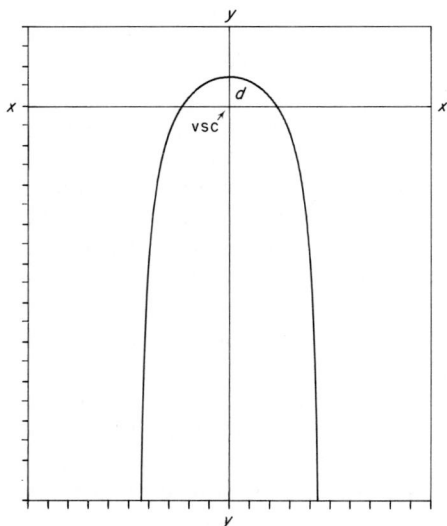

Fig. 3. Profile of a model hypha (*hyphoid*) plotted on an arbitrary scale from the equation

$$y = x \cot \frac{xV}{N}$$

d is the distance between the origin (VSC) and the extreme tip (apical pole)

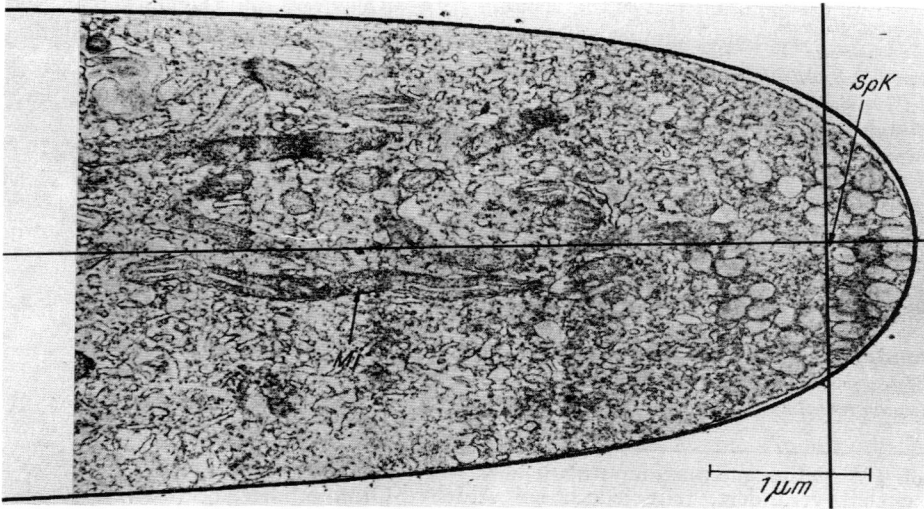

Fig. 4. Correspondence between curve predicted by Eq. 1 and the electron microscopic profile of a hypha with a Spitzenkörper (*Polystictus versicolor*). The curve plotted from the calculated value of *d* was superimposed on a photographic copy of Fig. 1 from Girbardt (1969). Note that the originally marked position of the Spitzenkörper (Spk) coincides with the position of the VSC in the model curve. Micrograph reproduced with publisher's permission

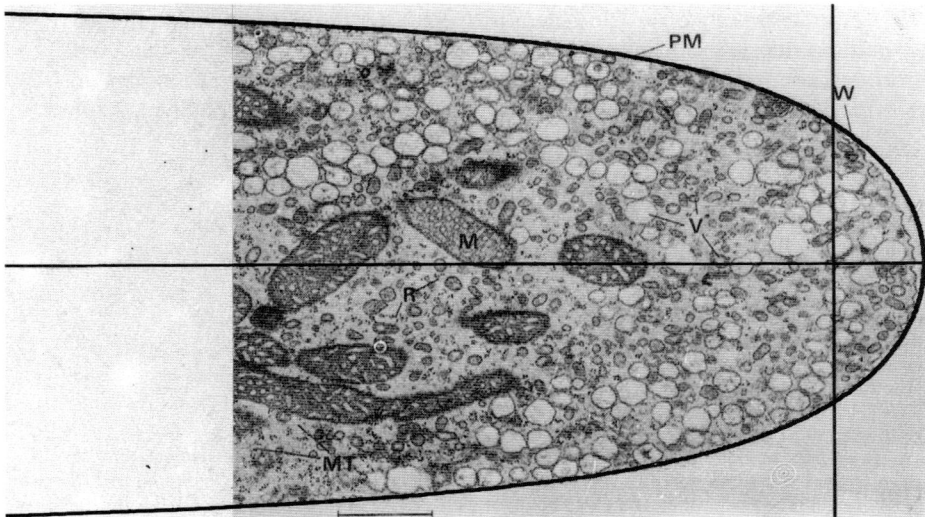

Fig. 5. Correspondence between shape predicted by Eq. 1 and the profile of a hypha from a fungus devoid of Spitzenkörper, *Pythium aphanidermatum*. Montage prepared as in previous figure for micrograph in Fig. 1 from Grove and Bracker (1970). Micrograph reproduced with author's permission

defines a key parameter: d, the distance between the VSC and the apical wall (Fig. 3). This single value defines the size of a hypha.

By calculating the d value for hyphae shown in published electron micrographs in median longitudinal section, one could draw curves using Eq. 1 that accurately match the profile of these hyphae (Figs. 4, 5)[1].

5.2 Hyphal Diameter

From Eq. 1 the diameter of a hypha can also be derived. Thus, since the first poles of Eq. 1 occur at

$$\frac{x V}{N} = \pm \pi$$

it follows that the maximum diameter (D) of the hyphae would be

$$D = 2\pi \frac{N}{V} \tag{3}$$

The position of the VSC in the apical dome (d) determines hyphal diameter; the closer the VSC lies to the apical wall, the narrower the tube produced. Since $d = N/V$, we can conclude that hyphal diameter will be determined by the ratio between the number of vesicles emitted per unit time and the rate of linear displacement of the VSC. For a given rate of vesicle generation, the faster the VSC advances, the narrower is the tube that is produced. Conversely, for a given rate of VSC displacement, the more vesicles produced per unit time, the wider the tube. Thus, a fast advancing VSC that produces relatively few vesicles would generate a very narrow hyphal tube.

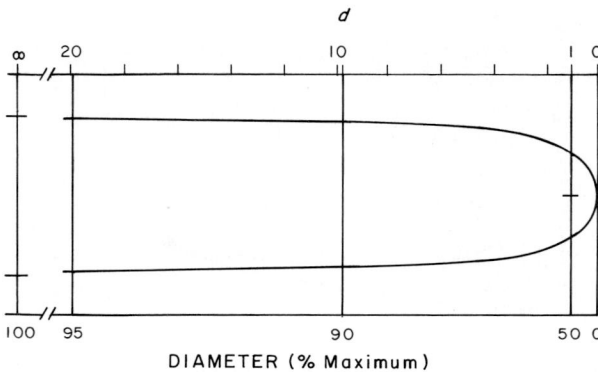

Fig. 6. Diameter of a model hypha (*hyphoid*) as a function of distance from tip. Relative distance calculated in units of d

---

[1] Values of d can be readily estimated from two simple measurements: 1) diameter of the hyphae at any given point and 2) distance of this point from the hyphal tip. Values of d were calculated from a numerical solution of Eq. 1.

According to Eq. 1, a hypha would never quite reach a maximum diameter, but for practical purposes we calculated that when the hyphal length is about $100 \times d$, the diameter is 99% of the theoretical maximum. At $10 \times d$, the diameter is approx. 90% of the maximum. Interestingly, at a distance from the tip equal to d, the diameter of the hypha is 50% of its theoretical maximum (Fig. 6).

6 Relationship of Spitzenkörper to VSC

One noteworthy correlation that emanates from the model is the remarkable correspondence between the position of the Spitzenkörper in a real hypha and the position of the VSC in the model (Fig. 4).

The Spitzenkörper is a remarkable structure found inside the tips of growing hyphae of higher fungi and always close to the apical wall (Brunswik 1924; Girbardt 1957). In living hyphae, it is seen as a dynamic spheroidal structure lacking a sharp boundary. With the advent of electron microscopy and ancillary techniques to study cell ultrastructure, the Spitzenkörper was identified as a conglomeration of vesicles (McClure et al. 1968; Girbardt 1969; Grove and Bracker 1970; Howard 1981). Because the Spitzenkörper appeared in actively growing tips and disappeared, or was absent from non-growing hyphae, a powerful argument could be made for a causal correlation between Spitzenkörper and tip growth (Girbardt 1957), particularly since changes in the position of the Spitzenkörper preceded changes in the direction of growth. Despite this strong circumstantial evidence, the exact role of the Spitzenkörper in apical growth has remained unsolved and, in view of the lack of an obvious Spitzenkörper in the hyphae of lower fungi (McClure et al. 1968; Grove and Bracker 1970), its necessity in apical growth has remained equivocal.

Our finding that the position of the Spitzenkörper coincides with the position of the VSC in our model ($= d$) may cast new light on the overall significance of the Spitzenkörper. The cytological evidence indicates that the Spitzenkörper is not a place for the generation of vesicles but an area of congregation of vesicles destined for the cell wall. Our model supports this concept by giving the Spitzenkörper a role as a delivery center for wall-destined vesicles.

For those fungi lacking a Spitzenkörper (McClure et al. 1968; Grove and Bracker 1970), our model suggests that they too have a functional equivalent of the Spitzenkörper, i.e., a place from which vesicles start on the final leg of their journey to the wall, but for some reason these transient vesicles do not accumulate in high enough numbers to produce a visible agglomeration. Even in fungi that normally have a visible Spitzenkörper, this may disappear during partial growth inhibition with anti-tubulin agents yet the hypha continues to elongate (Howard and Aist 1977).

7 The VSC Concept and Mechanisms for VSC Displacement

The main premise of the present model for hyphal growth, namely that vesicles are released randomly from a linearly advancing source (VSC), may be gross simplification but one that need not be in serious conflict with the more complex situation of a living hypha.

Since there is presently no obligatory structural correlate for the VSC in a fungal cell, it is best to regard the VSC solely as a functional feature of the vesicle generating apparatus of the cell: perhaps Golgi cisternae or equivalent vesicle generator plus accessory cytoskeletal elements. In increasing order of structural complexity, the VSC might represent 1) a single source of vesicles; 2) a geometric center for multiple sources of vesicles, or 3) the final release site of a vesicle transport mechanism that collects vesicles produced by an elaborate system of multiple vesicle-generating apparatus. The latter appears to be the case for hyphae. Present knowledge on hyphal ultrastructure suggests that vesicles are generated over a long span of the hyphal tube and then migrate to the tip (Grove and Bracker 1970; Barstow and Lovett 1974; Collinge and Rinci 1974; Howard and Aist 1979). Presumably, the VSC is the point of discharge of vesicles that may be traveling along cytoskeletal tracks to the tip and upon reaching the VSC are free to move to the cell surface. The observation that the apex is the preferred nucleation site for microtubules in hyphae of *Uromyces phaseoli* (Hoch and Staples 1985) gives us reason to speculate that the VSC of a hypha might be a microtubule organizing center, or a structure intimately associated with it.

In any of the above instances, any displacement of the VSC proposed in the model should be construed as a displacement of an *entire* vesicle-generating system.

The proposed linear displacement of the VSC is a crucial feature of the model. There are two different ways to generate the linear displacement of the VSC: pulling mechanism or pushing mechanism. The model would work with either mechanism, and both can be supported, in principle, by current cytological evidence.

7.1 Pulling Mechanism

In our first attempt to simulate hyphal growth, the VSC was simply linked to a point in the cell surface, that subsequently became the tip of the emerging tube. By the simple device of keeping the VSC at a constant distance from the wall, a tube was generated. The concept of a structural linkage between VSC and the fungal cell surface is not unreasonable; this might be established through microfilaments anchoring the VSC, plus its supporting structures, to the apical plasma membrane. Extensive arrays of actin microfilaments have been seen in hyphal tips (Hoch and Staples 1985; Anderson and Soll 1986; Runeberg and Raudaskoski 1986; Heath 1987), and other wall growing regions (Adams and Pringle 1984), and it has been proposed that they pull the cytoplasm towards the tip (McKerracher and Heath 1987).

7.2 Pushing Mechanism

In another simulation protocol, the VSC was programmed to advance at a constant rate while producing a constant number of vesicles. In fungal hyphae, this propulsion may result from a continuously extending 'scaffolding' of cytoskeleton pushing the entire vesicle-generating apparatus and its VSC in an apical direction. This is not unrealistic, as there is an elaborate system of cytoplasmic microtubules and microfilaments potentially responsible for organelle displacement inside the hypha (see review by Wessels 1986; McKerracher and Heath 1987).

8 Simulations of Morphogenetic Processes in Fungi

One virtue of the computer model is that one can simulate real morphogenetic transitions simply by adjusting the ratio of the two vesicle parameters N and V. When V is zero, a spherical shape would result, when V has a finite value, an elongated shape would be produced that may vary from a slightly prolate spheroid to an extremely

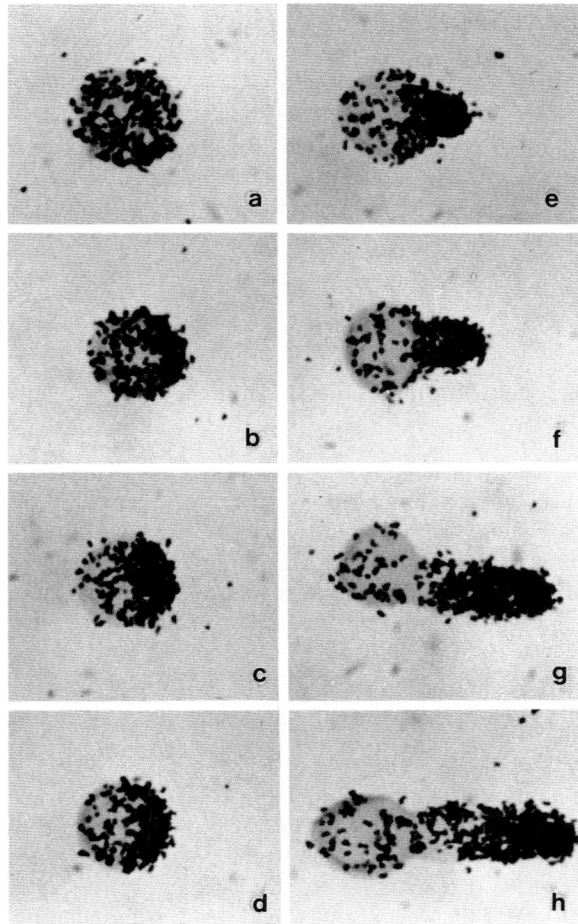

Fig. 7a–h. Reconstruction of the pattern of wall growth in germinating spores of *Mucor rouxii*. Autoradiographs show increasing polarization of wall synthesis (**b–d**) prior to germ tube emergence. Modified from Bartnicki-Garcia and Lippman (1977)

narrow and long tube. With these basic shapes the majority of fungal morphologies may be constructed. It is perhaps not coincidental that the common shapes of fungal cells oscillate between the limits predicted by the model.

8.1 Spore Germination and Hyphal Morphogenesis

The computer simulation described in Sect. 4 and Fig. 2 emulates the two-stage sequence of morphogenesis commonly observed during fungal spore germination (Bartnicki-Garcia et al. 1968; Bartnicki-Garcia 1981). The first stage of spherical growth is simulated by a stationary VSC; the second stage, the emergence of a germ tube from the germ sphere, is achieved by displacing the VSC toward the periphery. By changing the duration of the first stage and the ratio N/V, one can mimic most morphologies seen in actual cases of germination, variations depending mainly on the relative sizes of germ sphere and germ tube. Regarding the number of germ tubes per spore see Sect. 9.4.

Autoradiographic studies of wall growth during spore germination (Bartnicki-Garcia and Lippman 1977) show a gradual polarization of wall synthesis prior to emergence of germ tube (Fig. 7b−d). This is in accordance with the model predicting an increasing rate of vesicle discharge in the region towards which the VSC advances and where the germ tube will later emerge.

The model focuses attention on one event as the likely main cause for germ tube formation, namely the mechanism that triggers and/or regulates the displacement of the VSC.

8.2 Sporangium Development

A common morphology seen in reproductive structures of fungi consists of a long tube (sporangiophore) with a large spheroidal reservoir at one end (sporangium, gametangium). The process can be simulated by generating a long narrow hypha as above (Fig. 8a−c) and then stopping the linear displacement of the VSC ($V = 0$) without interrupting vesicle release (Fig. 8d−h).

8.3 Yeast Cell Morphogenesis

The production of ovoidal/ellipsoidal shapes typical of ordinary yeast cells can be simulated by invoking a relatively slow displacement of the VSC and high rate of vesicle generation. Variations in the N/V ratio produce buds of increasing cylindricity (Fig. 9). If there is no net displacement of the VSC, then the spherical buds typical of yeasts of *Mucor* spp. would result (simulated by sequence a−h in Fig. 2).

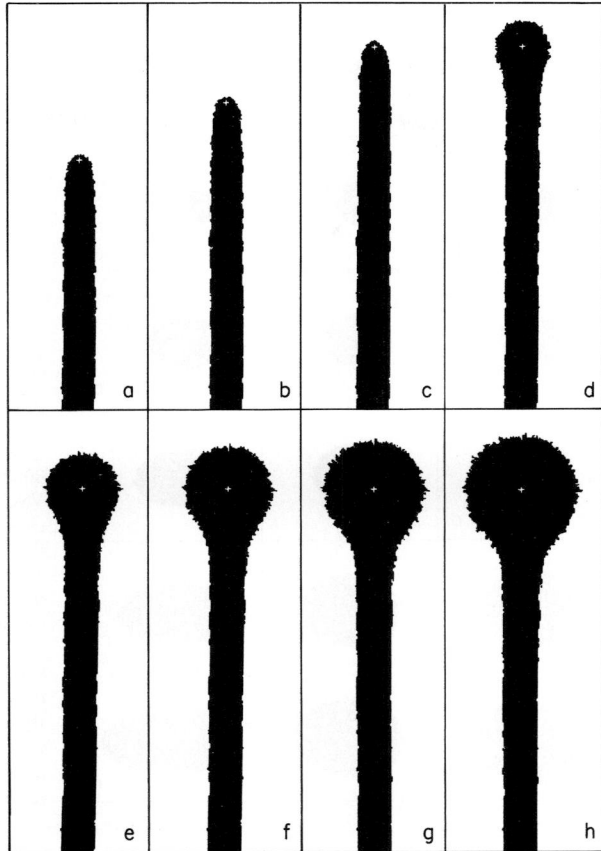

Fig. 8a–h. Computer simulation of sporulation. Consecutive phases of hyphal (sporangiophore) growth and spherical (sporangial) growth. Except for **a** each frame represents the growth from 8000 "vesicles" released from the VSC (+). In frames **a–c** a long tube was generated by advancing the VSC as in Fig. 2, except that the rate of advance was increased to generate a narrower tube. In frames **d–h** the linear rate of advance of the VSC was stopped but the release of vesicles was continued at the same rate. A spheroidal structure was then produced

8.4 Branching and Multiple Budding

Although we have not written a variant of the main computer program to simulate multiple growth processes from a single cell, e.g. multiple germ tubes from a germ sphere, or lateral branches from the main branch, this additional elaboration does not pose an insurmountable obstacle. In theory, each growth process would require creating a new VSC in a different location and/or moving in a different direction. The frequency and position of branching could be dictated by the same considerations stipulated by Prosser and Trinci (1979) in their vesicular model of mycelial development.

8.5 Mycelial-yeast Dimorphism

Another example of morphological variation explained by the model is the well known mycelial-yeast dimorphism of fungi (Bartnicki-Garcia 1963; Stewart and

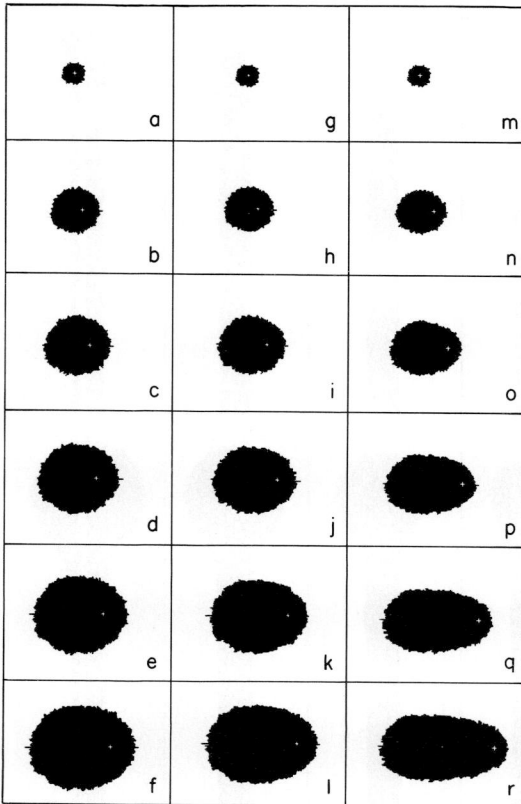

Fig. 9a–r. Computer simulation of three variations in yeast cell morphology. Each sequence started with a spherical bud generated by 4000 "vesicles". In all subsequent frames, 8000 vesicles were released from the VSC (+). A gradually expanding ovoidal/ellipsoidal shape results from the vesicles released by a slowly but continuously advancing VSC. Three different yeast shapes were generated by varying the rate of advance of the VSC. In frames **a–f** the VSC advances at the slowest of the three rates and produces a prolate spheroid. By increasing the rate of linear advance of the VSC 1.5- or 2.0-fold more elongated shapes are produced (**g–l** or **m–r**, respectively)

Rogers 1983). The capacity for dual vegetative morphology, whether the fungus grows spheroidal/ovoidal/ellipsoidal cells (yeast form) or whether it produces long branched tubes (mycelium), may revolve around a single decisive factor: the ability to displace its VSC during vegetative growth at two different rates; none or slow for yeast cells, fast for mycelium.

9 Limitations

The present model applies to surfaces or walls that grow isotropically (allometric coefficient of 1; see Green 1965; Prosser 1979), i.e. materials accrued from vesicle discharge expand the surface equally in transversal and longitudinal directions. It does not explain intercalary tubular growth (e.g. the subapical growth in elongating sporangiophores of *Phycomyces* during stage III) (Castle 1953). For such growth, additional parameters need to be considered that would restrict wall expansion in girth while allowing longitudinal extension, namely the anisotropy of wall expansion discussed by Green (1965, 1969).

As presently constructed, the model generates straight hyphae with a regular shape and size. Although living hyphae deviate to different extents from this highly idealized morphology, the present model provides a basis for explaining the more convoluted or irregular shapes seen in actual hyphae. Oscillations in the V/N ratio and/or drift in the path followed by the VSC might be the cause of fluctuations in the diameter of the hypha and/or its direction of growth.

10 Discussion and Conclusions

This study, which began as an attempt to develop a computer simulation of apical growth based on the vesicular theory of wall growth, has provided new insights into the basic mechanism of hyphal morphogenesis:

1. There is now a simpler mechanism to explain apical growth of hyphae than was previously anticipated. By simply advancing an existing wall-vesicle-generating apparatus in a continuous, linear fashion, the cell would automatically establish the polarized pattern of surface expansion that is so typical of fungal hyphae.
2. The observed or predicted gradients of wall properties in the hyphal apex, e.g. elasticity/rigidification (Robertson 1965; Saunders and Trinci 1979); plasticity (Bartnicki-Garcia 1973), or polymer cross-linking (Wessels and Sietsma 1981; Wessels 1986) are probably not the cause of hyphal morphogenesis but, rather, a reflection of the pattern of vesicle discharge which generates a graded distribution of biochemical/biophysical activities on the cell surface.
3. Likewise, our present model obviates the need for mechanisms to move vesicles to specific or preferential targets on the cell surface (apical pole). Seemingly, vesicles need only be endowed with the ability to move towards the cell surface in any random direction. (The massive polarized transport of vesicles from the subapical to the apical region should be viewed as an additional feature required not for morphogenesis *per se* but to provide an ample supply of vesicles to support the fast growth of fungal hyphae). Our present model is in accord with newer evidence (McGillviray and Gow 1987; Schreurs and Harold 1988) questioning whether the electric currents which flow through the tip of tubular cells play a determining role in morphogenesis (Jaffe 1968).
4. The simulation allowed us to predict which factors are critical in hyphal morphogenesis. Accordingly, the key to hyphal morphogenesis probably does not lie in the appex, as previously believed (Robertson 1965; Bartnicki-Garcia 1973; Gooday and Trinci 1980) but in the mechanisms that cause the linear displacement of the VSC. Conceivably, primary control of hyphal morphogenesis resides in the machinery that assembles elements of the cytoskeleton, a suggestion tentatively supported by experimental observations on the distortion of hyphal growth caused by inhibitors of microtubules (Howard and Aist 1977) or microfilaments (Betina et al. 1972; Grove and Sweigard 1980; Tucker et al. 1986).
5. The proposed equation for a hypha goes beyond earlier models which were mainly approximations of the shape of the apical dome. The present equation describes hyphal profile in its entirety. We therefore propose the name *hyphoid* for this

transcendental function, which apparently has not been previously applied to biological objects. (The cotangent function derived by Trinci and Saunders (1977) represents a totally different mathematical relationship.) In the hyphoid, there is no sharp boundary between the "dome" and the rest of the quasi-cylindrical portion. Hence, any presumed boundary between what is variously called "extension zone", "tapered zone", "apical dome", and the rest of the hypha would have to be purely arbitrary.

6. A wide spectrum of fungal morphogenesis (hyphal development, spore germination, sporangium formation, yeast cell growth) can be attained by a simple cytological device: the ability to displace the source, or sources, of randomly-released, wall-destined vesicles.

Acknowledgement. The experimental work that led to this model was supported in part by grants from the NIH (GM-33513) and NSF (INT-8413728). We thank Prof. C.E. Bracker for many valuable suggestions.

References

Adams AEM, Pringle JR (1984) Relationship of actin and tubulin distribution to bud growth in wild-type and morphogenetic-mutant *Saccharomyces cerevisiae.* J Cell Biol 98:934–945

Anderson JM, Soll DR (1986) Differences in actin localization during bud and hypha formation in the yeast *Candida albicans.* J Gen Microbiol 132:2035–2047

Barstow WE, Lovett JS (1974) Apical vesicles and microtubules in rhizoids of *Blastocladiella emersonii*: effects of actinomycin D and cycloheximide on development during germination. Protoplasma 82:103–117

Bartnicki-Garcia S (1963) Symposium on the biochemical bases of morphogenesis in fungi. III. Mold-yeast dimorphism of *Mucor.* Bacteriol Rev 27:293–304

Bartnicki-Garcia S (1973) Fundamental aspects of hyphal morphogenesis. In: Ashworth JM, Smith JE (eds) Microbial differentiation. Cambridge Univ Press, Cambridge, pp 245–267

Bartnicki-Garcia S (1981) Cell wall construction during spore germination in Phycomycetes. In: Turian G, Hohl HR (eds) The fungal spore: morphogenetic controls. Academic Press, London New York, pp 533–556

Bartnicki-Garcia S (1984) Kingdoms with walls. In: Dugger WM, Bartnicki-Garcia S (eds) Structure, function, and biosynthesis of plant cell walls. Am Soc Plant Physiol, Rockville, Maryland, pp 1–18

Bartnicki-Garcia S (1987) Chitosomes and chitin biogenesis. Food Hydrocolloids 1:353–358

Bartnicki-Garcia S, Lippman E (1969) Fungal morphogenesis: cell wall construction in *Mucor rouxii.* Science 165:302–304

Bartnicki-Garcia S, Lippman E (1977) Polarization of cell wall synthesis during spore germination of *Mucor rouxii.* Exp Mycol 1:230–240

Bartnicki-Garcia S, Nelson N, Cota-Robles E (1968) Electron microscopy of spore germination and cell wall formation in *Mucor rouxii.* Arch Microbiol 63:242–255

Betina V, Micekova D, Nemec P (1972) Antimicrobial properties of cytochalasins and their alteration of fungal morphology. J Gen Microbiol 71:343–349

Brunswik H (1924) Untersuchungen über Geschlechts- und Kernverhältnisse bei der Hymenonyzeten-gattung *Coprinus.* In: Goebel K (ed) Botanische Abhandungen, vol 5. Fischer, Jena, pp 1–152

Castle ES (1953) Problems of oriented growth and structure in *Phycomyces.* Q Rev Biol 28:364–372

Castle ES (1958) The topography of tip growth in a plant cell. J Gen Physiol 41:913–926

Collinge AJ, Trinci APJ (1974) Hyphal tips of wild type and spreading colonial mutants of *Neurospora crassa.* Arch Microbiol 99:353–368

Girbardt M (1957) Der Spitzenkörper von *Polystictus versicolor* (L.). Planta 50:47–59

Girbardt M (1969) Die Ultrastruktur der Apikalregion von Pilzhyphen. Protoplasma 67:413–441

Gooday GW (1971) An autoradiographic study of hyphal growth of some fungi. J Gen Microbiol 67:125–133

Gooday GW, Trinci APJ (1980) Wall structure and biosynthesis in fungi. In: Gooday GW, Lloyd D, Trinci APJ (eds) The eukaryotic microbial cell. Cambridge Univ Press, Cambridge, pp 207–251

Green PB (1965) Pathways of cellular morphogenesis. A diversity in *Nitella*. J Cell Biol 27: 343–363

Green PB (1969) Cell Morphogenesis. Annu Rev Plant Physiol 20:365–394

Green PB, King A (1966) A mechanism for the origin of specifically oriented textures in development with special reference to *Nitella* wall texture. Aust J Biol Sci 19:421–437

Grove SN, Bracker CE (1970) Protoplasmic organization of hyphal tips among fungi: vesicles and Spitzenkörper. J Bacteriol 104:989–1009

Grove SN, Sweigard JA (1980) Cytochalasin A inhibits spore germination and hyphal tip growth in *Gilbertella persicaria*. Exp Mycol 4:239–250

Grove SN, Bracker CE, Morre DJ (1970) An ultrastructural basis for hyphal tip growth in *Pythium ultimum*. Am J Bot 57:245–266

Heath IB (1987) Preservation of a labile cortical array of actin filaments in growing hyphal tips of the fungus *Saprolegnia ferax*. Eur J Cell Biol 44:10–16

Heath IB, Gay JL, Greenwood AD (1971) Cell wall formation in the saprolegniales: Cytoplasmic vesicles underlying developing walls. J Gen Microbiol 65:225–232

Hoch HC, Staples RC (1985) The microtubule cytoskeleton in hyphae of *Uromyces phaseoli* germlings: its relationship to the region of nucleation and to the F-actin cytoskeleton. Protoplasma 124:112–122

Howard RJ (1981) Ultrastructural analysis of hyphal tip cell growth in fungi: Spitzenkörper, cytoskeleton and endomembranes after freeze-substitution. J Cell Sci 48:89–103

Howard RJ, Aist JR (1977) Effects of MBC on hyphal tip organization, growth and mitosis of *Fusarium acuminatum*, and their antagonism by D_2O. Protoplasma 92:195–210

Howard RJ, Aist JR (1979) Hyphal tip cell ultrastructure of the fungus *Fusarium*: improved preservation by freeze substitution. J Ultrastruct Res 66: 224–234

Jaffe LF (1968) Localization in the developing *Fucus* egg and the general role of localizing currents. Adv Morphog 7:295–327

Koch AL (1982) The shape of the hyphal tips of fungi. J Gen Microbiol 128:947–951

McClure WK, Park D, Robinson PM (1968) Apical organization in the somatic hyphae of fungi. J Gen Microbiol 50:177–182

McGillviray AM, Gow NAR (1987) The transhyphal electrical current of *Neurospora crassa* is carried principally by protons. J Gen Microbiol 133:2875–2881

McKerracher LJ, Heath IB (1987) Cytoplasmic migration and intracellular organelle movements during tip growth of fungal hyphae. Exp Mycol 11:79–100

Picton JM, Steer MW (1982) A model for the mechanism of tip extension in pollen tubes. J Theor Biol 98:15–20

Prosser JI (1979) Mathematical modelling of mycelial growth. In: Burnett JH, Trinci APJ (eds) Fungal walls and hyphal growth. Cambridge Univ Press, Cambridge, pp 359–384

Prosser JI, Trinci APJ (1979) A model for hyphal growth and branching. J Gen Microbiol 111:153–164

Reinhardt MO (1892) Das Wachstum der Pilzhyphen. Jahrb Wiss Bot 23:479–566

Riva Ricci D da, Kendrick B (1972) Computer modelling of hyphal tip growth in fungi. Can J Bot 50:2455–2462

Robertson NF (1965) Presidential address: The fungal hypha. Trans Br Mycol Soc 48:1–8

Runeberg P, Raudaskoski M (1986) Cytoskeletal elements in the hyphae of the homobasidiomycete *Schizophyllum commune* visualized with indirect immunofluorescence and NBD phallacidin. Eur J Cell Biol 41:25–32

Saunders PT, Trinci APJ (1979) Determination of tip shape in fungal hyphae. J Gen Microbiol 110:469–473

Schreurs WJ, Harold FM (1988) Transcellular proton current in *Achlya bisexualis* hyphae: Relationship to polarized growth. Proc Natl Acad Sci USA 85:1534–1538

Stewart PR, Rogers PJ (1983) Fungal dimorphism. In: Smith JE (ed) Fungal Differentiation. Dekker, New York Basel, pp 267–313

Trinci APJ, Saunders PT (1977) Tip growth of fungal hyphae. J Gen Microbiol 103:243–248

Tucker BE, Hoch HC, Staples RC (1986) The Involvement of F Actin in *Uromyces* Cell Differentiation. The Effects of Cytochalasin E and Phalloidin. Protoplasma 135:88–101

Wessels JGH (1986) Cell wall synthesis in apical hyphal growth. Int Rev Cytol 104:37–79

Wessels JGH, Sietsma JH (1981) Cell wall synthesis and hyphal morphogenesis: a new model for apical growth. In: Robinson DG, Quader H (eds) Cell walls 1981. Wiss Verlagsges, Stuttgart, pp 135–142

Chapter 5

Inhibition of Chitin Metabolism

G. W. Gooday[1]

1 Chitin Metabolism as a Target for Antifungal Agents

Chitin, the $(1\rightarrow4)$-β-linked homopolymer of N-acetyl-D-glucosamine, is a character-istic component of the cell walls of nearly all zoopathogenic and phytopathogenic fungi, and also of the skeletal structures of most invertebrates. As vertebrates and higher plants do not produce this polysaccharide, chitin metabolism is an attractive target for antifungal agents and pesticides (Gooday 1977). A clear analogy is with the peptidoglycan of bacterial walls, the synthesis of which is inhibited by many an-tibacterial agents, notably the β-lactams, and the structure of which is attacked by the defensive enzyme, lysozyme. In the case of chitin we find that nature has exploited the potential of chitin as a target, with the antifungal antibiotics, polyoxins and nik-komycins, produced by streptomycetes isolated from soil, and with chitinases being used as defense enzymes by plants and animals. The antifungal actions to be discussed here, of the very specific polyoxins and nikkomycins and of purified chitinases, are ample proof of the essential nature of chitin in fungal growth.

2 Chitin Synthase

2.1 Properties of Chitin Synthase

Chitin synthase, UDP-N-acetamido-2-deoxy-D-glucose: chitin 4-β-acetamidodeoxy-D-glucosyl-transferase, EC 2.4.1.16, has now been characterised from many fungi and invertebrates (reviews by Cabib 1987; Gooday 1983; Kramer and Koga 1986). In all cases the substrate is the nucleotide sugar UDP-N-acetylglucosamine, UDP-GlcNAc, and the product is UDP:

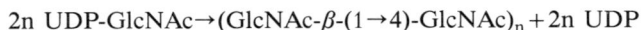

$$2n\ \text{UDP-GlcNAc} \rightarrow (\text{GlcNAc-}\beta\text{-}(1\rightarrow4)\text{-GlcNAc})_n + 2n\ \text{UDP}$$

No primer has been identified. A divalent cation is required, most probably Mg^{2+}, but Mn^{2+} or Co^{2+} can sometimes substitute. The Km value for UDP-GlcNAc is

[1] Department of Genetics and Microbiology, Marischal College, University of Aberdeen, Aberdeen, AB9 1AS, Scotland

Table 1. Representative effects of polyoxins and nikkomycins on chitin synthase and fungal growth

Organism	Km UDP-GlcNAc (µM)	Antibiotic	Ki (µM)	MIC (µM)	Effect	References
Chytridiomycete, *Allomyces macrogynus*	1200	Polyoxin D	–	25	Swelling and bursting	(1)
		Nikkomycin X/Z	–	0.01		
Zygomycete, *Mucor rouxii*	500	Polyoxin D	0.6, 2	19100	Swelling and bursting	(2)
		Polyoxin A	0.6	100		
		Nikkomycin X	0.5	0.5		
		Nikkomycin Z	3.5	0.5		
Ascomycete, *Neurospora crassa*	2100	Polyoxin D	1.4	190	Swelling and bursting	(3)
		Polyoxin B	32			
		Nikkomycin X/Z	2			
Hemiascomycete, *Saccharomyces cerevisiae*	(Chs1) 500, 800 (Chs2) 700,800	Polyoxin D	(Chs1) 1, 3 (Chs2) 1, 1.5	190	{ Exploded pairs, refringent pairs	(4)
Deuteromycete, *Candida albicans*	200, 2000	Polyoxin D	1.2, 0.75	20	Bulbous chains of cells, death	(5)
		Nikkomycin X	–	1		
Basidiomycete, *Coprinus cinereus*	900	Polyoxin D	3	0.5	No stipe elongation, autolysis	(6)

References (1) Porter and Jaworski (1966); J. Youatt, N.A.R. Gow, G.W. Gooday, unpublished results; (2) Bartnicki-Garcia and Lippman (1972); Müller et al. (1981); Furter and Rast (1985); (3) Endo et al. (1970b); Gow and Selitrennikoff (1984); (4) Bowers et al. (1974); Sburlati and Cabib (1986); Orlean (1987); (5) Chiew et al. (1980); J.C. Hardy, G.W. Gooday, unpublished results; Becker et al. 1983; Shenbagamurthi et al. (1983); Hector and Braun (1986); (6) Gooday et al. (1976).

usually of the order of 1 mM (Table 1), about the same as its intracellular concentration. The enzyme is activated by the monomer of chitin, N-acetylglucosamine, and its soluble oligomers, but at concentrations much higher than are likely to occur in the cell. The enzyme is usually at least in part found in a zymogenic form, both as cytoplasmic vesicles, chitosomes, and in the plasma membrane, in both cases being activatible by proteases (Bartnicki-Garcia et al. 1979; Cabib et al. 1979; see also Chapters 3 and 4). The active enzyme is an intrinsic membrane-bound protein, requiring phospholipids for activity. Thus its action is vectorial, accepting substrate, cations and other effectors from the cytoplasm, and feeding out macromolecular chitin chains to the wall.

The enzyme is inhibited by its product, UDP, competitively with the substrate UDP-GlcNAc. In *Coprinus cinereus* there is an accompanying nucleoside diphosphatase activity, hydrolysing the UDP to the less inhibitory UMP, and this enzyme may play a role in the endogenous regulation of chitin synthesis (Rousset-Hall and Gooday 1975; Gooday 1979a).

2.2 Inhibition by Polyoxins and Nikkomycins

2.2.1 Nature of the Antibiotics

Polyoxins and nikkomycins are two classes of closely related nucleoside-di- and -tripeptide antibiotics which are highly specific competitive inhibitors of chitin synthase, mimicking its substrate UDP-GlcNAc (Fig. 1).

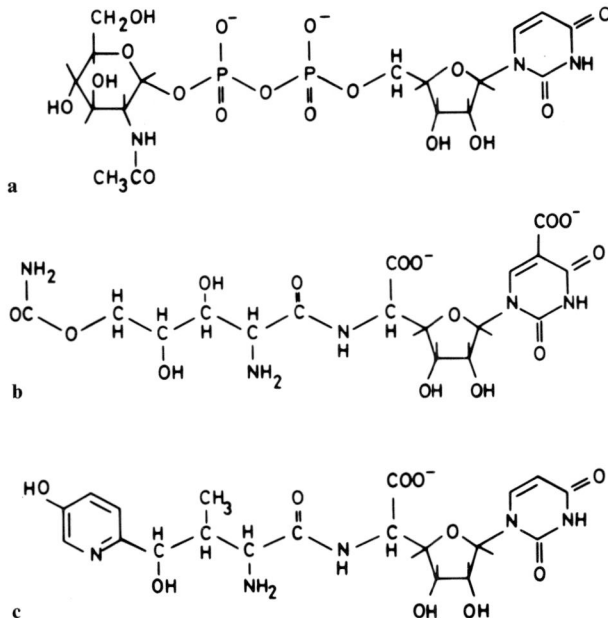

Fig. 1. a UDP-N-acetyl-glucosamine, **b** polyoxins D, **c** nikkomycin Z

The polyoxins were characterised in the 1960s as at least a dozen metabolites of *Streptomyces cacaoi* var. *asoensis* (Isono et al. 1965, 1969). Their action was elucidated soon afterwards (Endo and Misato 1969; Endo et al. 1970a, b). They inhibited incorporation of labelled glucosamine into chitin in fungal cells, resulting in an accumulation of labelled UDP-*N*-acetylglucosamine, and competitively inhibited cell-free chitin synthase preparations. This specificity has been amply confirmed in a wide range of other fungi. To date inhibition has not been reported of any related enzyme, such as UDP-*N*-acetylglucosamine pyrophosphorylase, glucan synthases, or glycoprotein synthesizing systems. Conversely, all cell-free chitin synthase systems to date have been shown to be susceptible, with Ki values typically in the order of a few micromolar, i.e. several-hundred-fold less than values for Km (Table 1; Gooday 1979b).

The nikkomycins were characterised in the 1970s as at least a dozen metabolites of *Streptomyces tendae* (Dähn et al. 1976; Hagenmaier et al. 1979, 1981; König et al. 1980; Fiedler et al. 1982). They have also been described as the neopolyoxins, by Kobinata et al. (1980) and Uramoto et al. (1980, 1982) from a new strain of *S. cacaoi* ssp. *asoensis*. The nikkomycins have a very similar action to polyoxins, acting as inhibitors of chitin synthase competitive with substrate UDP-*N*-acetylglucosamine (Müller et al. 1981; Furter and Rast 1985). Using a cell-free preparation of chitin synthase from *Neurospora crassa*, Gow and Selitrennikoff (1984) showed that polyoxin B, nikkomycin Z and UDP competed with each other and the substrate for binding to chitin synthase.

2.2.2 Uptake of Polyoxins and Nikkomycins

As polyoxins and nikkomycins are competitive inhibitors of chitin synthase which is an integral protein in the plasma membrane, accepting substrate from the cytosol, it is to be expected that these nucleoside-peptide antibiotics require transport into the cell for their antifungal activity. Evidence for this comes from observations that the enzyme chitin synthase from a wide range of fungi is approximately equally susceptible to inhibition, but that the intact fungi show a very wide range of susceptibility (Table 1). In general, fungi such as *Candida albicans* that are relatively resistant to polyoxin are more susceptible to nikkomycin, but both antibiotics are powerful inhibitors of the chitin synthase of the particular fungus. Bowers et al. (1974) showed that the action of polyoxin D on the relatively insensitive *Saccharomyces cerevisiae* was potentiated by non-lethal levels of the permeabilizing agents dimethylsulphoxide or amphotericin B. When polyoxins were used in the field as fungicides against black spot of pear, *Alternaria kikuchiana*, resistant strains emerged (Nishimura et al. 1973). Hori et al. (1974a) showed that the chitin synthase from resistant strains was still susceptible to polyoxin B. Thus impaired uptake of polyoxin was a likely cause of resistance. Mitani and Inoue (1968) had shown that the antifungal action of polyoxins against *Pellicularia sasakii* and *Rhizoctonia solani* was antagonised by the presence of some specific small peptides, such as glycyl-L-valine, in the medium, or by the presence of peptide-rich nutrients such as peptone, yeast extract or casein hydrolysate. Bowers et al. (1974) and Becker et al. (1983) demonstrated the same phenomenon for *S. cerevisiae* and *C. albicans*, respectively, both of which had previously been considered resistant to polyoxin. Hori et al. (1977) produced more direct evidence implicating a peptide uptake system in polyoxin transport by showing that the uptake of tritiated polyoxin A by *A. kikuchiana* was antagonised by peptides such as glycylglycine, in a competitive manner. Resistant strains of the fungus showed diminished uptake. The uptake of polyoxins and nikkomycins has been studied in detail for *C. albicans*. Payne and Shallow (1985) and McCarthy et al. (1985) showed that strains and mutants resistant to polyoxin showed cross-resistance to nikkomycin, and also to a toxic peptide, bacilysin. These authors and Yadan et al. (1984) showed that mutants resistant to nikkomycin had changed patterns of peptide uptake. McCarthy et al. (1985) showed that the uptake of two peptides that antagonise nikkomycin action, alanylalanine and leucylglycine, was impaired in resistant mutants. Yadan et al. (1984) and Gonneau et al. (1986) studied uptake of [^{3}H]-nikkomycin Z, and showed that this was antagonised by dimethionine and trimethionine, while uptake of [^{14}C]-dimethionine was antagonised by nikkomycin Z. A nikkomycin resistant mutant showed impaired uptake of both nikkomycin Z and dimethionine.

2.2.3 Synthetic and Semisynthetic Polyoxins and Nikkomycins

Efforts have been made to improve the antifungal potential of polyoxins and nikkomycins by producing semisynthetic or totally synthetic analogues. In particular the major aim has been to increase activity against human pathogenic fungi, especially *C. albicans*, where a major problem is the poor rate of uptake of the natural compounds by peptide permeases. The possibility of cleavage of peptide bonds by peptidases of pathogen or host must also be considered, as this could lead to inactivation (or possibly activation of a suicide substrate) according to the chemistry of the compound in question.

The polyoxins have uracil, 5-hydroxymethyl-uracil, uracil-5-carboxylic acid or thymine as the pyrimidine base. This forms a nucleoside, linked to the C1 of a unique 5-amino-uronic acid, which is in turn linked by a peptide bond through its amine group to an 5-amino acid, polyoximic acid. In polyoxins A, F, G, I, K the carboxyl group of the 5-amino-uronic acid is linked by a peptide bond to the amine group of polyoximix acid, to give a tripeptide structure (Isono et al. 1969). Synthesizing aminoacyl derivatives of 5'-amino-5'-deoxyuridine, Isono et al. (1971 a, b) showed that the 5' carboxyl group of the amino-uronic acid (or the carboxyl group of the polyoximic acid) is essential for biological activity, simulating the phosphate groups of UDP-GlcNAc.

Hori et al. (1971, 1974b) showed that uridine, 2'-deoxythymidine and 5'-amino-5'-deoxyuridine, which are basal structures of polyoxins, and several of their derivatives are competitive inhibitors of chitin synthase from *Pyricularia oryzae*. Cytidine and purine nucleosides however show no inhibitory activity. These authors conclude that a specific binding site to the uridine moiety of UDP-GlcNAc is present on the enzyme and that this is the site of competitive binding of uridine and its derivatives. Adams and Gooday (1983) found that a wide range of natural and synthetic derivatives of uridine inhibited the chitin synthase preparation from *C. cinereus* which they had developed as an assay for inhibitors (Adams and Gooday 1980). The natural products, uridine, UDP, polyoxin and nikkomycin, were the most inhibitory, but con-

siderable inhibition was observed with the synthetic derivatives 2'- and 5'-deoxy-
uridine, 5-fluorouridine, 5'-O-acetyluridine and 5'-O-octanoyluridine. Hori et al.
(1974b) investigated the inhibitory effects of a range of derivatives of polyoxin C and
thymine-polyoxin C on the chitin synthase preparation from P. oryzae, and expressed
them in terms of Ki and a calculated binding affinity (kcal·mol^{-1}, 25 °C). They concluded that the carbamoyl-polyoximic acid moiety of polyoxin helps to stabilize the
polyoxin-enzyme complex. Hori et al. (1974c) then investigated the effect of pH on
Ki values and concluded that the ionized amino group at the C-2'' position has a very
important role for the binding of polyoxins to the enzyme. Azuma et al. (1977) synthesized   derivatives   of   polyoxin   C   containing   the   hydrophobic   residue
(N-5-lauroyl)-ornithyl at the amino terminus and reported low activity towards a
range of phytopathogens. Boehm and Kingsbury (1986) report the synthesis of
polyoxin analogues with N-methylated peptide bonds, in order to confer resistance to
peptidase hydrolysis, but the products had only moderate activity on the enzyme. Emmer et al. (1985) also synthesized a range of peptide derivatives of polyoxin C. In
agreement with the results of Hori et al. (1974b) with chitin synthase from P. oryzae,
they found that some, in this case L-alanyl, L-citrullyl and L-threonyl, were moderate-
ly inhibitory to chitin synthase from C. albicans, but their 5'-epimers were inactive,
as were related compounds lacking the peptide bond. A novel dipeptide analogue of
nikkomycin, of uracil polyoxin C and an aromatic amino acid found in the antibiotic
echinocandin was a highly active inhibitor of the enzyme. This, however, had no inhibitory effect on cells of C. albicans, in contrast to the high activity of the L-citrullyl
derivative against growing cells of the plant pathogens A. kikuchiana, P. oryzae and
Cochliobolus miyabeanus (Isono et al. 1971b).

Shenbagamurthi et al. (1983) describe the synthesis of a range of polyoxin
analogues designed to penetrate the cell of C. albicans more readily than the natural
compounds. Using uridine as substrate they prepared peptide analogues of polyoxin
L. All compounds with a peptide bond inhibited the cell-free enzyme system. The
most active compounds had amino acid residues that were very hydrophobic. There
was a reasonable correlation between the ID$_{50}$ (concentration inhibiting enzyme reaction by 50%) and MEC (minimum concentration giving 5% morphologically abnormal cells), but not between ID$_{50}$ and viability (concentration at which 50% of cells
failed to form colonies after incubation with the compound). A possible explanation
was provided by their finding that most of their dipeptidyl analogues were rapidly
broken down to inactive metabolites by a cell-free extract from C. albicans, in contrast
to the naturally occurring polyoxin D. Thus they suggest that these analogues could
be taken up by the cells, result in morphological observations, but then be detoxified
by intracellular peptidases, allowing the cells to recover. Naider et al. (1983) synthesized tripeptidyl polyoxins and these gave similar results, showing some inhibition of
chitin synthase activity, of growth and of viability, but being readily hydrolysed by
cell-free extracts. Like the natural polyoxins, their activity on cells was antagonised
by peptides in the medium, suggesting that they are taken up by peptide permeases.
Only one analogue, leucyl-norleucyl-uracil polyoxin C, competed with trimethionine
uptake, which suggests that the others used another peptide uptake system. Smith et
al. (1986a, b) then synthesized hydrophobic polyoxin derivatives with the aim of producing compounds that would be taken up readily by the cells, be resistant to internal
hydrolysis and be inhibitors of chitin synthase activity. The most promising com-

pound combining these two latter properties was *N*-ε-(octanoyl)-lysyl uracil polyoxin C, but it appeared to be poorly taken up by cells, and was less active than polyoxin D against *C. albicans*.

2.2.4 Antifungal Action in Culture

All chitin synthase preparations are about equally susceptible to polyoxins and nikkomycins. Different fungi however show a very wide range of susceptibility in cultures (Table 1; Gooday 1979b). This is probably chiefly due to differences in uptake of the antibiotics, in part inherent in differences in specificities and kinetic properties of the peptide permeases, in part in some cases resulting from antagonism from components in the medium. Nevertheless, it is clear that any fungal morphogenetic process involving chitin deposition will be affected when polyoxins and nikkomycins have been applied in the appropriate concentration and conditions (Table 1).

Thus growing vegetative hyphae of *N. crassa* and *C. cinereus* respond to both polyoxin D and nikkomycin X/Z by ballooning of the apices, multiple apical branching, "beading" of the hyphae, and finally bursting (G.W. Gooday, unpublished results). The elongation of stipes of the mushroom *C. cinereus*, which involves the intercalary elongation of cells with much chitin synthesis, is completely inhibited by polyoxin D and nikkomycin X/Z, and the stipe tissue eventually autolyses (Gooday et al. 1976; G.W. Gooday, unpublished results). J. Youatt, N.A.R. Gow and G.W. Gooday (unpublished results) have investigated the effects of polyoxin and nikkomycin on zoospores of *Allomyces macrogynus*. Ungerminated zoospores (which lack chitin) swam for up to two days in the presence of both antibiotics, emphasizing their specificity for chitin synthesis, but germination and outgrowth of spores were inhibited. With polyoxin, growth was normal at 12.5 μM, but abnormal swellings occurred at the tip and base of the cyst and in the rhizoids after 4.5 h at 25 μM and after 3 h at 100 μM. Eventually the cells lost their contents. Nikkomycin was more potent, having effects more quickly and at lower concentrations. Growth was normal at 5 nM, but abnormal swellings occurred at 10 nM. If nikkomycin was present when germination was induced, the spores still rounded up but soon disintegrated. At later stages swelling and bursting of rhizoids was observed. At still later stages apical swellings of hyphae were observed, often with bursting.

Both chitin synthases characterised from *S. cerevisiae*, Chs 1 (with no apparent synthetic role in cell growth) and Chs 2 (active during budding), are sensitive to polyoxin D (Table 1; Sburlati and Cabib 1986; Orlean 1987; see also Chapter 3). Treatment of growing cells with polyoxin D results in two responses, "exploded pairs" with bursting at the site of budding, and partially lysed "refringent pairs", in both cases of unseparated mother cells and full-sized buds (Bowers et al. 1974). When strains with a disrupted Chs 1 were constructed, they were found to be more sensitive to polyoxin D than the parental strain, with a higher proportion of refringent cells (Bulawa et al. 1986). Polyoxin D also inhibits septation and separation of buds of yeast cells of *C. albicans*, leading to cell death, but four different strains tested showed very different sensitivities (Becker et al. 1983). Treatment of both yeast cells and mycelia of this dimorphic fungus with polyoxin D led to a chain of bulbous cells (Becker et al. 1983; Hilenski et al. 1986). The polyoxins and, much more powerfully, nikkomycins X and Z, are active against yeast cells and mycelium of the dimorphic fungus *Mucor rouxii* (Bartnicki-Garcia and Lippman 1982; Müller et al. 1981; Furter and Rast 1985). The antibiotics caused abnormal swellings and the bursting of both growth forms. Sporangiospores were more sensitive to nikkomycins than yeast or mycelium. They showed enhanced swelling, and cell death at the presumptive time of germ tube emergence (Furter and Rast 1985).

Although in all systems examined so far the direct effect of polyoxins and nikkomycins is solely to inhibit chitin synthesis, there is an interesting secondary effect in the inhibition of accumulation of alkali-insoluble $(1\rightarrow3)$-β-glucan in *Schizophyllum commune* (Sonnenberg et al. 1982). These authors ascribe this to the prevention of the formation of the glucan/chitin complex, as there is a concomitant accumulation of water-soluble glucan, the precursor of the complex (see also Chapter 6).

2.2.5 Antifungal Action In Vivo

As well as these activities on fungi growing in vitro, the polyoxins and nikkomycins also have activities on pathogenic fungi in vivo. Polyoxins are formulated as agricultural sprays, and used widely in Japan against plant pathogens, especially *A. kikuchiana*, black spot of pear; *P. sasakii*, rice sheath blight; and *C. miyabeanus* (Misato and Kakiki 1977; Misato et al. 1977). Nikkomycins also are active in the field, against *Botrytis cinerea* and *Uromyces phaseoli* on beans, and *Puccinia recondita* on wheat (Fiedler et al. 1982). Becker et al. (1988) report the effects of administering polyoxin or nikkomycin on survival of mice experimentally infected with *C. albicans* to give systemic candidiasis. In their regime, polyoxin D gave no protection, while treatment with nikkomycin increased the survival times of the mice. In their experiments no mouse died of candidal infection while receiving nikkomycin, but stopping the drug treatment resulted in re-establishment of the infection and death of the mice.

2.2.6 Effect on Other Organisms

Neither polyoxins nor nikkomycins have any reported significant toxicities to vertebrates or higher plants (Isono et al. 1967; Fiedler et al. 1982; Adams and Gooday 1983), reflecting their specificity for chitin synthase and reinforcing the appeal of chitin synthesis as a target for fungal control.

Against microbes other than fungi, polyoxins and nikkomycins prevent synthesis of the β-chitin flotation spines of centric diatoms (Gooday et al. 1985), and of chitinous cyst formation in amoebae (Avron et al. 1982). Nikkomycin inhibits the production of microfilariae by the nematode *Brugia pahangi* in culture (R. Howells, L. J. Brydon, L. H. Chappell, G. W. Gooday, unpublished results). Both polyoxins and nikkomycins are active against a wide range of chitin-synthesizing systems in insects, mites and other invertebrates (Fiedler et al. 1982; Kramer and Koga 1986).

2.3 Effects of Other Compounds on Chitin Synthase

To date polyoxins and nikkomycins are the only reported antifungal agents with precise specificity for chitin synthesis. Many other compounds inhibit chitin synthase preparations, but also affect other cell processes. Thus tunicamycin, a potent inhibitor of glycoprotein lipid-intermediate biosynthesis, is a weak competitive inhibitor of chitin synthase preparations from *N. crassa* (Selitrennikoff 1979), *C. cinereus* and *C. albicans* (D. J. Adams, J. C. Hardy, G. W. Gooday, unpublished observations). The membrane-active polyene antibiotics, amphotericin B, nystatin and filipin, are non-competitive inhibitors of chitin synthase preparations from *M. rouxii, Agaricus bisporus* and *C. albicans* (Rast and Bartnicki-Garcia 1981; Hänseler et al. 1983; Nozawa et al. 1985). Ergosterol, the characteristic sterol of fungal membranes, inhibits a chitin synthase preparation from *C. albicans* (Chiew et al. 1982), and unsaturated fatty acids inhibit chitin synthase from *S. cerevisiae* (Duran and Cabib 1978). Novobiocin, an antibiotic interacting with magnesium ions and strongly inhibiting DNA metabolism, is a non-competitive inhibitor of chitin synthase from *C. cinereus*, giving 50% inhibition at 140 µM (D. J. Adams and G. W. Gooday, unpublished results). Calcofluor White M2R, a fluorescent brightener with a strong affinity for chitin, non-competitively inhibits chitin synthase from *N. crassa* (Selitrennikoff 1984), *Geotrichum lactis* (Roncero and Duran 1985) and *C. cinereus* (P. Reilly and G. W. Gooday, unpublished results). This dye also has the effect of disrupting the assembly of chitin chains into microfibrils, but this results in increasing accumulations of chitin cells treated with it (Roncero and Duran 1985). Another fluorescent dye, primulin, also non-competitively inhibits chitin synthase from *N. crassa*, though Congo red, which has similar effects on cells to Calcofluor White M2R, does not (Selitrennikoff 1985).

The effect of Calcofluor White to give abnormal accumulations of chitin in walls of *C. albicans* (Rico et al. 1985) is also shown by monensin, an ionophore (Poli et al. 1986), but probably by an unrelated mechanism, and both probably not directly by interacting directly with chitin synthase. Also indirect is the effect of the organophosphorus fungicide, Kitazin P (IBP), which inhibits chitin synthesis in *P. oryzae* via inhibiting phosphatidylcholine biosynthesis (Kodama et al. 1979; see also Chapter 17).

The benzoylaryl ureas, very successful insecticides, specifically inhibit chitin biosynthesis during insect development, but not by direct interaction with either chitin synthase or chitinases (Willems et al. 1986). Rather, Mitsui (1986) suggests that they inhibit transport of UDP-GlcNAc within the insect tissues. Neither the benzoylaryl ureas, nor the unrelated inhibitor of insect chitin synthesis, buprofezin (Izawa et al. 1985) have any appreciable effect on fungal growth or chitin synthesis (G. W. Gooday, unpublished results). Effects of a wide range of other insecticides on a chitin synthase system from *Phycomyces blakesleeanus* are described by Marks et al. (1982), but no other compounds have noteworthy specificity.

3 Glutamine-Fructose-6-P-Aminotransferase

This enzyme, EC 2.6.1.16 (formerly glucosamine-6-phosphate synthetase EC 5.3.1.19), catalyses the first step of the biosynthetic pathway leading to UDP-GlcNAc:

L-Glutamine + D-Fructose-6-P→D-Glucosamine-6-P + L-Glutamate.

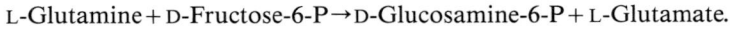

In *N. crassa* this enzyme is subject to a feedback inhibition by UDP-GlcNAc, so that nucleotide sugar does not accumulate when cells are treated with polyoxin D (Endo et al. 1970a). The peptide antibiotic tetaine (bacilysin, bacillin) is a powerful inhibitor of growth of yeast and especially mycelium of *C. albicans*, of the germination of the yeast cells and of chitin and mannoprotein biosynthesis, by an interesting route (Milewski et al. 1983, 1986a, b). It is taken up by a different peptide uptake system for each cell type, and when in the cell is cleaved by peptidases to give the antibiotic anticapsin, L-β-(2,3-epoxycyclo-hexanono-4)-alanine, an analogue of glutamine and a powerful inhibitor of the glutamine-fructose-6-P-aminotransferase from *C. albicans*. It is competitive with respect to L-glutamine and uncompetitive with respect to D-fructose-6-phosphate, and incubation of the enzyme in the absence of glutamine leads to the formation of an inactive enzyme, irreversibly modified. It thus inhibits the supply of UDP-GlcNAc for chitin and glycoprotein biosynthesis. It is also antibacterial, inhibiting peptidoglycan and lipopolysaccharide biosynthesis. Milewski et al. (1984) report the synthesis of other glutamine analogues that also inhibit the glutamine-fructose-6-P-aminotransferase from *C. albicans*, and clearly this is a worthwhile route to pursue further.

4 Chitinase

4.1 Roles for Chitinases in Fungi

Chitin-containing fungi that have been investigated all produce chitinases (EC 3.2.1.14). These may have an autolytic role, most obviously during gross autolysis during spore release in puff-balls, *Lycoperdon* spp. and ink caps, *Coprinus* spp. (Tracey 1955; Iten and Matile 1970). They may have a nutritional role, in saprophytes such as *Mortierella* spp. and *Aspergillus* spp. (Gray and Baxby 1968; Reyes et al. 1988), and in pathogens of other fungi and invertebrates such as *Aphanocladium album* and *Beauvaria bassiana*, respectively (Srivastava et al. 1985; Coudron et al. 1984). What concerns us here, however, is the possibility that they have morphogenetic roles during fungal growth. Most obviously they could have a role during branching of hyphae or budding of chitinous yeast cells, by locally weakening the wall to allow establishment of a new hyphal tip or bud, and an analogous role in germ tube emergence during spore germination. Mahadevan and Mahadkar (1970), Polachek and Rosenberger (1978) and Rosenberger (1979) describe autolytic enzymes including chitinases that are bound to sub-apical walls of *N. crassa* and *Aspergillus nidulans* and suggest that these are associated with hyphal branching rather than autolytic wall turnover. Bartnicki-Garcia (1973) presents the unitary model of cell wall growth in which lytic enzymes play a vital role in maintaining a balance between wall synthesis and wall lysis during hyphal apical growth, maintaining the apex in a plastic state and allowing insertion of new chitin into the wall.

Chitinases also however could have more positive roles in wall growth. The first could be regulating the formation of crystalline-chitin in its fully hydrogen-bonded form. Minke and Blackwell (1978) have shown by X-ray crystallography that α-chitin is composed of adjacent chains arranged in antiparallel form, and this clearly requires an orderly assembly of individual chains after synthesis. Vermeulen and Wessels (1984, 1986; see also Chapter 6) have shown that synthesis and crystallisation of chitin chains are two distinct processes, separated in space and time.

The second role could be in the modelling of the resultant microfibrils so that they take the form characteristic of the particular construction of the wall in that particular fungus. Thus Gow and Gooday (1983) show that some dimorphic fungi characteristically have short stubby chitin microfibrils in their hyphal walls compared to long microfibrils in filamentous fungi, for example being an average of 33 nm long in the dimorphic *C. albicans*, and very much longer in the mycelial *C. cinereus* and *Mucor mucedo*.

The third role for morphogenetic chitinases could be cross-linking of chitin to other wall components through glycosidic linkages. Chitinases, like other polysaccharases, can have transglycosylase activities (Usai et al. 1987). Covalent linkages between chitin and glucans have been described for a variety of fungal walls by Sietsma and Wessels (1979, 1981) and Mol and Wessels (1987; see also Chapter 6). They suggest that the linkages are through lysine and/or citrulline, possibly through amino groups of glucosamine (via deacetylation of *N*-acetylglucosamine), but do not rule out the possibility of other types of linkages. Whatever the chemistry of the linking, they implicate it in the rigidification of the hyphal wall which is essential for the transition between the plastic apical wall and the rigid hyphal tube. Young regenerating protoplasts not yet fully equipped to produce all wall components lay down long crystalline microfibrillar chitin unlike that of mature walls, and Wessels and Sietsma (1981) suggest that this may reflect the absence of enzymes which could cross-link the chitin chains to other wall polymers. In *S. cerevisiae* part of the chitinase activity is in the periplasmic space (Elango et al. 1982). This enzyme is secreted to the medium by protoplasts incubated in growth medium. In growing cells, the enzyme activity is highest during logarithmic phase of growth, and these authors suggest it plays a role during septum formation and cell separation (see also Chapter 3).

Evidence for the association of chitinases with chitin synthase comes from parallel behaviour of the two activities during spore germination in *M. mucedo* (Gooday et al. 1986), and during exponential growth in *M. rouxii* (D. M. Rast, R. Furter and G. W. Gooday, unpublished results) and *C. albicans* (Barrett-Bee and Hamilton 1984), and from the finding of a membrane-bound chitinase activity in the same cell fraction as chitin synthase in *M. mucedo* (Humphreys and Gooday 1984a, b). Herrera-Estrella and Ruiz-Herrera (1983), studying the light growth response of stage I sporangiophores of *P. blakesleeanus*, showed that exposure to white light resulted in increases in growth, chitin synthase activity, apical staining with fluorescein isothiocyanate-labelled wheat germ agglutinin (signifying accessible chitin), and free reducing groups in the chitin (as measured by reaction with sodium boro[^{3}H]hydride). They interpret this last observation as indicating the action of an endo-chitinase loosening the chitin in the wall as part of the process of the light-induced cell elongation, and correlate it with the finding of Ortega et al. (1975) of a transient mechanical weakening of the wall in response to light.

4.2 Allosamidins

Thus chitinase activities could provide a target for antifungal compounds. The first antibiotics specifically active against chitinases have recently been described. These are the allosamidins, produced by *Streptomyces* sp. no. 1713. They were identified in a screen to detect compounds with potential insecticidal activity by testing for inhibition of chitinase from silkworm pupae *Bombyx mori*, using the chromogenic substrate γ-chitin red (Sakuda et al. 1987a). Allosamidin proved to be insecticidal by preventing ecdysis in vivo. Independently Somers et al. (1987) isolated the same metabolite, designated as A82516, from *Streptomyces* strain A82516, using chitinase from *Streptomyces griseus* as the screen and the chromogenic substrate chitin azure. They found that it did not affect egg hatching of the housefly *Musca domestica*, but completely prevented development from larvae to pupae.

The allosamidins are pseudotrisaccharides, consisting of a disaccharide of *N*-acetylallosamine linked to a novel aminocylitol derivative, named as allosamizoline (Fig. 2) (Sakuda et al. 1986a, b; 1987b). Methylallosamidin, methylated at the 6-hydroxyl group of the non-reducing end of the disaccharide, is a co-metabolite.

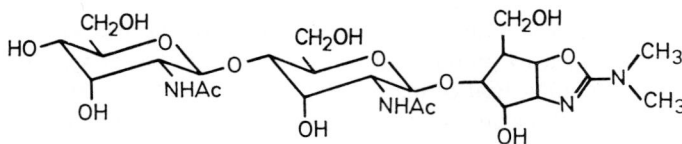

Fig. 2. Allosamidin

Table 2. Effect of allosamidin on chitinases from different sources

Organism	Assay[a]	IC$_{50}$ (nM)[b]	Reference
Nematode, *Onchocerca gibsoni*	A	0.20	Gooday et al. (1988)
Fish, *Scophthalmus maximus*, plasma	A	1.45	F. D. C. Manson and G. W. Gooday (unpublished)
Insect, *Bombyx mori*, pupal gut	B	386	Koga et al. (1987)
Fungus, *Neurospora crassa*	A	500	R. McNab, L. A. Glover and G. W. Gooday (unpublished)
Fish, *Scophthalmus maximus*, stomach	A	2000	F. D. C. Manson and G. W. Gooday (unpublished)
Bacteria, *Streptomyces griseus*	C	3700	Somers et al. (1987)
Streptomyces griseus	B	>53570	Koga et al. (1987)
Serratia marcescens	B	>53570	Koga et al. (1987)
Plant, yam	D	No inhibition[c]	Koga et al. (1987)

[a] A) Assayed with 45 μM 3,4-dinitrophenyl-tetra-*N*-acetylchitotetraoside; B) with 80 μM GlcNAc$_5$; C) with chitin azure (1.2 mg ml^{-1}); D) with 160 μM GlcNAc$_5$ and 0.4% colloidal chitin.
[b] Concentration giving 50% inhibition.
[c] Tested up to 100 μM. Also no activity against lysozymes or β-*N*-acetylglucosaminidases.

Koga et al. (1987) report interesting differential activity of allosamidin on chitinases from different sources, with good activity against the insect chitinase from *Bombyx mori*, poorer activities against bacterial chitinases from *S. griseus* and *Serratia marcescens* and no detectable activity against the plant chitinase from yam, lysozymes from egg white or human urine, or insect β-N-acetyl-glucosaminidase from *B. mori* (Table 2). We have shown that the allosamidin is active against secreted chitinases from the fungi *N. crassa* and *A. nidulans*, and very powerfully active against the chitinase of the nematode *Onchocerca gibsoni* (Table 2; Gooday et al. 1988). An endochitinase from *S. cerevisiae* is also inhibited by allosamidin (see Chapter 3). Interestingly it had very different activities against two different chitinase enzymes from the fish, turbot, *Scophthalmus maximus*. The enzyme in the blood plasma (probably a defence enzyme) was much more sensitive than that from the stomach (a food-processing enzyme making chitinous food accessible to digestive enzymes) (Table 2).

4.3 Other Inhibitors of Chitinase

The allosamidins are to date the only antibiotics reported that specifically inhibit chitinases. A report by Calcott and Fatig (1984) that avermectins, antibiotics with strong insecticidal, acaricidal and antihelminthic activities, were also fungicidal and inhibitors of chitinase has been shown to be incorrect when re-examined with purified compounds (Onishi and Miller 1985; Gordiner et al. 1987; D. M. Rast and G. W. Gooday, unpublished results).

There is however a report of an endogenous macromolecular inhibitor of chitinase which could play a role in regulating hyphal growth. Prick and Diekmann (1979) describe a lectin produced in hyphal walls of *N. crassa* that has a high affinity for chitin and *N*-acetylglucosamine, and that inhibits chitinase. Exogenously applied wheat germ agglutinin, a plant lectin with a high affinity for chitin, has been reported to be fungistatic for a range of fungi, perhaps by inhibiting selective lysis in the hyphal tip (Mirelman et al. 1975; Barkai-Golan et al. 1979).

4.4 Antifungal Activity of Chitinase Itself

High levels of chitinase activity are induced in plants in response to viral, bacterial and fungal infections, wounding, ethylene treatment and leaf abscission (Pegg and Young 1982; Metraux and Boller 1986; Kelly et al. 1987). This chitinase is clearly implicated in defense against potential pathogens, and plant chitinases have been shown to be inhibitors of fungal growth (Schlumbaum et al. 1986; Roberts and Selitrennikoff 1988). Chitinase activities with a potentially similar defensive role are found in the serum of some vertebrates, such as ruminants (Lundblad et al. 1979) and fish (Lindsay 1986; F. D. C. Manson and G. W. Gooday, unpublished results). Davies and Pope (1978) and Pope and Davies (1979) present results pointing to a possible therapeutic use for chitinase, since in particular when combined with polyene antibiotics nystatin or amphotericin B, injections of the enzyme extracts from the ink cap *Coprinus com-atus*, the puff-ball *Lycoperdon pyriforme* or the slime mould *Physarum poly-cephalum* prolonged survival of mice infected with *Aspergillus fumigatus* or *C. albicans*. Oranusi and Trinci (1985) have shown that chitinase from the bacterium *Cytophaga* sp. potentiated the antifungal activity of amphotericin B against *Candida pseudotropicalis* in vitro. The finding of Roberts and Selitrennikoff (1988) of an-tifungal activities of plant chitinases from the grains of wheat, barley and maize, but not those from the bacteria *Serratia marcescens, S. griseus* and *Pseudomonas stutzeri*, shows that the source of a chitinase has to be considered in any experimental ap-proach to controlling fungal pathogens of animals and plants.

5 Conclusions

From the foregoing it should be clear that chitin metabolism is a worthwhile target for antifungal agents, but that the ideal compounds remain elusive. The polyoxins, despite being used in large quantities in the field for many years, have the problems of a relatively narrow range of susceptible fungi and relatively rapid acquisition and spread of resistance. It is to be expected that resistance to nikkomycin would also arise relatively easily by changes in uptake specificities. Nevertheless, the promise of these compounds, and the recent discovery of the allosamidins, should encourage us to con-tinue to study in this area.

References

Adams DJ, Gooday GW (1980) A rapid chitin synthase preparation for the assay of potential fungicides and insecticides. Biotech Lett 2:75–78

Adams DJ, Gooday GW (1983) Chitin synthesis as a target – current progress. In: Lyr H, Polter C (eds) VI Int Symp Systemische Fungizide und antifungale Verbindungen. Abh Akad Wiss DDR 1982 Nr 1N. Akademie-Verlag, Berlin, pp 39–45

Avron B, Deutsch RM, Mirelman D (1982) Chitin synthesis inhibitors prevent cyst formation by *Entamoeba* trophozoites. Biochem Biophys Res Commun 108:815–821

Azuma T, Saita Y, Isoni K (1977) Polyoxin analogs. II. Synthesis and biological activity of aminoacyl derivatives of polyoxin C and L. Chem Pharm Bull 25:1740–1748

Barkai-Golan R, Mirelman D, Sharon N (1979) Studies on growth inhibition by lectins of Penicillia and Aspergilli. Arch Microbiol 116:119–124

Barrett-Bee K, Hamilton M (1984) The detection and analysis of chitinase activity from the yeast form of *Candida albicans*. J Gen Microbiol 130:1857–1861

Bartnicki-Garcia S (1973) Fundamental aspects of hyphal morphogenesis. In: Ashworth JM, Smith JE (eds) Microbial differentiation. Symp Soc Gen Microbiol, vol 23. Cambridge Univ Press, Cambridge, pp 245–268

Bartnicki-Garcia S, Lippman E (1972) Inhibition of *Mucor rouxii* by polyoxin D: effects on chitin syn-thetase and morphological development. J Gen Microbiol 71:301–309

Bartnicki-Garcia S, Ruiz-Herrera J, Bracker CE (1979) Chitosomes and chitin synthesis. In: Burnett JH, Trinci APJ (eds) Fungal walls and hyphal growth. Symp Br Mycol Soc, vol 2. Cambridge Univ Press, Cambridge, pp 149–168

Becker JM, Covert NL, Shenbagamurthi P, Steinfeld AS, Naider F (1983) Polyoxin D inhibits growth of zoopathogenic fungi. Antimicrob Agents Chemother 23:926–929

Becker JM, Marcus S, Tullock J, Miller D, Krainer E, Khare RK, Naider F (1988) Use of chitin-synthesis inhibitor nikkomycin to treat disseminated candidiosis in mice. J Infect Dis 157:212–214

Boehm JC, Kingsbury WD (1986) Rapid and convenient syntheses of polyoxin peptides containing N-methylated peptide bonds. J Org Chem 51:2307–2314

Bowers B, Levin G, Cabib E (1974) Effect of polyoxin D on chitin synthesis and septum formation in *Saccharomyces cerevisiae*. J Bacteriol 119:564–575

Bulawa CE, Slater M, Cabib E, Au-Young J, Sburlati A, Adair WC, Robbins PW (1986) The *S. cerevisiae* structural gene for chitin synthase is not required for chitin synthesis in vivo. Cell 46:213–225

Cabib E (1987) The synthesis and degradation of chitin. Adv Enzymol 59:59–101

Cabib E, Duran A, Bowers B (1979) Localized activation of chitin synthase in the initiation of yeast septum formation. In: Burnett JH, Trinci APJ (eds) Fungal walls and hyphal growth. Symp Br Mycol Soc, vol 2. Cambridge Univ Press, Cambridge, pp 189–201

Calcott PH, Fatig RO (1984) Inhibition of chitin metabolism by avermectin in susceptible organisms. J Antibiot 37:253–259

Chiew YY, Shepherd MG, Sullivan PA (1980) Regulation of chitin synthesis during germ-tube formation in *Candida albicans*. Arch Microbiol 125:97–104

Chiew YY, Sullivan PA, Shepherd MG (1982) The effects of ergosterol and alcohols on germ-tube formation and chitin synthase in *Candida albicans*. Can J Biochem 60:15–20

Coudron TA, Kroha MJ, Ignoffo CM (1984) Levels of chitinolytic activity during development of three entomopathogenic fungi. Comp Biochem Physiol 79B:339–348

Dähn U, Hagenmaier H, Höhne H, König WA, Wolf G, Zähner H (1976) Stoffwechselprodukte von Mikroorganismen. 154. Mitteilung. Nikkomycin, ein neuer Hemmstoff der Chitin Synthese bei Pilzen. Arch Microbiol 107:143–160

Davies DAL, Pope AMS (1978) Mycolase, a new kind of systemic antimycotic. Nature (London) 273:235–236

Duran A, Cabib E (1978) Solubilization and partial purification of yeast chitin synthetase. J Biol Chem 235:4419–4425

Elango N, Correa JV, Cabib E (1982) Secretory nature of yeast chitinase. J Biol Chem 257:1398–1400

Emmer G, Ryder NS, Grassberger MA (1985) Synthesis of new polyoxin derivatives and their activity against chitin synthase from *Candida albicans*. J Med Chem 28:278–281

Endo A, Misato T (1969) Polyoxin D, a competitive inhibitor of UDP-*N*-acetylglucosamine: chitin *N*-acetylglucosaminyltransferase in *Neurospora crassa*. Biochem Biophys Res Commun 37:718–722

Endo A, Kakiki K, Misato T (1970a) Feedback inhibition of L glutamine D fructose-6 phosphate amido transferase by UDP *N* acetylglucosamine in *Neurospora crassa*. J Bacteriol 103:588–594

Endo A, Kakiki K, Misato T (1970b) Mechanism of action of the antifungal agent polyoxin D. J Bacteriol 104:189–196

Fiedler HP, Kurth R, Langhärig J, Delzer J, Zähner H (1982) Nikkomycins: microbial inhibitors of chitin synthetase. J Chem Tech Biotechnol 32:271–280

Furter R, Rast DM (1985) A comparison of the chitin synthase-inhibitory and antifungal efficacy of nucleoside-peptide antibiotics: structure: activity relationships. FEMS Microbiol Lett 28:205–211

Gonneau M, Yadan JC, Sarthou P, Le Goffic F (1986) Nikkomycin X as inhibitor of *Candida albicans* growth. In: Muzzarelli RAA, Jeuniaux C, Gooday GW (eds) Chitin in nature and technology. Plenum Press, New York London, pp 203–205

Gooday GW (1977) Biosynthesis of the fungal wall – mechanisms and implications. The first Fleming lecture. J Gen Microbiol 99:1–11

Gooday GW (1979a) Chitin synthesis and differentiation in *Coprinus cinereus*. In: Burnett JH, Trinci APJ (eds) Fungal walls and hyphal growth. Symp Br Mycol Soc, vol 2. Cambridge Univ Press, Cambridge, pp 204–223

Gooday GW (1979b) The action of polyoxin on fungi. In: Lyr H, Polter C (eds) V Int Symp Systemfungizide. Abh Akad Wiss DDR 1979 Nr 2N. Akademie-Verlag, Berlin, pp 159–168

Gooday GW (1983) The microbial synthesis of cellulase, chitin and chitosan. In: Bushell ME (ed) Microbial polysaccharides. Prog Indust Microbiol, vol 18. Elsevier, Amsterdam, pp 85–127

Gooday GW, Rousset-Hall A de, Hunsley D (1976) The effect of polyoxin D on chitin synthesis in *Coprinus cinereus*. Trans Br Mycol Soc 67:77–83

Gooday GW, Woodman J, Casson EA, Browne CA (1985) Effect of nikkomycin on chitin spine formation in the diatom *Thalassiosira fluviatilis*, and observations on its peptide uptake. FEMS Microbiol Lett 28:335—340

Gooday GW, Humphreys AM, McIntosh WH (1986) Roles of chitinases in fungal growth. In: Muzzarelli RAA, Jeuniaux C, Gooday GW (eds) Chitin in nature and technology. Plenum Press, New York London, pp 83—91

Gooday GW, Brydon LJ, Chappell LH (1988) Chitinase in female *Onchocerca gibsoni* and its inhibition by allosamidin. Mol Biochem Parasitol 29:223—225

Gordiner PM, Brezner J, Tanenbaum SW (1987) Chitin metabolism: not a target of avermectin/milbemycin activity in insects. J Antibiot 40:110—112

Gow LA, Selitrennikoff CP (1984) Chitin synthetase of *Neurospora crassa*: inhibition by nikkomycin, polyoxin and UDP. Curr Microbiol 11:211—216

Gow NAR, Gooday GW (1983) Ultrastructure of chitin in hyphae of *Candida albicans* and other dimorphic and mycelia fungi. Protoplasma 115:52—58

Gray TRG, Baxby P (1968) Chitin decomposition in soil. II. The ecology of chitinoclastic microorganisms in forest soil. Trans Br Mycol Soc 51:293—309

Hänseler E, Nyhlen LE, Rast DM (1983) Isolation and properties of chitin synthetase from *Agaricus bisporus* mycelium. Exp Mycol 7:17—30

Hagenmaier H, Keckeisen A, Zähner H, König WA (1979) Stoffwechselprodukte von Mikroorganismen, 182. Aufklärung der Struktur des Nukleosidantibiotikums Nikkomycin X. Liebigs Ann Chem 1979:1494—1502

Hagenmaier H, Keckeisen A, Dehler W, Fiedler H, Zähner H, König WA (1981) Stoffwechselprodukte von Mikroorganismen, 199. Konstitutionsaufklärung der Nikkomycine I, J, M und N. Liebigs Ann Chem 1981:1018—1024

Hector RF, Braun PC (1986) Synergistic action of nikkomycins X and Z with papulacandin B on whole cells and regenerating protoplasts of *Candida albicans*. Antimicrob Agents Chemother 29:389—394

Herrera-Estrella L, Ruiz-Herrera L (1983) Light response in *Phycomyces blakesleeanus*: Evidence for roles of chitin biosynthesis and breakdown. Exp Mycol 7:362—369

Hilenski LKL, Naider F, Becker JM (1986) Polyoxin D inhibits colloidal gold-wheat germ agglutinin labelling of chitin in dimorphic forms of *Candida albicans*. J Gen Microbiol 132:1441—1457

Hori M, Kakiki K, Suzuki S, Misato T (1971) Studies on the mode of action of polyoxins. Part III. Relation of polyoxin structure to chitin synthetase inhibition. Agric Biol Chem 35:1280—1291

Hori M, Eguchi J, Kakiki K, Misato T (1974a) Studies on the mode of action of polyoxins. VI. Effect of polyoxin B on chitin synthesis in polyoxin-sensitive and resistant strains of *Alternaria kikuchiana*. J Antibiot 27:26—266

Hori M, Kakiki K, Misato T (1974b) Further study on the relation of polyoxin structure to chitin synthetase inhibition. Agric Biol Chem 38:691—698

Hori M, Kakiki K, Misato T (1974c) Interaction between polyoxin and active center of chitin synthetase. Agric Biol Chem 38:699—705

Hori M, Kakiki K, Misato T (1977) Antagonistic effect of dipeptides on the uptake of polyoxin A by *Alternaria kikuchiana*. J Pestic Sci 2:139—150

Humphreys AM, Gooday GW (1984a) Properties of chitinase activities from *Mucor mucedo*: evidence for a membrane-bound zymogenic form. J Gen Microbiol 130:1359—1366

Humphreys AM, Gooday GW (1984b) Phospholipid requirement of microsomal chitinase from *Mucor mucedo*. Curr Microbiol 11:187—190

Isono K, Nagatsu J, Kawashiwa Y, Suzuki S (1965) Studies on polyoxins, antifungal antibiotics. Part 1. Isolation and characterisation of polyoxins A and B. Agric Biol Chem 29:848—854

Isono K, Nagutsu J, Kobinata K, Sasaki K, Suzuki S (1967) Studies of polyoxins, antifungal antibiotics. Part V. Isolation and characterisation of polyoxins C, D, E, F, G, H and I. Agric Biol Chem 31:190—199

Isono K, Asahi K, Suzuki A (1969) Studies on polyoxins, antifungal antibiotics. XIII. The structure of polyoxins. J Am Chem Soc 91:7490—7505

Isono K, Azuma T, Suzuki S (1971a) Polyoxin analogs. I. Synthesis of aminoacyl derivatives of 5′-amino-5′-deoxyuridine. Chem Pharm Bull 19:505—512

Isoni K, Suzuki S, Azuma T (1971b) Semi-synthetic polyoxins: aminoacyl derivatives of polyoxin C and their in vitro activity. Agric Biol Chem 35:1986—1989

Iten W, Matile P (1970) Role of chitinase and other lysosomal enzymes of *Coprinus lagopus* in the autolysis of fruiting bodies. J Gen Microbiol 61:301 – 309

Izawa Y, Uchida M, Sugimoto T, Asai T (1985) Inhibition of chitin biosynthesis by buprofezin analogs in relation to their activity controlling *Nilaparvata lugens* Stal. Pestic Biochem Physiol 24:343 – 347

Kelly P, Trewavas AJ, Lewis LN, Durbin ML, Sexton R (1987) Translatable messenger RNA changes in ethylene induced abscission zones of *Phaseolus vulgaris* (Red Kidney). Plant Cell Environ 10:11 – 16

Kobinata K, Oramoto M, Nishii M, Kusakabe H, Nakamura G, Isono K (1980) Neopolyoxin A, neopolyoxin B and neopolyoxin C, new chitin synthetase inhibitors. Agric Biol Chem 44:1709 – 1722

Kodama O, Yamada H, Akatsuka T (1979) Kitazin P, inhibitor of phosphatidylcholine biosynthesis in *Pyricularia oryzae*. Agric Biol Chem 43:1719 – 1725

König WA, Hass W, Dehler W, Fiedler HP, Zähner H (1980) Stoffwechselprodukte von Mikroorganismen, 189. Strukturaufklärung und Partialsynthese des Nukleosidantibiotikums Nikkomycin B. Liebigs Ann Chem 1980:622 – 628

Koga D, Isogai A, Sakuda S, Matsumoto S, Suzuki A, Kimura S, Ide A (1987) Specific inhibition of *Bombyx mori* chitinase by allosamidin. Agric Biol Chem 51:471 – 476

Kramer KJ, Koga D (1986) Insect chitin. Physical state, synthesis, degradation and metabolic regulation. Insect Biochem 16:851 – 877

Lindsay GJH (1986) The significance of chitinolytic enzymes and lysozyme in rainbow trout (*Salmo gairdneri*) defence. Aquaculture 51:169 – 173

Lunblad G, Elander M, Lind J, Slettengren K (1979) Bovine serum chitinase. Eur J Biochem 100:455 – 460

Mahadevan PR, Mahadkar UR (1970) Role of enzymes in growth and morphology of *Neurospora crassa*: cell wall bound enzymes and their possible role in branching. J Bacteriol 101:941 – 947

Marks EP, Leighton T, Leighton F (1982) Modes of action of chitin synthesis inhibitors. In: Coats JR (ed) Insecticide mode of action. Academic Press, London New York, pp 281 – 313

McCarthy P, Troke PF, Gull K (1985) Mechanism of action of nikkomycin and the peptide transport system of *Candida albicans*. J Gen Microbiol 131:775 – 780

Metraux JP, Boller T (1986) Local and systemic induction of chitinase EC 3.2.1.14 in cucumber *Cucumis sativus* cultivar Wisconsin plants in response to viral, bacterial and fungal infections. Physiol Mol Plant Pathol 28:161 – 170

Milewski S, Chmara H, Borowski E (1983) Growth inhibitory effect of antibiotic tetaine on yeast and mycelial forms of *Candida albicans*. Arch Microbiol 135:130 – 136

Milewski S, Chmara H, Andruszkiewicz R, Borowski E (1984) Synthetic derivatives of ^{3}N-fumaroyl-L-2,3-diaminopropanoic acid inactivate glucosamine synthetase from *Candida albicans*. Biochem Biophys Acta 828:247 – 254

Milewski S, Chmara H, Borowski E (1986a) Antibiotic tetaine – a selective inhibitor of chitin and mannoprotein biosynthesis in *Candida albicans*. Arch Microbiol 145:234 – 240

Milewski S, Chmara H, Borowski E (1986b) Anticapsin: an active site directed inhibitor of glucosamine-6-phosphate synthetase from *Candida albicans*. Drugs Exp Clin Res 12:577 – 583

Minke R, Blackwell J (1978) The structure of α-chitin. J Mol Biol 120:167 – 181

Mirelman D, Galun E, Sharon N, Lotan R (1975) Inhibition of fungal growth by wheat germ agglutinin. Nature (London) 256:414 – 416

Misato T, Kakiki K (1977) Inhibition of fungal cell wall synthesis and cell membrane function. In: Siegal MR, Sisler HD (eds) Antifungal compounds, vol 2. Dekker, New York Basel, pp 277 – 300

Misato T, Ko K, Yamaguchi U (1977) Use of antibiotics in agriculture. Adv Appl Microbiol 21:53 – 88

Mitani M, Inoue Y (1968) Antagonists of antifungal substance polyoxin. J Antibiot 21:492 – 496

Mitsui T (1986) Mode of inhibition of chitin synthesis by diflubenzuron. In: Muzzarelli RAA, Jeuniaux C, Gooday GW (eds) Chitin in nature and technology. Plenum Press, New York London, pp 193 – 196

Mol PC, Wessels JGH (1987) Linkages between glucosaminoglycan and glucan determine alkali-insolubility of the glucan in walls of *Saccharomyces cerevisiae*. FEMS Microbiol Lett 41: 91 – 99

Müller H, Furter R, Zähner H, Rast DM (1981) Metabolic products of microorganisms 203. Inhibition of chitosomal chitin synthetase and growth of *Mucor rouxii* by nikkomycin Z, nikkomycin X and polyoxin A: a comparison. Arch Microbiol 130:195 – 197

Naider F, Shenbagamurthi P, Steinfeld AS, Smith HA, Boney C, Becker JM (1983) Synthesis and biological activity of tripeptidyl polyoxins as antifungal agents. Antimicrob Agents Chemother 24:787–796

Nishimura M, Kohmoto K, Udagawa H (1973) Field emergence of fungicide tolerant strains of *Alternaria kikuchiana*. Rep Tottori Mycol Inst 10:677–686

Nozawa Y, Yaginuma H, Ozeki K (1985) Advances in the concept of molecular mechanism for antifungal action by amphotericin B. In: Arai T (ed) Filamentous microorganisms. Biomedical aspects. Jpn Sci Soc Press, Tokyo, pp 345–354

Onishi JC, Miller TW (1985) The lack of antifungal activity by avermectin B1a. J Antibiot 38:1568–1572

Oranusi NA, Trinci APJ (1985) Growth of bacteria on chitin cell walls and fungal biomass, and the effect of extracellular enzyme produced by these cultures on the antifungal activity of amphotericin B. Microbios 43:17–30

Orlean P (1987) Two chitin synthases in *Saccharomyces cerevisiae*. J Biol Chem 262:5732–5739

Ortega JKE, Gamow RI, Ahlquist CN (1975) *Phycomyces*: A change in mechanical properties after a light stimulus. Plant Physiol 55:333–337

Payne JW, Shallow DA (1985) Studies on drug targeting in the pathogenic fungus *Candida albicans*: peptide transport mutants resistant to polyoxins, nikkomycins and bacilysin. FEMS Microbiol Lett 28:55–60

Pegg GF, Young DH (1982) Purification and characterisation of chitinase enzymes from healthy and *Verticillium albo-atrum*-infected tomato plants and from *V. albo-atrum*. Physiol Plant Pathol 21:389–409

Polachek Y, Rosenberger RF (1978) The distribution of autolysins in hyphae of *Aspergillus nidulans*: existence of a lipid mediated attachment to hyphal walls. J Bacteriol 135:741–754

Poli F, Pancaldi S, Vannini GL (1986) The effect of monesin on chitin synthesis in *Candida albicans* blastospores. Eur J Cell Biol 42:79–83

Pope AMS, Davies DAL (1979) The influence of carbohydrates on the growth of fungal pathogens in vitro and in vivo. Postgrad Med J 55:674–676

Porter CA, Jaworski EG (1966) The synthesis of chitin by particulate preparations of *Allomyces macrogynus*. Biochemistry 5:1149–1154

Prick G, Diekmann H (1979) A chitin-binding lectin in *Neurospora crassa*. FEMS Microbiol Lett 6:427–429

Rast DM, Bartnicki-Garcia S (1981) Effects of amphotericin B, nystatin, and other polyene antibiotics on chitin synthase. Proc Natl Acad Sci USA 78:1233–1236

Reyes F, Calatayud J, Martinez MJ (1988) Chitinolytic activity in the autolysis of *Aspergillus nidulans*. FEMS Microbiol Lett 49:239–243

Rico H, Miragall F, Sentandreau R (1985) Abnormal formation of *Candida albicans* produced by Calcofluor White: an ultrastructural and stereologic study. Exp Mycol 9:241–253

Roberts WK, Selitrennikoff CP (1988) Plant and bacterial chitinases differ in antifungal activity. J Gen Microbiol 134:169–178

Roncero C, Duran A (1985) Effect of Calcofluor White and Congo Red on fungal cell wall morphogenesis: in vivo activation of chitin polymerization. J Bacteriol 163:1180–1185

Rosenberger RF (1979) Endogenous lytic enzymes and wall metabolism. In: Burnett JH, Trinci APJ (eds) Fungal walls and hyphal growth. Symp Br Mycol Soc, vol 2. Cambridge Univ Press, Cambridge, pp 265–278

Rousset-Hall A de, Gooday GW (1975) A kinetic study of a solubilized chitin synthetase preparation from *Coprinus cinereus*. J Gen Microbiol 89:146–154

Sakuda S, Isogai A, Makita T, Matsumoto S, Suzuki AM, Koga A, Koseki K, Kodama H, Noma M (1986a) Structures of allosamidins, novel insect chitinase inhibitors, produced by microorganisms. Tennen Yuki Kagobutsu Toronkai Koen Yoshishu 28:65–72

Sakuda S, Isogai A, Matsumoto S, Suzuki A, Koseki K (1986b) The structure of allosamidin, a novel insect chitinase inhibitor, produced by *Streptomyces* sp. Tetrahedron Lett 27:2475–2478

Sakuda S, Isogai A, Matsumoto S, Suzuki A (1987a) Search for microbial insect growth regulators. II. Allosamidin, a novel insect chitinase inhibitor. J Antibiot 40:296–300

Sakuda S, Isogai A, Makita T, Matsumoto S, Koseki K, Kodama H, Suzuki A (1987b) Structures of Allosamidins, novel insect chitinase inhibitors, produced by actinomycetes. Agric Biol Chem 51:3251–3259

Sburlati A, Cabib E (1986) Chitin synthetase 2, a presumptive participant in septum formation in *Saccharomyces cerevisiae*. J Biol Chem 261:15147–15152

Schlumbaum A, Mauch F, Vögeli U, Boller T (1986) Plant chitinases are potent inhibitors of fungal growth. Nature (London) 324:365–367

Selitrennikoff CP (1979) Competitive inhibition of *Neurospora crassa* chitin synthetase by tunicamycin. Arch Biochem Biophys 195:243–244

Selitrennikoff CP (1984) Calcofluor White inhibits *Neurospora* chitin synthetase activity. Exp Mycol 8:269–272

Selitrennikoff CP (1985) Chitin synthetase activity of *Neurospora crassa*: effect of Primulin and Congo red. Exp Mycol 9:179–182

Shenbagamurthi P, Smith HA, Becker JM, Steinfeld A, Naider F (1983) Design of anticandidal agents: synthesis and biological properties of analogues of polyoxin L. J Med Chem 26:1518–1522

Sietsma JH, Wessels JGH (1979) Evidence for covalent linkages between chitin and β-glucan in a fungal wall. J Gen Microbiol 114:99–108

Sietsma JH, Wessels JGH (1981) Solubility of (1–3)-β-D-(1–6)-β-D-glucan in fungal walls: importance of presumed linkage between glucan and chitin. J Gen Microbiol 125:209–212

Smith HA, Shenbagamurthi P, Naider F, Kundu B, Becker JM (1986a) Hydrophobic polyoxins are resistant to intracellular degradation in *Candida albicans*. Antimicrob Agents Chemother 29:33–39

Smith HA, Shenbagamurthi P, Naider F, Kundu B, Becker JM (1986b) New synthetic polyoxin analogs for chitin synthesis inhibition. In: Muzzarelli RAA, Jeuniaux C, Gooday GW (eds) Chitin in nature and technology. Plenum Press, New York London, pp 197–202

Somers PJB, Yao RC, Doolin LR, McGowan MJ, Fakuda DS, Mynderse JS (1987) Method for the detection and quantitation of chitinase inhibitors in fermentation broths; isolation and insect life cycle. Effect of A82516. J Antibiot 40:1751–1756

Sonnenberg ASM, Sietsma JH, Wessels JGH (1982) Biosynthesis of alkali-insoluble cell-wall glucan in *Schizophyllum commune* protoplasts. J Gen Microbiol 128:2667–2674

Srivastava AK, Defago G, Boller T (1985) Secretion of chitinase by *Aphanocladium album*, a hyperparasite of wheat. Experientia 41:1612–1613

Tracey MV (1955) Chitinase in some Basidiomycetes. Biochem J 61:579–586

Uramoto M, Kobinata K, Isono K, Higashijima T, Miyazawa T, Jenkins EE, McCloskey JA (1980) Structures of neopolyoxins A, B, and C. Tetrahedron Lett 21:3395–3398

Uramoto M, Kobinata K, Isono K, Higashijima T, Miyazawa T, Jenkins EE, McCloskey JA (1982) Chemistry of the neopolyoxins, pyrimidine and imidazoline nucleoside peptide antibiotics. Tetrahedron 38:1559–1608

Usai T, Hayashi Y, Nanjo F, Sakai K, Ishido Y (1987) Transglycosylation reaction of a chitinase purified from *Nocardia orientalis*. Biochem Biophys Acta 923:302–309

Vermeulen CA, Wessels JGH (1984) Ultrastructural differences between wall apices of growing and non-growing hyphae of *Schizophyllum commune*. Protoplasma 12:123–131

Vermeulen CA, Wessels JGH (1986) Chitin biosynthesis by a fungal membrane preparation. Evidence for a transient non-crystalline state of chitin. Eur J Biochem 158:411–415

Wessels JGH, Sietsma JH (1981) Fungal cell walls: a survey. In: Tanner W, Loewus FA (eds) Plant carbohydrates. II. Extracellular carbohydrates. Encycl Plant Physiol, N Ser, vol 13 B. Springer, Berlin Heidelberg New York, pp 352–394

Willems AGH, Brouwer MS, Jongsma B (1986) Benzoylaryl ureas: insecticidal compounds interfering with chitin synthesis. In: Muzzarelli RAA, Jeuniaux C, Gooday GW (eds) Chitin in nature and technology. Plenum Press, New York London, pp 131–138

Yadan J, Gonneau M, Sarthou P, Le Goffic F (1984) Sensitivity to nikkomycin in *Candida albicans*: role of peptide permeases. J Bacteriol 160:884–888

Chapter 6

Wall Structure, Wall Growth, and Fungal Cell Morphogenesis

J. G. H. WESSELS, P. C. MOL, J. H. SIETSMA, and C. A. VERMEULEN[1]

1 Introduction

Because of its rigidity, the wall maintains the shape of the fungal hypha and various arguments have indicated chitin microfibrils embedded in an amorphous matrix as the major polymeric assemblage conferring rigidity to the wall (Bartnicki-Garcia 1973; Burnett 1979; Gooday and Trinci 1980; Wessels and Sietsma 1981). This rigidity contrasts with the need for expansion of the wall when growth occurs either apically, as in all vegetatively growing hyphae, or by diffuse extension, as in hyphae in some fruit bodies. To accommodate these contrasting needs, two basic hypotheses have been advanced. Either the wall as synthesized is inherently rigid and must be continuously loosened by lysins in order to expand and to allow for intercalation of new polymers. Alternatively, the newly synthesized wall is inherently visco-elastic and expandable and gradually develops rigidity to fulfill its role in the mature wall.

Many observations on shadowed extracted wall preparations have shown the occurrence of a microfibrillar network in the wall covering the hyphal apex (Bartnicki-Garcia 1973; Burnett 1979), although the general occurrence of microfibrils in growing apical walls has recently been challenged (Vermeulen and Wessels 1984). This previously assumed similarity in ultrastructure between the extending apical wall and the non-extending longitudinal wall may have contributed to a strong adherence to the first hypothesis, implying an important role for lysins both in apical extension (Bartnicki-Garcia 1973; Gooday and Trinci 1980) and in diffuse extension of walls in stipes of fruit bodies (Gooday 1979). Burnett (1979) in his introductory address to a meeting on 'Fungal Walls and Hyphal Growth' expressed his doubts on the concept of lysins playing a role in apical wall growth. In his words: "teleologically speaking, the apex would seem to be a most dangerous location for a lytic entity". A number of conceptual difficulties related to the involvement of lysins in apical growth have also been raised (Wessels 1984, 1986). On the basis of measurements of densities and dimensions of microfibrils over the apical dome and in the mature wall, Burnett (1979) arrived at the suggestion that during apical growth, slippage of the microfibrils occurs and that gradual stiffening of the wall is the result of interlocking and aggregation of microfibrils. He thus adhered more closely to the second hypothesis stating that

[1] Department of Plant Biology, Biological Centre, University of Groningen, Kerklaan 30, 9751 NN Haren, The Netherlands

the newly synthesized wall is inherently plastic but develops rigidity. However, although one would expect slipping microfibrils to become re-orientated during elongation, this was not observed. Also during diffuse extension growth of walls in stipes of fruit bodies, re-orientation of microfibrils was not observed (Gooday 1979).

Our own work during the last decade on the structure and biogenesis of the wall of vegetatively growing hyphae of *Schizophyllum commune* has provided data which definitively support the second hypothesis of wall growth. The evidence leading to the formulation of the so-called steady-state model for apical wall growth has recently been reviewed (Wessels 1986, 1988). In essence, this model considers the newly-synthesized wall at the apex as essentially consisting of a mixture of non-crystalline chitin and β-glucan with visco-elastic properties. Molecular interactions between and among these polymers, occurring within the domain of the wall, then gradually convert this mixture into a rigid composite which resists turgor at the base of the extension zone. Below we will summarize basic observations which led to the formulation of this theory. The theory will be extended to formulate an hypothesis on determinate growth as occurring in budding yeast cells. In addition we will report on some observations on walls in fruit bodies of *Agaricus bisporus* which suggest that delay of rigidification of an originally plastic wall may account for diffuse extension growth.

2 Alkali-Insoluble β-Glucan in the Mature Wall Forms Part of a Glucosaminoglycan-Glucan Complex

Traditionally the alkali-insoluble $(1-3)$-β-D/$(1-6)$-β-D-glucan in the wall of ascomycetes and basidiomycetes has been considered as an amorphous matrix material in which chitin microfibrils are physically embedded (Bartnicki-Garcia 1973; Burnett 1979; Gooday and Trinci 1980; Wessels and Sietsma 1981; Farkaš 1985). Some evidence that the glucan may actually be covalently bonded to glucosaminoglycan was obtained by Stagg and Feather (1973) who found that the glucan in the alkali-insoluble wall complex of *Aspergillus niger* could be solubilised completely in dimethylsulfoxide and partly in alkali after treatment with nitrous acid. Such a treatment specifically hydrolyzes glucosaminoglycans at sites where non-acetylated amino residues occur (Shively and Conrad 1970). However, it is not known whether the low pH during the nitrous acid treatment contributed to the change in solubility of the glucan. Treatments with acid, even very mild ones, may render insoluble β-glucans soluble in dimethylsulfoxide and alkali (Bacon et al. 1969), probably by breaking acid-labile bonds. Sietsma and Wessels (1979) showed that depolymerization of the glucosaminoglycan with chitinase resulted in solubilisation of most of the β-glucan from the alkali-insoluble wall residue of *Schizophyllum commune*. Recently this has also been demonstrated for the alkali-insoluble wall residues of *Agaricus bisporus*, *Aspergillus nidulans*, and *Neurospora crassa* (Mol et al. 1988). Even in *Saccharomyces cerevisiae* which contains extremely little glucosaminoglycan in its wall, all of the β-glucan became soluble in alkali after a chitinase treatment (Mol and Wessels 1987). In the ascomycetes mentioned above no evidence for the occurrence of nitrous-acid

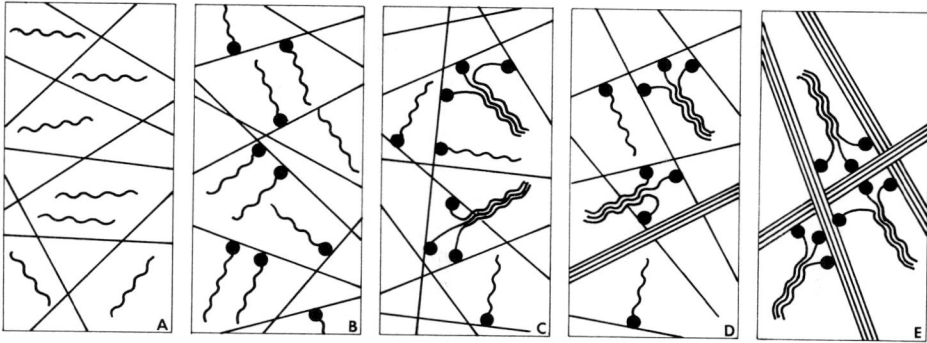

Fig. 1A–E. Possible interactions between glucosaminoglycan (chitin) chains (*straight lines*) and (1−3)-β-glucan chains (*wavy lines*). **A** Immediately after synthesis and excretion into the wall the two polymers are assumed to be present as individual polymers. **B** Linkage of glucan chains to the alkali-insoluble glucosaminoglycan chains makes the former insoluble in alkali. Glucan not bound remains in the wall as alkali-/water-soluble chains. **C** Triple helices formed by the bound glucan chains inter-connect glucosaminoglycan chains. **D** Hydrogen-bonding between unsubstituted glucosaminoglycan chains may produce chitin microfibrils. **E** Hydrogen-bonding of substituted glucosaminoglycan chains to chitin microfibrils would produce a stiff network of interconnected microfibrils. In the mature wall all these arrangements may be present at the same time. Not shown are (1−6)-β-linked glucose branch-es and (1−6)-β-linked glucan branches on the (1−3)-β-linked glucan chains

labile bonds was found before alkali treatment (which caused some deacetylation). However, such bonds were clearly present in *S. commune* and particularly in *A. bi-sporus* where treatment with nitrous acid led to considerable depolymerization of glucosaminoglycan and solubilisation of β-glucan (Mol et al. 1988).

The evidence for the existence of covalent linkages between glucan and glucosaminoglycan is corroborated by results of kinetic experiments using labelled acetylglucosamine and glucose as precursors for the individual polymers (Sonnenberg et al. 1985; Wessels et al. 1983). Whereas acetylglucosamine was immediately incorpo-rated into an alkali-insoluble glucosaminoglycan (chitin), glucose first appeared in water-/alkali-soluble (1−3)-β-linked glucan chains which were subsequently converted into an alkali-insoluble form from which they could again be released by a chitinase treatment. Specific inhibition by polyoxin D of chitin synthesis in regenerating pro-toplasts of *S. commune* prevented insolubilisation of the glucan and led to its accu-mulation in a water-/alkali-soluble form (de Vries and Wessels 1975; Sonnenberg et al. 1982). This has also been observed in regenerating protoplasts of *Candida albicans* after inhibition of chitin synthesis with nikkomycin (Elorza et al. 1987). A highly schematic view depicting the insolubilisation of β-glucan chains by linkage to alkali-insoluble glucosaminoglycan is given in Fig. 1A, B.

The nature of the covalent linkage(s) between glucan and glucosaminoglycan re-mains a matter of conjecture. Similarly, the nature of presumed inter-polymer link-ages in the chitin-protein complexes of animals have not been resolved (Poulicek et al. 1985). Linkage of the protein through non-acetylated amino groups of the glucosaminoglycan are thought to occur in the squid *Loligo australis* (Hackman and Goldberg 1965) and have also been surmised to occur in the chitin-protein complex of the zygomycete *Mucor mucedo* (Datema et al. 1977). With respect to the inter-polymer linkages in chitin-glucan complexes, direct links between free amino groups in the glucosaminoglycan and the reducing end of the glucan chains are one possibility. A disaccharide containing both glucose and glucosamine was isolated after partial degradation of the glucosaminoglycan-glucan complexes of *Schizophyllum commune* (Sietsma and Wessels 1979) and *Saccharomyces cerevisiae* (E. Cabib, personal communication). A Schiff reaction between free amino groups in the glucosaminoglycan and aldehyde groups in substituents can easily lead to a branched glucosaminoglycan (Kurita 1985). However, amino acids were found in the glucosaminoglycan-glucan complex of *S. commune*, 50% of which were lysine. After enzymic removal of the glucan these amino acids remained attached to the glucosaminoglycan. Subsequent degradation with chitinase produced a small fragment containing glucosamine, glucose, lysine, and an amino acid tentatively identified as citrulline (Sietsma and Wessels 1979). High lysine contents have also been found in chitinous residues of *Verticillium alboatrum* (Wang and Bartnicki-Garcia 1970) and *Chaetomium globosum* (Maret 1972). The presence of lysine attached to the glucosaminoglycan may suggest a Schiff reaction between lysine and the reducing end of glucans as postulated in age-related browning of human collagen (Monnier et al. 1984). Also the formation of highly reactive aldehyde groups by lysyl oxidase as thought to occur in the cross-linking of collagen helices (Hay 1981) can be envisaged. Importantly, none of these suggested mechanisms require direct enzymic catalysis during actual formation of the crosslinks which would be a very unlikely event in reactions between pre-formed polymers.

3 Secondary Structures in the Glucosaminoglycan-Glucan Wall Complex

There is little evidence from X-ray diffraction studies that in native hydrated or dried walls of hyphae glucosaminoglycans and β-glucans occur as highly ordered crystallites (cf. Wessels and Sietsma 1981). Even non-extracted mycelial walls and alkali-extracted residues of *Agaricus bisporus* containing 34% and 42% glucosaminoglycan, respectively, do not reveal clear X-ray reflections corresponding to those of chitin, chitosan or hydroglucan (Mol 1989). On the other hand, chemical treatments involving acids generally result in the appearance of sharp X-ray reflections of α-chitin and in some cases of hydroglucan (Wessels and Sietsma 1981). Also enzymic removal of glucan from the glucosaminoglycan-glucan complex can result in the generation of crystalline chitin. A few cases can be cited in which crystalline and microfibrillar polymers were observed without any treatments. Crystalline microfibrillar α-chitin was formed during early regeneration of *S. commune* protoplasts (van der Valk and Wessels 1976) and during in vitro synthesis by chitosomes (Ruiz-Herrera and Bartnicki-Garcia 1974) and isolated plasma membranes (Vermeulen et al. 1979). Similarly, $(1-3)$-β-glucan synthesized on regenerating protoplasts of *Saccharomyces cerevisiae* (Kreger and Kopecká 1976) or in vitro by a mixed membrane preparation of *Phytophthora cinnamomi* (Wang and Bartnicki-Garcia 1976) showed clear crystallinity and microfibrils. In all these cases the polymers were synthesized in isolation without any cross-linking or branching occurring. On the one hand, the paucity of crystallites of chitin and β-glucan in hyphal walls could be due to the fact that these polymers are excreted simultaneously in the wall to form a mixture, reducing the change of homologous chains to form regular crystal lattices. On the other hand, branching and cross-linking of the polymers would also be expected to hamper their crystallization. The absence of X-ray reflections from the glucosaminoglycan-glucan complex [native walls do show crystallinity of other wall components, e.g. the unbranched alkali-soluble $(1-3)$-α-D-glucan (Wessels et al. 1972)] does not, of course, preclude the occurrence of hydrogen bonds between homologous chains. Such bonds can be thought of as being similar to those occurring in α-chitin and hydroglucan but with a much more irregular distribution and lower frequency than in perfect crystallites. For instance, about 40% of the glucan chains in the glucosaminoglycan-glucan complex of S. commune are $(1-3)$-β linked with frequent $(1-6)$-β-linked glucose branches along the chains (Sietsma and Wessels 1977, 1979). These chains closely resemble a glucan freely occurring in the wall and in the extracellular medium (Wessels et al. 1972). This extracellular glucan has gel-like properties because it forms interconnected triple helices (Sato et al. 1981) similar to those suggested for unbranched $(1-3)$-β-glucan (Jelsma and Kreger 1975; Marchessault and Deslandes 1979). Linkage of such triple helices to the glucosaminoglycan could thus result in structures as depicted in Fig. 1 C, D, E. The water-soluble β-glucan chains occurring in the wall and surrounding medium (variously called slime, mucilage, and schizophyllan) may thus simply represent chains that have escaped linkage to the glucosaminoglycan. One would also expect at least some hydrogen-bonding to occur between substituted and unsubstituted glucosaminoglycan chains. Cooperation between all these bonds could then result in a highly cross-linked glucosaminoglycan-glucan complex of high rigidity as schematically shown in Fig. 1 E. However, all the structures shown in Fig. 1, and probably others, may actually exist in the mature wall at the same time.

Do the interactions between glucosaminoglycan chains lead to the formation of chitin microfibrils in the native wall, notwithstanding the fact that clear reflections of crystalline α-chitin cannot be seen? At the inner surface of dried and shadowed non-extracted walls imprints of microfibrils can often be seen but clear microfibrils only appear after chemical and enzymic extractions which also induce crystallinity (van der Valk et al. 1977; Gow and Gooday 1983). Such treatments remove the glucan but also up to 30% of the glucosaminoglycan from the wall (Sietsma and Wessels 1977; Mol 1989). Apparently, the treatments used to visualize the microfibrils facilitate the formation of regular crystal lattices and possibly lead to crystallization of originally dispersed glucosaminoglycan chains. It is therefore uncertain to what extent the microfibrillar nets seen in hyphal walls after rigorous extractions reflect their actual occurrence in the native wall. A further complication arises from the fact that various microfibril morphologies can be discerned although the X-ray diffraction patterns and infrared spectra of the chitin contained in these microfibrils are indistinguishable (Gow and Gooday 1987).

An important point that relates to the existence of wall structures as shown in Fig. 1 is that the natural tendency of glucosaminoglycan chains to form α-chitin crystallites must be sufficiently slow to allow for interspersion of glucan chains between the glucosaminoglycan chains and for cross-linking to occur. Immediately after synthesis in vivo (Vermeulen and Wessels 1984) and in vitro (Molano et al. 1979; Lopez-Romero et al. 1982; Vermeulen and Wessels 1986) chitin is much more sensitive to chitinase than some time after formation. Because one expects chitinase to be unable to penetrate the crystal lattice, this suggests a time lag (in the order of minutes) between polymerization and crystallization of the chitin chains. Vermeulen and Wessels (1986) showed by X-ray diffraction that Calcofluor White M2R, which binds to chitin chains, when added during synthesis of chitin by a membrane preparation of *S. commune* could indefinitely delay crystallization, provided the wet state was maintained. Using dried preparations, Elorza et al. (1983) demonstrated that Calcofluor interfered with microfibril formation on the surface of regenerating protoplasts of *Candida albicans*.

With respect to the morphogenetic importance of the formation of the glucosaminoglycan-glucan complex it is relevant to note that interference with either the β-glucan or the glucosaminoglycan component prevents the formation of hyphae by regenerating protoplasts of *S. commune* (Sonnenberg et al. 1982). Unfortunately, no agents are known that specifically block the covalent cross-linking between the polymers. Conversely, regenerating protoplasts produce normal hyphae when the synthesis of another quantitatively important wall polymer, $(1-3)-\alpha$-glucan, is specifically inhibited by 2-deoxyglucose (Sietsma and Wessels 1988).

4 Apical Biosynthesis and Morphogenesis of the Hyphal Wall

It is generally accepted that alkali-insoluble glucosaminoglycan (chitin) is exclusively generated outside the plasma membrane. Electron microscopy and high resolution autoradiography have shown that the glucosaminoglycan is polymerised at or just outside the plasma membrane in the inner layer of the wall (van der Valk and Wessels 1977; Gooday 1982). If the structure of the glucosaminoglycan-glucan complex as outlined in the previous section is accepted, it follows that the covalent linkage of glucan to glucosaminoglycan chains and the hydrogen bonding between homologous chains can only occur after the two different polymers have been excreted into the wall. It is hypothesized that these processes, occurring within the domain of the wall, change the physical properties of the wall, e.g. from a visco-elastic wall to a wall displaying increased rigidity with maturation.

On the basis of this consideration the steady-state growth theory of apical wall growth was originally formulated (Wessels et al. 1983). A highly schematic view of this theory is presented in Fig. 2. The rates of polymerization and excretion of the two wall polymers, glucosaminoglycan and β-glucan, decreasing in subapical direction, are thought to be essentially independent of the rate of cross-linking and the development of rigidity. During linear growth of the hypha the rate of formation of visco-elastic and expanding wall material is equal to the rate of withdrawal of rigidified wall material at the base of the extension zone, hence the reference to a steady state. With respect to the stretching of visco-elastic wall segments until further stretching is prevented by rigidification there is similarity to a scheme proposed by Green (1974) on the basis of biophysical considerations. On the mechanisms involved Green (1974) could only speculate but he proposed that the maximum degree of stretch could be a function of previous stretch, i.e. an inbuilt property of the wall. In a general sense

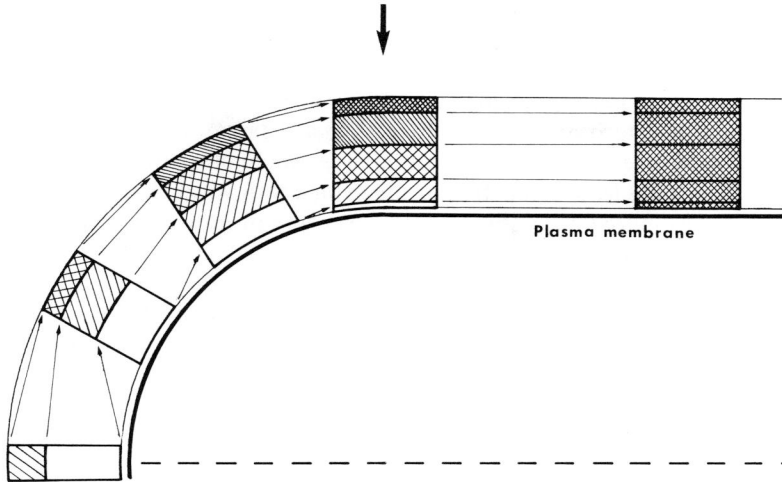

Fig. 2. The steady-state theory of apical wall extension. Wall material consisting of a visco-elastic mixture of individual polymers, e.g. $(1-3)$-β-glucans and $(1-4)$-β-glucosaminoglycans, is added to the wall at the apex by apposition according to a gradient symbolised by the height of the open wall segments. Because the wall maintains constant thickness during expansion this can only be done if previously deposited wall material undergoes stretching and moves to outer regions of the wall (note that segments of wall connected by *arrows* have the same volume). With increasing time of residence in the wall the individual polymers become more cross-linked (*heavier shading, cf.* Fig. 1) increasing the rigidity of the wall. At a certain distance from the apex (*thick arrow*) the average cross-linking in the wall has proceeded to the point that the wall no longer yields to internal pressure; this marks the base of the extension zone. Cross-linking, however, may continue and reach its maximum distally in the cylindrical part of the hypha. When expansion ceases, irrespective of continued apposition of new wall material, cross-linking and rigidification would be expected to proceed over the apex. Note that in reality infinitely small wall segments and a smooth gradient of cross-linking should be considered. Therefore real expansion of wall material is much larger than indicated. The thickness of the wall is also not drawn to scale

the scheme also fits ideas of Robertson (1965) on 'unset' and 'set' wall formulated on the basis of experiments with living hyphae.

In its simplest form the steady-state growth theory assumes the formation at the apex of an expandable mixture of two polymers, A and B, which are continuously removed at the base of the extension zone as a rigid complex AB arising by interactions between and among the two polymers. If it is assumed that the formation and excretion of the wall polymers A and B is governed by zero order rate constants k_A^0 and k_B^0 and the formation of the complex AB by a second order rate constant k_{AB}^2, then in the steady state the rate of formation of expandable wall material composed of A and B is determined by k_A^0 and k_B^0 and will be equal to its rate of conversion into rigid wall material, i.e. k_{AB}^2 [A] [B]. The product of the steady state concentrations, [A] [B], will thus equal $k_A^0 \cdot k_B^0 / k_{AB}^2$. Increased excretion of the two wall polymers would thus lead to an increase in the amount of expandable wall material and consequently to an increase in length and width of the extension zone provided wall thickness is maintained. Indeed, an increase in length and width of the extension zone with increased extension rate is generally observed (Steele and Trinci 1975). On the other hand, if elongation ceases for some reason, irrespective of continued excretion of the two wall polymers, the covalently-linked complex will form over the whole apex and the steady state will be interrupted by rigidification of the apical wall. This agrees with observations of Robertson (1958) who showed irreversible blockage of elongation after interrupting growth for less than 1 minute. Failure to excrete one of the two polymers, e.g. by inhibiting its synthesis, would block the process of rigidification and thus could lead to bursting of the hyphal wall when exposed to full turgor pressure. It has indeed been observed that exposure of growing hyphae to polyoxin D, a specific inhibitor of chitin synthesis (see Chapter 5), can result in bursting at hyphal apices (Bartnicki-Garcia and Lippman 1972).

Observations supporting the steady-state growth theory of apical wall growth obtained with germlings of *Schizophyllum commune* can be summarized as follows (Fig. 3):

At extending hyphal apices, glucans are not yet present in an alkali-insoluble complex; they become alkali-insoluble during subapical displacement and after cessation

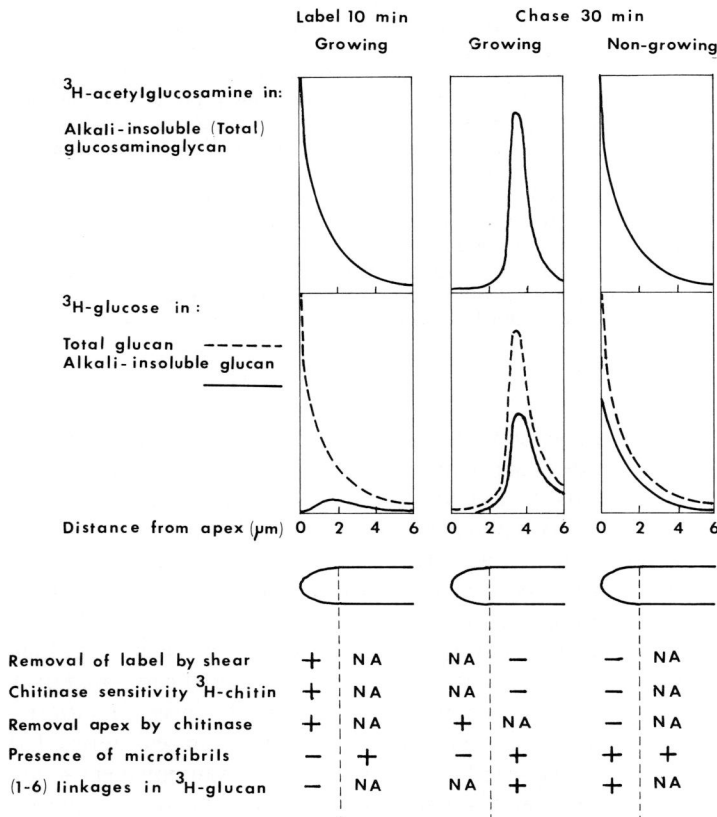

Fig. 3. Experimental evidence supporting the steady-state growth theory of apical wall expansion. Patterns of wall labelling and tabulated results are from Wessels et al. 1983; Vermeulen and Wessels 1984; and Sietsma et al. 1985. NA: not applicable

of hyphal extension (Wessels et al. 1983). Pulse-labelling (10 min) with [^{3}H]acetylglu-
cosamine resulted in labelling of glucosaminoglycan, nearly all alkali-insoluble, maximal at the extreme apex and rapidly decreasing subapically. [^{3}H]glucose labelled total
glucan in the wall in a similar pattern. It is important to note that within the short
period of labelling [^{3}H]glucose did not label glucosaminoglycan (Sietsma et al.
1985). However, at the apex very little label appeared in alkali-insoluble glucan; nearly
all label was present in water- and alkali-soluble glucans, indicating a paucity of linkage of glucan to glucosaminoglycan at the apex. After a chase of radioactivity and
continued incubation in the presence of unlabelled precursors, two patterns of labelling were observed (Fig. 3). In a portion of the hyphae the label was displaced
subapically showing these hyphae to grow at a rate of 7 μm h^{-1}. Other hyphae ceas-
ed growth due to mechanical disturbance by the chase procedure and retained all label
in their apices. In both cases, however, during the incubation in unlabelled medium
a considerable part of the label in glucan appeared in the alkali-insoluble fraction at
the expense of label in the water-/alkali-soluble glucan fraction. It thus appeared that
in both cases cross-linking of glucan to alkali-insoluble glucosaminoglycan had
occurred.

  *Immediately after labelling wall polymers can be removed from the apex by me-
chanical shear* (Wessels et al. 1983). After pulse-labelling with [^{3}H]acetylglucosamine
or [^{3}H]glucose labelling patterns as shown in Fig. 3 were only observed after chemical removal of cytoplasm (with ethanolic KOH). Mechanical breakage of the hyphae
to remove cytoplasm resulted in disappearance of labelled apices: labelled glucan was
solubilised, labelled glucosaminoglycan was fragmented. After a chase, however, all
label present subapically in growing walls and apically in non-growing walls was resistant to the shear produced by mechanical breakage. This indicates transfer of label
from mechanically fragile to rigid wall structures.

  *Immediately after its synthesis glucosaminoglycan is very sensitive to chitinase
and dilute mineral acid* (Vermeulen and Wessels 1984). Immediately after pulse-labelling (5 min), incubation of the alkali-insoluble wall residue with chitinase (0.5 mg
ml^{-1}, 1 h, 30 °C) or 0.5 M HCl (1 h, 80 °C) effected dissolution of 75% and 50%, respectively, of the label incorporated into glucosaminoglycan. After a chase of 60 min
these percentages dropped to 5 and 20, respectively. This may reflect the gap between
polymerization and crystallization of chitin (Vermeulen and Wessels 1986) but also
the attachment of glucan chains making the glucosaminoglycan chains less accessible
to chitinase.

  Growing hyphae contain non-fibrillar chitinase-sensitive chitin in their apices
(Vermeulen and Wessels 1984). Germlings were briefly labelled with [^{3}H]acetylgluco-
samine, followed by a chase, to distinguish between extending (label displaced and
non-extending (label still apical) hyphae during the chase period. The location of the
label was monitored by autoradiography of shadowed wall preparations. Extending
hyphae extracted with hot water and alkali did not show microfibrils over their apices,
although the chemical treatments removed all the glucan from these apices. These
apices were rapidly disintegrated by treatment with chitinase, indicating the presence
of chitin which was very susceptible to chitinase. Non-extending apices were resistant
to chitinase but treatments with water and alkali were not expected to uncover
microfibrils because of the presence of alkali-insoluble glucan in these apical walls.
A subsequent treatment with 0.5 M HCl (1 h, 80 °C) did, however, reveal the presence

of long microfibrils in these apices whereas extending apices simply disintegrated. We therefore suspect that all records of microfibrillar chitin nets occurring in apical wall fragments concern observations on wall fragments from non-extending apices which are sufficiently robust to survive the drastic cleaning and extraction procedures employed.

(1−6)-β linkages in wall glucan arise subapically (Sietsma et al. 1985). Using glucose labelled with [³H] at the C_3 position and localizing the label by autoradiography before and after treatment with periodate, the distribution of (1−3)- and (1−6)-linkages in the alkali-insoluble wall glucan could be observed. The alkali-insoluble glucan in the most apical region was found to contain few (1−6)-linkages in accordance with the finding that a water-/alkali-soluble unbranched (1−3)-β-glucan is the immediate precursor of the alkali-insoluble glucan (Wessels et al. 1983). Subapically, the number of (1−6)-linkages rose rapidly, probably by attachment of glucose by (1−6)-β-linkage to the insolubilized (1−3)-β-linked glucan chains. Such a mechanism involving branching enzymes operating within the wall was also suggested by kinetic experiments (Sonnenberg et al. 1985). Such an activity within the wall may be partly responsible for the low level of labelling of the wall observed beyond the extension zone as is particularly evident when [³H]glucose is used as precursor. Whatever the mechanism, it was significant that the (1−6)-linkages also appeared in the alkali-insoluble glucan covering non-extending apices so that also in this respect the wall over these apices became very similar to that of the longitudinal walls.

5 Mycelium-Yeast Transitions

The fact that in the yeast *Saccharomyces cerevisiae* cross-linking between β-glucans and glucosaminoglycans in the wall seems to occur (Mol and Wessels 1987) points to the possibility that expansion of the wall in yeasts is also based on initial visco-elasticity of the wall followed by rigidification due to crosslinking. Recently a careful analysis was made of wall expansion in the dimorphic fungus *Candida albicans* (Staebell and Soll 1985; Soll et al. 1985). In growing buds there was an apical and a general component in wall expansion, the first component predominating in the earlier phases of bud growth but only the latter component remained after the bud had attained approximately half its final size (Staebell and Soll 1985).

Possibly the general component of wall expansion in yeast, i.e. expansion of the whole surface area, results from a low rate of cross-linking and rigidification. Because wall addition is less polarized than in hyphal growth, or at least becomes so during growth of the bud, cross-linking and rigidification would be expected to spread slowly over the whole bud wall, eventually checking further expansion. In other words, steady-state conditions for wall addition, wall expansion, and wall rigidification as envisaged for hyphal tip growth (Fig. 2) would never be realized during bud growth; expansion ends with rigidification of the whole bud wall, i.e. determinate wall growth. Further proliferation would then be possible only by local wall softening and formation of new buds. Such a mechanism would fit observations of Soll and coworkers on the pattern of yeast-hypha transitions (Soll et al. 1985). Environmental conditions conducive to hyphal development caused young buds to become transformed into hyphal apices but after the buds had attained approximately half their final size this was no longer possible; the buds grew to their final size but hyphae arose only from new evaginations.

6 Diffuse Extension of Hyphal Walls

The best known example of diffuse extension of hyphal walls is that occurring during rapid expansion of mushrooms. The rapid elongation of stipes of agaric fruit bodies has received particular attention. The upper part of the stipe expands to a much greater extent than the base and this has been correlated with a parallel increase in cell length in *Coprinus radiatus* (Haffner and Thielke 1970; Eilers 1974) or with cell elongation followed by septation, maintaining a unit cell length, in *Agaricus bisporus* (Craig et al. 1977). Although during elongation there may be some decrease in dry weight of the walls per unit stipe length (Kamada and Takemaru 1977) there is continuous addition of wall materials along the whole lengths of the hyphae. This was shown by autoradiography both for chitin (Craig et al. 1977; Gooday 1982) and glucan (Mol et al. 1989).

Kamada and Takemaru (1977) have investigated the mechanical properties of hyphal walls in elongating stipes of *Coprinus macrorhizus* (= *cinereus*). They found that the osmotic potential of the cytoplasm remained constant but that there was a positive correlation between elasticity of the walls (measured as shrinkage after plasmolysis), extensibility and minimum stress-relaxation time of the walls (measured by mechanically stretching the wall) and the rate of elongation these walls could sustain in the stipe.

Apart from being visco-elastic, during elongation the hyphal walls in stipes must expand more in a longitudinal direction than in a tangential direction, notwithstanding the fact that the stresses in the wall due to turgor are expected twice as large in tangential than in longitudinal direction. In elongating hyphae which do not maintain constant cell lengths by septation the mechanism for promoting longitudinal elongation versus increase in width may therefore be found in an anisotropic deposition of wall polymers. Indeed, in *Coprinus cinereus* stipes Gooday (1979) observed (by polarization microscopy) a strong anisotropic component in the walls while extracted walls showed predominantly transversely oriented microfibrils in the electron microscope, before and after elongation. In an elongationless mutant of *C. macrorhizus* (*cinereus*) the stipe hyphae increased in volume as much as in the elongating wild-type strain but the walls expanded more in width than in length (Kamada and Takemaru 1977). It would be of interest to establish whether the anisotropic component of the wall is missing in this mutant. In *Agaricus bisporus* (Mol et al. 1989) we observed a very weak anisotropic component by polarization microscopy in stipe hyphae, but we could not visualize transversely running microfibrils. At the inside of alkali-extracted walls from elongating hyphae we did, however, see fine transverse striations suggesting a preferred orientation of wall polymers. It may be relevant that in this species the observed intercalary formation of septa maintaining a unit cell length (Craig et al. 1977) could also greatly contribute to the maintenance of cylindrical shape during elongation. A volume bounded by a visco-elastic wall can be drawn into a cylinder with a length of approximately three times its diameter if the cylinder is supported at its ends by rigid material, as originally pointed out by d'Arcy Thompson (c.f. Koch (1983) for an application of this principle to bacterial wall growth).

Our findings on apical wall growth, indicating the excretion of an inherently visco-elastic wall at the apex which subsequently gains rigidity by cross-linking, has suggested to us the possibility that the visco-elastic properties of the walls of elongating stipe hyphae may be due to a delay in complete rigidification of the wall. A relationship between mechanisms involved in apical wall extension and diffuse wall extension is suggested by results of Kamada et al. (1984) who found that many temperature-sensitive mutants of *Coprinus cinereus* restricted in apical extension also exhibited diminished stipe elongation. We thus hypothesise that walls of elongating hyphae in the stipe of fruit bodies do not proceed through all the secondary modifications that lead to rigidification of the wall in vegetative substrate hyphae but maintain a degree of plasticity. While being stretched by turgor pressure the apposition of similar visco-elastic wall material could compensate for thinning of the wall. Cessation of elongation could occur either by completion of the cross-linking process throughout the wall or by apposition of a final wall layer in which cross-linking occurs as in the walls of vegetative substrate hyphae.

Comparison of the structure of hyphal walls from elongating and non-elongating hyphae from stipes and from the apically growing substrate hyphae of *Agaricus bisporus* (Mol 1989) gives some support to the hypothesis outlined above. Contrary to the situation encountered in the wall covering growing apices we found that the walls of hyphae throughout the stipe do contain an alkali-insoluble glucosaminoglycan-glucan complex. From this complex the glucan could be solubilised by specific depolymerization of the glucosaminoglycan by nitrous acid and chitinase. The presence of such a complex has also been shown to occur in walls from stipes of *Coprinus cinereus* (Kamada and Takemaru 1983). However, we found that in the elongating region of stipes of *A. bisporus* the glucosaminoglycan-glucan complex is extremely sensitive to chitinase and $(1-3)$-β-glucanase; this sensitivity decreased towards the base of the stipe but remained higher than that of the complex from substrate hyphae. This suggests that in the diffusely elongating walls there are few hydrogen bonds between the glucosaminoglycan chains and between the glucan chains. Such bonds were considered an essential component of the non-expandable wall of substrate hyphae (Sect. 3). Hyphal wall fragments from substrate hyphae and from hyphae from the basal non-elongating part of stipes, obtained by mechanical breakage of the hyphae, showed a net of interwoven chitin microfibrils after treatment with alkali followed by $(1-3)$-β-glucanase. However, such treatments when applied to hyphal wall fragments from the elongating part of the stipe resulted in complete disintegration of most of the fragments leaving dispersed chitin often as aggregates of short fibrils. It thus appears that the glucosaminoglycan in these elongating walls (accounting for approximately 40% of the dry weight) is not organized into distinct chitin microfibrils but occurs in a rather dispersed condition. When hyphae from the elongating part of the stipe were not subjected to mechanical shear but treated in situ with alkali to remove alkali-soluble wall polymers, a subsequent $(1-3)$-β glucanase treatment caused up to 30% longitudinal contraction of the walls. Possibly, the removal of interspersed

$(1-3)$-β-glucan caused aggregation of transversely oriented glucosaminoglycan chains resulting in longitudinal contraction. The remaining wall residues still maintained the hyphal shape but easily disintegrated when subjected to mechanical disturbances such as sonication. Perhaps these observations which show that the wall of diffusely extending hyphae is structurally unique are not surprising because, to our knowledge, the rigidified wall of substrate hyphae can never become engaged in diffuse extension growth.

7 Conclusions

Experimental evidence is presented suggesting that apical wall growth in hyphal morphogenesis is based on polarized extrusion of visco-elastic wall material which is stretched and rigidified while extension of the hypha proceeds. This results in a steady-state amount of visco-elastic wall material at the growing hyphal tip but the tip wall rigidifies when extension ceases. It is hypothesized that during budding of yeast cells the extrusion of wall material is less polarized and that rigidification is slow, leading to determinate growth of daughter cells. Branching of hyphae and budding of yeast cells are thought to depend on evaginations of the wall caused by local wall softening followed by reinitiation of steady-state growth and determinate growth in mycelium and yeast, respectively. Delayed rigidification coupled to anisotropic wall deposition, sometimes accompanied by regular intercalary formation of septa, are suggested to be responsible for diffuse cylindrical extension of walls as occurring in fruit-body stipes.

References

Bacon JSD, Farmer VC, Jones D, Taylor IF (1969) The glucan components of the cell wall of baker's yeast (*Saccharomyces cerevisiae*) considered in relation to its ultrastructure. Biochem J 114: 557–567
Bartnicki-Garcia S (1973) Fundamental aspects of hyphal morphogenesis. In: Microbial differentiation. 23rd Symp Soc Gen Microbiol. Cambridge Univ Press, Cambridge, pp 245–267
Bartnicki-Garcia S, Lippman E (1972) The bursting tendency of hyphal tips of fungi: presumptive evidence for a delicate balance between wall synthesis and wall lysis in apical growth. J Gen Microbiol 73:487–500
Burnett JH (1979) Aspects of the structure and growth of hyphal walls. In: Burnett JH, Trinci APJ (eds) Fungal walls and hyphal growth. Cambridge Univ Press, Cambridge, pp 1–25
Craig GD, Gull K, Wood DA (1977) Stipe elongation in *Agaricus bisporus*. J Gen Microbiol 102:337–347
Datema R, Wessels JGH, Ende H van den (1977) The hyphal wall of *Mucor mucedo*. 2. Hexosamine-containing polymers. Eur J Biochem 80:621–626
Eilers FI (1974) Growth regulation in *Coprinus radiatus*. Arch Microbiol 96:353–364
Elorza MV, Rico H, Sentandreu R (1983) Calcofluor white alters the assembly of chitin fibrils in *Saccharomyces cerevisiae* and *Candida albicans* cells. J Gen Microbiol 129:1577–1582
Elorza MV, Murgui A, Rico H, Miragall F, Sentandreu R (1987) Formation of a new cell wall by protoplasts of *Candida albicans*: effect of papulacandin B, tunicamycin and nikkomycin. J Gen Microbiol 133:2315–2325

Farkaš V (1985) The fungal cell wall. In: Peberdy JF, Ferenczy L (eds) Fungal protoplasts. Dekker, New York Basel, pp 3–29

Gooday GW (1979) Chitin synthesis and differentiation in *Coprinus cinereus*. In: Burnett JH, Trinci APJ (eds) Fungal walls and hyphal growth. Cambridge Univ Press, Cambridge, pp 203–223

Gooday GW (1982) Metabolic control of fruitbody morphogenesis in *Coprinus cinereus*. In: Wells K, Wells EK (eds) Basidium and basidiocarp. Springer, Berlin Heidelberg New York, pp 157–173

Gooday GW, Trinci APJ (1980) Wall structure and biosynthesis in fungi. In: Gooday GW, Lloyd D, Trinci APJ (eds) The eukaryotic microbial cell. 30th Symp Soc Gen Microbiol. Cambridge Univ Press, Cambridge, pp 208–251

Gow NAR, Gooday GW (1983) Ultrastructure of chitin in hyphae of *Candida albicans* and other dimorphic and mycelial fungi. Protoplasma 115:52–58

Gow NAR, Gooday GW (1987) Infrared and X-ray diffraction data in chitins of variable structure. Carbohydr Res 165:105–110

Green PB (1974) Morphogenesis of the cell and organ axis – biophysical models. Brookhaven Symp Biol 25:166–199

Hackman RH, Goldberg M (1965) Studies on chitin. Aust J Biol Sci 18:935–946

Haffner L, Thielke C (1970) Kernzahl und Zellgrösse im Fruchtkörperstiel von *Coprinus radiatus* (Solt) Fr. Ber Dtsch Bot Ges 83:27–31

Hay ED (1981) Extracellular matrix. J Cell Biol Suppl 91:205–223

Jelsma J, Kreger DR (1975) Ultrastructural observations on (1→3)-β-D-glucan from fungal cell walls. Carbohydr Res 43:200–203

Kamada T, Takemaru T (1977) Stipe elongation during basidiocarp maturation in *Coprinus macrorhizus*: Mechanical properties of stipe cell wall. Plant Cell Physiol 18:831–840

Kamada T, Takemaru T (1983) Modification of cell-wall polysaccharides during stipe elongation in the basidiomycete *Coprinus cinereus*. J Gen Microbiol 129:703–709

Kamada T, Katsuda H, Takemaru T (1984) Temperature-sensitive mutants of *Coprinus cinereus* defective in hyphal growth and stipe elongation. Curr Microbiol 11:309–312

Koch AL (1983) The surface stress theory of microbial morphogenesis. Adv Microbiol Physiol 24:301–366

Kreger DR, Kopecká M (1976) On the nature and formation of the fibrillar nets produced by protoplasts of *Saccharomyces cerevisiae* in liquid media: an electronmicroscopic, X-ray diffraction and chemical study. J Gen Microbiol 92:207–220

Kurita K (1985) Chemical modifications of chitin and chitosan. In: Muzarelli R, Jeuniaux Ch, Gooday GW (eds) Chitin in nature and technology. Plenum, New York London, pp 287–293

Lopez-Romero E, Ruiz-Herrera J, Bartnicki-Garcia S (1982) The inhibitory protein of chitin synthase from *Mucor rouxii* is a chitinase. Biochem Biophys Acta 702:233–236

Marchessault RH, Deslandes Y (1979) Fine structure of (1→3)-β-D-glucans: curdlan and paramylon. Carbohydr Res 75:231–242

Maret R (1972) Chimie et morphologie submicroscopique des parois cellulaires de l'Ascomycète *Chaetomium globosum*. Arch Mikrobiol 81:68–90

Mol PC (1989) Hyphal wall elongation during expansion growth of the common mushroom (*Agaricus bisporus*). PhD Thesis, Univ Groningen

Mol PC, Wessels JGH (1987) Linkages between glucosaminoglycan and glucan determine alkali-insolubility of the glucan in walls of *Saccharomyces cerevisiae*. FEMS Microbiol Lett 41:95–99

Mol PC, Vermeulen CA, Wessels JGH (1988) Glucan-glucosaminoglycan linkages in fungal walls. Acta Bot Neerl 37:17–21

Molano J, Polacheck I, Durán A, Cabib E (1979) An endochitinase from wheat germ; activity on nascent and preformed chitin. J Biol Chem 254:4901–4907

Monnier VM, Kohn RR, Cerami A (1984) Accelerated age-related browning of human collagen in diabetes mellites. Proc Natl Acad Sci USA 81:583–587

Poulicek M, Voss-Foucart MF, Jeuniaux Ch (1985) Chitinoproteic complexes and mineralization in mollusk skeletal structures. In: Muzarelli R, Jeuniaux Ch, Gooday GW (eds) Chitin in nature and technology. Plenum, New York London, p 7–12

Robertson NF (1958) Observations on the effect of water on the hyphal apices of *Fusarium oxysporum*. Ann Bot (London) 22:159–173

Robertson NF (1965) The fungal hypha. Trans Br Mycol Soc 48:1–8

Ruiz-Herrera J, Bartnicki-Garcia S (1974) Synthesis of cell wall microfibrils in vitro by a "soluble" chitin synthase from *Mucor rouxii*. Science 186:357–359

Sato T, Norisuye T, Fujita H (1981) Melting behaviour of *Schizophyllum commune* polysaccharides in mixtures of water and dimethyl sulfoxide. Carbohydr Res 95:195–204

Shively JE, Conrad HE (1970) Stoichiometry of the nitrous acid deaminative cleavage of model aminoglucan glucosides and glycosaminoglycuronans. Biochemistry 9:33–43

Sietsma JH, Wessels JGH (1977) Chemical analysis of the hyphal wall of *Schizophyllum commune*. Biochem Biophys Acta 496:225–239

Sietsma JH, Wessels JGH (1979) Evidence for covalent linkages between chitin and β-glucan in a fungal wall. J Gen Microbiol 114:99–108

Sietsma JH, Wessels JGH (1988) Total inhibition of wall synthesis by 2-deoxyglucose and polyoxin D in protoplasts of *Schizophyllum commune*. Acta Bot Neerl 37:23–29

Sietsma JH, Sonnenberg ASM, Wessels JGH (1985) Localization by autoradiography of synthesis of (1→3)-β and (1→6)-β linkages in a wall glucan during hyphal growth of *Schizophyllum commune*. J Gen Microbiol 131:1331–1337

Soll DR, Herman MA, Staebell MA (1985) The involvement of cell wall expansion in the two modes of mycelium formation of *Candida albicans*. J Gen Microbiol 131:2367–2375

Sonnenberg ASM, Sietsma JH, Wessels JGH (1982) Biosynthesis of alkali-insoluble cell-wall glucan in *Schizophyllum commune* protoplasts. J Gen Microbiol 128:2667–2674

Sonnenberg ASM, Sietsma JH, Wessels JGH (1985) Spatial and temporal differences in the synthesis of (1→3)-β linkages in a wall glucan of *Schizophyllum commune*. Exp Mycol 9:141–148

Staebell M, Soll DR (1985) Temporal and spatial differences in cell wall expansion during bud and mycelium formation in *Candida albicans*. J Gen Microbiol 131:1467–1480

Stagg CM, Feather MS (1973) The characterization of a chitin-associated D-glucan from the cell walls of *Aspergillus niger*. BioChem Biophys Acta 320:64–72

Steele GC, Trinci APJ (1975) The extension zone of mycelial hyphae. New Phytol 75:583–587

Valk P van der, Wessels JGH (1976) Ultrastructure and localization of wall polymers during regeneration and reversion of protoplasts of *Schizophyllum commune*. Protoplasma 90:65–87

Valk P van der, Wessels JGH (1977) Light and electron microscopic autoradiography of cell-wall regeneration by *Schizophyllum commune* protoplasts. Acta Bot Neerl 26:43–52

Valk P van der, Marchant R, Wessels JGH (1977) Ultrastructural localization of polysaccharides in the wall and septum of the basidiomycete *Schizophyllum commune*. Exp Mycol 1:69–82

Vermeulen CA, Wessels JGH (1984) Ultrastructural differences between wall apices of growing and non-growing hyphae of *Schizophyllum commune*. Protoplasma 120:123–131

Vermeulen CA, Wessels JGH (1986) Chitin biosynthesis by a fungal membrane preparation. Evidence for a transient non-crystalline state of chitin. Eur J Biochem 158:411–415

Vermeulen CA, Raeven MBJM, Wessels JGH (1979) Localization of chitin synthase activity in subcellular fractions of *Schizophyllum commune* protoplasts. J Gen Microbiol 114:87–97

Vries OMH de, Wessels JGH (1975) Chemical analysis of cell wall regeneration and reversion of protoplasts from *Schizophyllum commune*. Arch Microbiol 102:209–218

Wang MC, Bartnicki-Garcia S (1970) Structure and composition of walls of the yeast form of *Verticillium alboatrum*. J Gen Microbiol 64:41–54

Wang MC, Bartnicki-Garcia S (1976) Synthesis of β-1,3-glucan microfibrils by a cell-free extract from *Phytophthora cinnamomi*. Arch Bioch Biophys 175:351–354

Wessels JGH (1984) Apical wall extension. Do lytic enzymes play a role? In: Nombela C (ed) Microbial cell wall synthesis and autolysis. Elsevier, Amsterdam, pp 31–42

Wessels JGH (1986) Cell wall synthesis in apical hyphal growth. Int Rev Cytol 104:37–79

Wessels JGH (1988) A steady state model for apical wall growth. Acta Bot Neerl 37:3–16

Wessels JGH, Sietsma JH (1981) Fungal cell walls: a survey. In: Tanner W, Loewus FA (eds) Plant carbohydrates II. Encycl Plant Physiol, N Ser 13B. Springer, Berlin Heidelberg New York, pp 352–394

Wessels JGH, Kreger DR, Marchant R, Regensburg BA, Vries OMH de (1972) Chemical and morphological characterization of the hyphal wall surface of the basidiomycete *Schizophyllum commune*. Biochim Biophys Acta 273:346–358

Wessels JGH, Sietsma JH, Sonnenberg ASM (1983) Wall synthesis and assembly during hyphal morphogenesis in *Schizophyllum commune*. J Gen Microbiol 129:1607–1616

Chapter 7

Cellulose and β-Glucan Synthesis in *Saprolegnia*

M. Fèvre, V. Girard, and P. Nodet [1]

1 Introduction

The biological and physiological properties of fungal cell walls reside in their chemical composition and also in the mode of spatial arrangement of the individual polymers in the wall. The general organization is that skeletal microfibrillar wall components are embedded in an amorphous matrix (Gooday and Trinci 1980). In *Saprolegnia monoica*, as in other Oomycetes, microcrystalline cellulose is overlayed by an outer layer of glucans containing $\beta(1 \to 3)$ and $\beta(1 \to 6)$linkages (Sietsma 1969; Hunsley and Burnett 1970). During normal hyphal growth, synthesis and deposition of cell wall components must be highly co-ordinated in order to provide and maintain the cell wall architecture. However, the morphogenetic roles of the cell wall polysaccharides are of differential importance. Amorphous glucans have probably little morphogenetic significance. In contrast, microfibrillar polysaccharide synthesis seems to have a primary role. The importance of chitin in cell wall maintenance was explored using chitin deficient mutants of *Aspergillus nidulans* (Katz and Rosenberger 1971) and by inhibiting chitin synthesis. Polyoxin D, a potent inhibitor of chitin synthase, prevented primary septum formation in yeast (Bowers et al. 1974), inhibited stipe elongation of *Coprinus* fruit-bodies (Gooday 1972), and induced abnormal hyphal growth of *Trichoderma* (Benitez et al. 1976). Therefore, microfibrillar chitin synthesis plays a fundamental role in the ontogeny of the mycelial wall. Chitin crystallization and its cross-linking to β-glucan chains would lead to normal hyphal wall extension (Wessels 1984; see also Chapter 6).

Chitin synthesis by fungi is by far the best characterized enzyme system catalyzing the formation of insoluble polysaccharide (Cabib 1981; Bartnicki-Garcia and Bracker 1984; Ruiz-Herrera 1984; see also Chapter 3).

Biochemical studies on the mechanism of cellulose synthesis have proved difficult. Cellulosic organisms (some fungi and green plants) possess at least two β-glucosyl transferase activities. These enzyme systems are capable of using as substrate the same sugar nucleotide, UDP-glucose, to produce $\beta(1 \to 4)$ or $\beta(1 \to 3)$glucans (Maclachlan and Fèvre 1982; Delmer 1987). β-glucan synthases are complex enzyme systems; their composition and properties may largely depend on the type of organism but their expression in vitro depends also on the method of isolation and assay conditions.

[1] Laboratoire de Différenciation Fongique-UMR-CNRS 106, Université Claude Bernard, Lyon I Bat 405, 69622 Villeurbanne Cedex, France

2 Synthesis and Deposition of Cell Wall Glucans in *Saprolegnia* hyphae

2.1 Cell Wall Synthesis During Apical Growth

The sites of polysaccharide synthesis and deposition in the hyphal cell wall were determined by pulse-labelling mycelium with [^{3}H]glucose. Light microscopy autoradiography showed that incorporation of glucose was very high in the apical zones and decreased rapidly in subapical parts (Fèvre 1979 a; Fèvre and Rougier 1980). This pattern of incorporation corresponds to the classical gradient of wall synthesis characterizing apical growth (Gooday and Trinci 1980).

Electron microscope autoradiography confirmed that cell wall growth is due to polar addition of cell wall material and located the site of glucose polymerization to the cell wall/plasma membrane region of the apex (Fèvre and Rougier 1982). Very little label was seen in any other region of the hyphae, and dictyosomes and apical vesicles did not accumulate radioactivity. The same distribution pattern was observed when hyphae were labelled for 3, 5, 10 or 15 min or labelled for 15 min and chased for 10 min. The absence of [^{3}H]glucose labelling indicates that dictyosome and Golgi vesicles do not transport long polymerized cell wall polysaccharides. Cytochemical staining of (1 → 4)-linked polysaccharides showed that cell walls and plasma membrane were heavily stained, while vesicles and dictyosomes exhibited a very low reaction (Fèvre and Rougier 1980). In *Saprolegnia*, therefore, cell wall glucan synthesis does not appear to occur in cytoplasmic structures but solely in the cell wall or at the plasmalemma level.

2.2 Cell Wall Synthesis During Protoplast Regeneration

Regeneration of protoplasts from *Saprolegnia* provides a simple system to examine the activity of the cell wall synthesizing enzymes (Girard and Fèvre 1984 a, b). Immediately after their isolation, 80% to 90% of the protoplasts are capable of producing a cell wall as revealed by autoradiography of [^{3}H]glucose incorporation (Girard et al. 1984). Cell wall regeneration is characterized by the synthesis of a network of cellulose microfibrils deposited around the spherical cells and is clearly visible after 2 h in culture. While the rate of cell wall synthesis, measured by [^{14}C]glucose incorporation was linear over 6 to 7 h, glucose incorporation into cellulose reached a plateau after 3 h. Thus, wall biogenesis by protoplasts started by the deposition of a microfibrillar skeleton with the subsequent deposition of other polysaccharides (Fèvre and Girard 1985).

Electron microscopy studies indicate that the early stages of regeneration are independent of the endomembrane system. After numerous observations there has been no evidence to show restoration of a vesicular transport system through the Golgi flow. Therefore, cellulose and other cell wall polysaccharides are only produced by plasma membrane enzymes which remain active during protoplast formation. This confirms that in the fungal cell, polysaccharide polymerization does not occur in secretory vesicles but at the cell surface only.

3 Cellulose Synthesis and Hyphal Morphogenesis

The apical growth of fungal hyphae is closely related to cell wall synthesis and alterations of polysaccharide polymerization lead to abnormal morphogenesis.

Congo red is a dye which interacts with various polysaccharides and exhibits a strong affinity for β-glucans (Wood 1980). As a consequence of dye interaction, the glucan chains are prevented from crystallizing to form microfibrils in *Acetobacter* (Haigler and Benziman 1982) and in algae (Quader et al. 1983).

Congo red and Calcofluor white modify morphogenesis and cell wall structure of yeast and chitinous fungi (Elorza et al. 1983; Pancaldi et al. 1984). In our studies, Congo red was used to investigate cellulose synthesis and its relation to apical growth of *Saprolegnia* hyphae (Nodet et al. 1986; Nodet 1987).

3.1 Morphogenetic Effect of Congo Red

When actively growing colonies of *Saprolegnia* were transferred to liquid medium containing $25 \, \mu g \cdot ml^{-1}$ of Congo red, hyphal growth was completely disturbed. Hyphal tips started to swell producing large bulges after 1 to 3 h incubation. Very intense coloured cell walls were observed at the base of the swollen apices and along the old part of the hyphae, indicating thickened cell walls. After 6 to 8 days treatment, high concentrations of Congo red lead to the complete disorganization of hyphal growth. Mycelia produced large bulges (up to 1 mm diameter) which detached from old mycelia. These isolated bulges were able to produce hyphae when placed in control nutrition media (Nodet 1987). Light microscopy autoradiography showed that cell wall polymer deposition was completely modified. After 5 min labelling of Congo red treated hyphae, [³H]glucose incorporation was very high at the base of the swollen apices and, in some places, along the tubular part of the hyphae. The normal gradient of cell wall synthesis was destroyed by Congo red and deposition of cell wall polysaccharides was unco-ordinated (Nodet et al. 1986b). Hyphae labelled for 1 min with [³H]glucose were chased for 3 h in nutrient medium containing Congo red. Autoradiography revealed that the bases of the swollen apices were labelled, while the cell walls deposited during hyphal tip swelling were unlabelled. This showed that Congo red prevented hyphal elongation and induced an immediate tip expansion leading to large bulges (Nodet 1987).

Electron microscopy studies revealed that the typical polarity of elongating apices was lost following Congo red treatment. Bulges contained numerous vacuoles: nuclei and mitochondria had an irregular distribution. The cell walls, 5 to 6 times thicker than in normal hyphae were composed of multiple layers. This 'lens-shaped' thickening of the walls, also observed along the hyphae, corresponded to the zones of [³H] glucose incorporation and intensively stained zones.

3.2 Effects of Congo Red on Cellulose Synthesis

The rate of cell wall polymer synthesis was measured by incorporation of glucose after 3 h treatment at different Congo red concentrations. Total incorporation was increased by 40% at low concentration (25 $\mu g \cdot ml^{-1}$) then decreased by increasing dye concentrations. Cellulose synthesis, estimated as the radioactivity insoluble in acetic-nitric acid reagent (Updegraff 1969), was stimulated at low Congo red concentration, while a hot water soluble polysaccharide was largely produced at high dye concentration.

The stimulation of cellulose synthesis by Congo red was immediate and was observed from the beginning of incubation in the dye, increasing by 40% after 6 h treatment compared to controls (Nodet 1987).

Electron microscopy of shadow cast preparations of cell wall ghosts extracted by the Updegraff reagent (1969) showed that Congo red affects cellulose crystallization. The cellulose of control cell walls remaining after acetic-nitric acid extraction appeared as interwowen microfibrils. Cell wall cellulose from Congo red-treated hyphae showed different states of crystallization. Short microfibrils embedded in an amorphous material were observed but in thickened cell walls the cellulose fraction was in an amorphous state.

Thus, in the presence of Congo red, cellulose lost its characteristic microfibrillar organization but was synthesized in significant quantitites in an amorphous state (Nodet et al. 1986b).

Congo red effects showed that in *Saprolegnia*, like in other organisms, cellulose synthesis and its crystallization into microfibrils are two distinct processes which can be dissociated. Crytallization represent the limiting step of cellulose synthesis (Haigler and Benziman 1982). By preventing microfibril formation, Congo red would allow the polymerization of a higher amount of amorphous cellulose. One can suggests that this kind of cellulose is not able to assume the rigidity of the cell wall. As a result the soft cell wall would expand and lead to swollen apices. These results clearly show that the normal hyphal tip growth of *Saprolegnia* depends on the presence of a cellulose microfibrillar skeleton conferring rigidity. The presence of very thick, amorphous cell walls is not sufficient to ensure apical elongation.

4 β-Glucan Synthases from *Saprolegnia*

4.1 Enzymatic Synthesis of β-Glucans

Cell free extracts of numerous fungi contain enzyme systems transferring glucose from UDP glucose to produce β-glucans. In yeast and chitinous fungi, these enzymes synthesize $\beta(1 \rightarrow 3)$glucans (Shematek et al. 1980; Lopez-Romero and Ruiz-Herrera 1978; Quigley and Selitrennikoff 1984; Larriba et al. 1981).

In *Saprolegnia*, as in green plants, different glucans are produced from UDP-glucose depending on the assay conditions (Fèvre 1979a; Maclachlan and Fèvre

1982). The relative proportions of $\beta(1 \to 3)$ and $\beta(1 \to 4)$polysaccharides synthesized by particulate enzymes vary according to concentrations of UDP-glucose and $MgCl_2$ and do not involve glycolipid intermediates (Fèvre 1983a).

The products formed at high substrate (UDP-Glc) concentrations were hydrolyzed by purified exo $\beta(1 \to 3)$glucanase from Basidiomycete Qm 806 which released 90% of the radioactivity. Only cellulase was active on glucans produced at low substrate concentrations in the presence of $MgCl_2$. The presence of glucose and cellobiose on paper chromatography of the hydrolysates confirmed that the polymer was $\beta(1 \to 4)$ linked (Fèvre and Rougier 1981).

$\beta(1 \to 3)$glucans produced by particulate enzymes collected by centrifugation and then observed by electron microscopy are constituted of single microfibrils of 10 nm diameter (Fèvre and Rougier 1981). The synthesis of $\beta(1 \to 3)$ glucans is also achieved at a high rate by different fungal enzymes. Microfibrillar glucans have also been produced by cell-free extracts from other filamentous fungi and from yeast (Wang and Bartnicki-Garcia 1976; Larriba et al. 1981). No cellulose microfibrils were observed when glycosyl transferases were assayed in conditions of $\beta(1 \to 4)$glucans synthesis (Fèvre and Rougier 1981). A convincing demonstration of cellulose biosynthesis must meet several criteria such as insolubility of the unbranched polymer, $\beta(1 \to 4)$linkages only, and chain length in the range of natural cellulose (Maclachlan and Fèvre 1982). In fact, although fungal or green plant systems are capable of producing $\beta(1 \to 4)$ glucans there have been no reports showing evidence for the synthesis of authentic cellulose in vitro.

4.2 β-Glucan Synthases Have a Transmembrane Orientation in the Plasma Membrane

Particulate enzymes isolated from mycelia or protoplasts and separated by isopycnic centrifugation are mainly associated with membranes exhibiting ATPase activity at the density $1.16 \, g \cdot cm^{-1}$ (Girard and Fèvre 1984a). Glucan synthases are also associated with internal membranes, endoplasmic reticulum, dictyosomes and Golgi vesicles (Fèvre 1979a; Fèvre and Rougier 1981; Girard and Fèvre 1984a). Radioactive ConA was used as a specific marker to label protoplast plasma membranes which were isolated by isopycnic centrifugation (Girard and Fèvre 1984a). $\beta(1 \to 3)$ and $\beta(1 \to 4)$glucan synthases sedimented as a single peak, corresponding to that of vanadate-sensitive ATPase and [³H]concanavalin A indicating that both synthases are associated with the plasma membrane.

The action of external inactivating agents (proteases, glutaraldehyde) on intact protoplasts reduced $\beta(1 \to 3)$ and $\beta(1 \to 4)$glucan synthase activities suggesting that the enzymes face the outer surface of the plasma membrane (Girard and Fèvre 1984a). Regenerating protoplasts are unable to use external UDP-glucose to produce cell wall polysaccharides which indicates that the substrate binding sites of the enzymes face the inside of the cell (Girard et al. 1984).

The transmembrane orientation of the synthases in the plasma membrane was confirmed by the comparison of the effect of Congo red on cellulose synthesis and glucan synthase activities. Particulate glucan synthases isolated from mycelia are inhibited in vitro by Congo red. The apparent Ki values were $40\,\mu g \cdot ml^{-1}$ and $100\,\mu g \cdot ml^{-1}$ for the $\beta(1 \rightarrow 3)$ and $\beta(1 \rightarrow 4)$glucan synthases, respectively (Nodet et al. 1986b). The addition of Congo red during the course of incubation produced an immediate inhibition of glucose incorporation, i.e. of glucan elongation. In vitro, Congo red can reach the catalytic site of the isolated enzymes inhibiting polymerization. In contrast, in vivo, Congo red binds to the glucan, preventing crystallization, but not synthesis. Glucan synthases of the plasma membrane should have a transmembrane orientation as their catalytic sites cannot be reached from the outside (Nodet 1987). Glucan synthesis would be vectorial characterized by the polymerization on the inner face or in the plasma membrane and the extension of polymerized glucans to the outside of the cell.

4.3 Regulation of Glucan Synthases

Glycosyl transferase activities are modulated by divalent cations and cellobiose, but different exogenous compounds can modify glucan synthesis.

$\beta(1 \rightarrow 3)$glucan synthases are stimulated by the presence of trypsin but inhibited by other proteases, pepsin, papain, and acid protease (Fèvre 1979b). This effect was observed on membrane-bound enzymes and on digitonin-solubilized enzymes (Fèvre and Rougier 1981). Stimulation occurs from the beginning of incubation in the presence of the protease but prolonged action of trypsin leads to inactivation of the glycosyl transferases. $\beta(1 \rightarrow 3)$glucan synthases must, therefore, exist in an inactive state (zymogen) which can be activated by moderate proteolysis. Such regulation, which also appears to modulate green plant glycosyl transferases (Girard and Maclachlan 1987), characterizes the chitin synthase system of various fungi (Cabib 1981).

In contrast, $\beta(1 \rightarrow 4)$glucan synthases from *Saprolegnia* are inactivated by trypsin, but stimulated in the presence of certain nucleotides. Of the various nucleotides tested, only ATP and GTP produced strong stimulation of glucan synthesis (Fèvre 1983b). These nucleotides exhibit the same pattern and level of stimulation, comparable Km values, and pH and $MgCl_2$ dependency (Fèvre 1983b, 1984). This stimulation seems independent of the nature of the nucleoside portion of the activator (guanosine or adenosine) but the integrity of the three phosphate groups is essential. Nucleoside diphosphate and nucleotide analogues in which the β-oxygen have been substituted, are inefficient, while nucleoside thiotriphosphates are activators (Fèvre 1984). Thus, stimulation might occur by a phosphorylation process. However, ATP and GTP do not stimulate digitonin-solubilized enzymes, indicating that these stimulators do no act directly on the enzymes (Fèvre 1984). It is possible that a compound associated with the enzymes, lost or destroyed during enzyme solubilization, may be the target of the nucleotides and the true activator. Aloni et al. (1982) have shown that solubilized cellulose synthase from *Acetobacter xylinum* is stimulated by the co-operative action of GTP and a soluble protein factor. $\beta(1 \rightarrow 3)$glucan synthase of yeast seems to be modulated by a proteinaceous component binding nucleotides (Kang and Cabib 1986). The enzyme system of *Saprolegnia* is more like that from *A. xylinum* since GTP and ATP stimulate synthesis of $\beta(1 \rightarrow 4)$glucans, unlike yeast and chitinous filamentous fungi where $\beta(1 \to 3)$glucan synthesis is stimulated by nucleotides (Fèvre 1984; Ross et al. 1986; Szaniszlo et al. 1985).

If exogenous compounds such as protease and nucleotides modulate the glucan synthase activities of *Saprolegnia*, some membrane-bound constituents are also able to regulate enzyme activities. A heat stable effector associated with internal membranes has been characterized (Girard and Fèvre 1988). Solubilization of particulate enzymes using the zwitterionic detergent 3-[3-(cholamidopropyl)dimethylammonio]-1-propane-sulphonate (CHAPS) allows recovery of a compound in the soluble fraction that gives five-fold stimulation of $\beta(1 \to 3)$glucan synthase activity. Preliminary characterization has shown this membrane-bound effector to be heat stable, protease-insensitive and non-dialyzable. It increased the V_{max} of the enzyme without affecting the apparent Km for UDP-glucose. As glucan synthases bound to nitrocellulose can fix and retain this effector, the stimulation seems to be due to direct interaction of the effector with the enzymes. This effector may represent a natural effector which modulates $\beta(1 \to 3)$glucan synthase activity.

All the results show that glucan synthase activity can be regulated in vitro by various compounds. Exogenous protease or nucleotides may not have physiological significance, but one cannot exclude the possibility that the types of modulation produced by these activators could be induced, in vivo, by other compounds. The characterization of a membrane-bound stimulator of $\beta(1 \to 3)$glucan synthase from *Saprolegnia* may indicate one of the natural effectors.

5 Purification of Glucan Synthases from *Saprolegnia*

In spite of numerous studies on bacteria, fungi and green plants, glucan synthase systems are not fully understood (Maclachlan and Fèvre 1982; Delmer 1987). Enzymes from numerous systems have been solubilized but their assay conditions for maximal activity are still to be ascertained. One major reason for this lack of information is that these membrane-bound enzymes are difficult to purify.

Density-gradient centrifugation has recently been used to assist in the identification of the proteins implicated in glucan synthesis (Girard, Bulone and Fèvre, unpublished results). Monoclonal antibodies to intracellular membranes of *Saprolegnia* were produced in order to select hybrid myeloma lines secreting antibodies to the β-glucan synthases (Nodet et al. 1986 b).

5.1 Monoclonal Antibodies to Glucan Synthases

Mice were immunized by injection of particulate or solubilized enzymes and their splenic lymphocytes fused with myeloma cells. Hybridoma secreting antibodies to β-glucan synthases were identified in a direct antigen-binding assay in which glucan synthase activities, retained by the antibodies previously fixed on nitrocellulose, were measured (Nodet 1987; Nodet et al. 1988). By using this screening assay, 30 polyclonal cultures were isolated from 528 clones tested. Three lines were subcloned and 15 monoclonal lines selected.

The antibodies selected were capable of precipitating $\beta(1 \rightarrow 4)$ and $\beta(1 \rightarrow 3)$glucan synthase activities, indicating that both enzyme systems have identical immunodominant epitopes. [^{35}S]methionine-labelled proteins were immunoprecipitated then analyzed in SDS-PAGE. Fluorography revealed a major band corresponding to a MW of 45000 together with some minor bands.

Radioactive antibodies produced by growing hybridoma in the presence of [^{3}H]lysine were used to study the localization of glucan synthases in protoplasts. Autoradiographic studies showed that most antibodies were strongly fixed to the protoplast cell surface while one antibody was weakly fixed (Nodet 1987). Since the antibodies are directed against the same enzymes, this indicates that the epitopes are more or less accessible from the external face of the plasma membrane confirming the transmembrane orientation of glucan synthases.

Quantitative analysis of antibody fixation on the protoplast surface showed that radioactive antibodies were preferentially fixed on protoplasts originating from the apical zones of hyphae compared to protoplasts produced from the older part of the hyphae (Nodet 1987). This confirms the polarized distribution of glucan synthases along the hyphae in relation to apical growth (Girard and Fèvre 1984b).

These antibodies provide the basis for the localization, at the electron microscope level, of the glucan synthases and their characterization and purification by immunoaffinity chromatography.

5.2 Isolation of $\beta(1 \rightarrow 3)$ Glucan Synthases

Solubilized enzymes were separated by zonal centrifugation in glycerol gradients. The bulk of protein was recovered at the top of the gradients, while $\beta(1 \rightarrow 3)$ and $\beta(1 \rightarrow 4)$glucan synthase activities entered the gradients. Maximal activity of these enzymes was separated and recovered at densities of 1.06 and 1.08 for $\beta(1 \rightarrow 3)$ and $\beta(1 \rightarrow 4)$glucan synthases (Girard and Bulone, unpublished results). Analysis of the glucan synthase fractions by sodium dodecyl sulphate-polyacrylamide gel electrophoresis (SDS-PAGE) showed differences in the protein subunit composition. $\beta(1 \rightarrow 3)$glucan synthases were enriched in proteins with mol wts varying from 40000 to 60000 while $\beta(1 \rightarrow 4)$glucan synthases contained, in addition, proteins with mol wts higher than 10000.

The purification of $\beta(1 \rightarrow 3)$glucan synthase was improved by product entrapment of the enzymes as described for chitin synthase (Kang et al. 1986) followed by glycerol gradient centrifugation. By this procedure, only $\beta(1 \rightarrow 3)$glucan synthase activity was recovered in the gradient with a purification approaching 1000-fold. The purified preparation was examined by SDS-PAGE. Silver staining of the proteins revealed a major band with a mol wt of 40000 and several minor bands with mol wts from 30000 to 70000 (Girard and Bulone, unpublished results). This indicated that the $\beta(1 \rightarrow 3)$glucan synthase complex which, in its native form, has a mol wt higher than 500000 (Girard, unpublished results) contained subunits with different mol wts. The remaining problem is the identification of the band representing the active glucan synthase. The availability of antibodies will allow elucidation of this point.

6 Conclusions

Glucan and cellulose synthesis occur in the plasma membrane of hyphal tips or protoplasts as revealed by [^{3}H]glucose labelling and cytological studies. $\beta(1 \rightarrow 3)$ and $\beta(1 \rightarrow 4)$glucan synthases have a transmembrane orientation in the plasmalemma leading to a vectorial synthesis of the cell wall polysaccharides. These enzymes may have a common structure or organization, as revealed by preliminary immunological studies, but they are different systems which can be separated by glycerol-gradient centrifugation.

One can speculate that these enzymatic complexes are transmembrane structures composed of several integral proteins in a model resembling the green plant plasma membrane rosettes involved in cellulose synthesis (Müller and Brown 1980; Montezinos 1982). Some proteins, sensitive to proteases or capable of reacting with nucleotides, would be involved in the regulatory processes. Other proteins would be involved in UDP-glucose binding c.f. such an enzyme seems to exist in green plants (Delmer 1987).

Cellulose synthases may have a more complicated organization than $\beta(1 \rightarrow 3)$glucan synthases. $\beta(1 \rightarrow 4)$glucan synthase activities of cell free extracts are always much lower than $\beta(1 \rightarrow 3)$glucan synthase activities. It is probable that these enzymes require a specific factor that is lost in the course of isolation.

The opposite behaviour of the synthases towards protease and nucleotides, and the presence of a membrane bound activator of $\beta(1 \rightarrow 3)$ glucan synthase, may indicate a difference in the regulation of their activities. This would have implications in cell wall assembly where the deposition of the different polysaccharides during apical growth is co-ordinated in time and space.

References

Aloni Y, Delmer DP, Benziman M (1982) Achievement of high rates of in vitro synthesis of 1,4-β-D-glucan: activation by cooperative interaction of the *Acetobacter xylinum* enzyme system with GTP, polyethylene glycol and a protein factor. Proc Natl Acad Sci USA 79:6448–6452

Bartnicki-Garcia S, Bracker CE (1984) Unique properties of chitosomes. In: Nombela C (ed) Microbial cell wall synthesis and autolysis. Elsevier, Amsterdam, pp 101–112

Benitez T, Villa TG, Garcia-Acha I (1976) Effect of polyoxin D on germination, morphological development and biosynthesis of the cell wall of *Trichoderma viride*. Arch Microbiol 108:183–188

Bowers B, Levin G, Cabib E (1974) Effect of polyoxin D on chitin synthesis and septum formation in *Saccharomyces cerevisiae*. J Bacteriol 119:564–575

Cabib E (1981) Chitin: structure metabolism, and regulation of biosynthesis. In: Tanner W, Loewus TA (eds) Plant carbohydrates II. Encyclo Plant Physiol, vol 13 B. Springer, Berlin Heidelberg New York, pp 395–415.

Delmer DP (1987) Cellulose biosynthesis. In: Briggs WR, Jones RL, Walbot V (eds). Annu Rev Plant Physiol, Palo Alto, pp 259–290

Elorza MV, Rico H, Sentandreu R (1983) Calcofluor white alters the assembly of chitin fibrils in *Saccharomyces cerevisiae* and *Candida albicans* cells. J Gen Microbiol 129:1577–1582

Fèvre M (1979a) Glucanases, glucan synthases and wall growth in *Saprolegnia monoica*. In: Burnett JH, Trinci APJ (eds) Fungal walls and hyphal growth. Br Mycol Soc Symp, vol 2. Cambridge Univ Press, Cambridge, pp 225–264

Fèvre M (1979b) Digitonin solubilization and protease stimulation of β glucan synthetases of *Saprolegnia*. Z Pflanz Physiol 95:129–140

Fèvre M (1983a) Inhibitors of synthesis of lipid-linked Saccharides also inhibit β-glucans synthesis by cell free extracts of the fungus *Saprolegnia monoica*. J Gen Microbiol 129:3007–3013

Fèvre M (1983b) Nucleotide effects on glucan synthesis activities of particulate enzyme from *Saprolegnia*. Planta 159:130–135

Fèvre M (1984) ATP and GTP stimulate membrane bound but not digitonin-solubilized β-glucan synthases from *Saprolegnia monoica*. J Gen Microbiol 130:3279–3284

Fèvre M, Girard V (1985) Glucan synthases and cell-wall regeneration in fungal protoplasts. In: Pilet PE (ed) The physiological properties of plant protoplasts. Springer, Berlin Heidelberg New York Tokyo, pp 184–190

Fèvre M, Rougier M (1980) Hyphal morphogenesis of *Saprolegnia*: cytological and biochemical effects of coumarin and glucono lactone. Exp Mycol 4: 343–361

Fèvre M, Rougier M (1981) $\beta(1 \rightarrow 3)$ and $\beta(1 \rightarrow 4)$glucan synthesis by membrane fractions from the fungus *Saprolegnia*. Planta 151:232–241

Fèvre M, Rougier M (1982) Autoradiographic study of hyphal cell wall synthesis of *Saprolegnia*. Arch Microbiol 131:212–215

Girard V, Fèvre M (1984a) $\beta(1 \rightarrow 4)$ and $\beta(1 \rightarrow 3)$ glucan synthases are associated with the plasma membrane of the fungus *Saprolegnia*. Planta 160:400–406

Girard V, Fèvre M (1984b) Distribution of $(1 \rightarrow 3)\beta$ and $(1 \rightarrow 4)\beta$-glucan synthases along the hyphae of *Saprolegnia monoica*. J Gen Microbiol 130:1557–1562

Girard V, Fèvre M (1989) Solubilization and partial characterization of a membrane bound stimulator of $\beta(1 \rightarrow 3)$ glucan synthase from *Saprolegnia*. J Gen Microbiol (submitted)

Girard V, Maclachlan G (1987) Modulation of Pea membrane β-glucan synthase activity by calcium, polycation, endogenous protease, and protease inhibitor. Plant Physiol 85:131–136

Girard V, Capellano A, Fèvre M (1984) β-glucan synthesis and glucan synthase activities during early stages of cell wall regeneration by protoplasts from *Saprolegnia monoica*. J Gen Microbiol 130:2367–2374

Gooday GW (1972) The effect of Polyoxin D on morphogenesis of *Coprinus cinereus*. Biochem J 129:17p

Gooday GW, Trinci APJ (1980) Wall structure and biosynthesis in fungi. In: Gooday GW, Trinci APJ (eds) The eukaryotic microbial cell. Soc Gen Microbiol Symp, vol 30. Cambridge Univ Press, Cambridge, pp 207–251

Haigler CH, Benziman M (1982) Biogenesis of cellulose microfibrils occurs by cell-directed self-assembly in *Acetobacter xylinum*. In: Malcolm, Brown R Jr (ed) Cellulose and other natural polymer systems. Plenum, New York London, pp 273–297

Hunsley D, Burnett JH (1970) The ultrastructural architecture of the walls of some fungi. J Gen Microbiol 62:203–218

Kang MS, Cabib E (1986) Regulation of fungal cell wall growth: a guanine nucleotide binding, proteinaceous component required for activity for $(1 \rightarrow 3)\beta$-glucan synthase. Biochemistry 83:5808–5812

Kang MS, Elango N, Mattia E, Au-Young J, Robbins PW, Cabib E (1984) Isolation of chitin synthetase from *Saccharomyces cerevisiae*. Purification of an enzyme by entrapment in the reaction product. J Biol Chem 259:14966–14972

Katz D, Rosenberger RF (1971) Lysis of an *Aspergillus nidulans* mutant blocked in chitin synthesis and its relation to wall assembly and wall metabolism. Arch Microbiol 80:284–292

Larriba G, Morales M, Ruiz-Herrera J (1981) Biosynthesis of β-glucan microfibrils by cell-free extracts from *Saccharomyces cerevisiae*. J Gen Microbiol 124:375–383

Lopez-Romero E, Ruiz-Herrera J (1978) Properties of β glucan synthetase from *Saccharomyces cerevisiae*. Antonie van Laeuwenhoek 44:329–339

Maclachlan G, Fèvre M (1982) An overview of cell wall synthesis. In: Lloyd CW (ed) The cytoskeleton in plant growth and development. Academic Press, London New York, pp 127–146

Montezinos B (1982) The role of the plasma membrane in cellulose microfibril assembly. In: Lloyd CW (ed) The cytoskeleton in plant growth and development. Academic Press, London New York, pp 147–162

Müller SC, Brown RM Jr (1980) Evidence for an intramembrane component associated with a cellulose microfibril synthesizing complex in higher plants. J Cell Biol 84:315–326

Nodet P (1987) Fonctionnement et caractérisation immunologique des glucane synthases de *Saprolegnia monoica*. Thesis, Univ Claude Bernard, Lyon I

Nodet P, Capellano A, Fèvre M (1986a) Effects of Congo red on polysaccharide synthesis and glucan synthase activities of the fungus *Saprolegnia*. In: Vian B, Reiss B, Goldberg R (1986b) Cell wall 86. Proc 4th Cell Wall Meet, Paris, pp 346−347

Nodet P, Grange J, Fèvre M (eds) Monoclonal antibodies to glucan synthases from the fungus *Saprolegnia*. In: Vian B, Reiss B, Goldberg R (eds) Cell wall 86. Proc 4th Cell Wall Meet, Paris, pp 348−349

Nodet P, Grange J, Fèvre M (1988) $\beta(1 \rightarrow 3)$ and $\beta(1 \rightarrow 4)$glucan synthase dot blot assays and their use as a direct antigen binding assay to screen monoclonal antibodies. Anal Biochem 174:662−665

Pancaldi S, Poli F, Dall'Olio G, Vannini GL (1984) Morphological anomalies induced by Congo red in *Aspergillus niger*. Arch Microbiol 137:185−187

Quader H, Robinson DG, Kempen R Van (1983) Cell wall development in *Oocystis solitaria* in the presence of polysaccharide binding dyes. Planta 157:317−323

Quigley DR, Selitrennikoff CP (1984) $\beta(1 \rightarrow 3)$ glucan synthase activity of *Neurospora crassa*: Stabilization and partial characterization. Exp Mycol 8:202−204

Ross P, Aloni Y, Weinhouse H, Michaeli D, Weinberger-Ohana P, Mayer R, Benziman M (1986) Control of cellulose synthesis in *Acetobacter xylinum*. An unique guanyl oligonucleotide is the immediate activator of the cellulose synthase. Carbohydr Res 149:101−117

Ruiz-Herrera J (1984) The role of chitosomes in the apical growth of fungi. In: Nombela C (ed) Microbial cell wall synthesis and autolysis. Elsevier, Amsterdam, pp 113−120

Shematek EM, Braatz JA, Cabib E (1980) Biosynthesis of the yeast cell wall. Preparation and properties of a $\beta(1 \rightarrow 3)$ glucan synthetase. J Biol Chem 255: 888−894

Sietsma JH (1969) Protoplast formation and cell wall composition of some Oomycete species. Thesis, Amsterdam

Szaniszlo PJ, Mohinder SK, Cabib E (1985) Stimulation of $\beta(1 \rightarrow 3)$ glucan synthetase of various fungi by nucleotide triphosphates: Generalized regulatory mechanism for cell wall biosynthesis. J Bacteriol 161: 1188−1194

Updegraff DM (1969) Semi-micro determination of cellulose in biological material. Anal Biochem 32:420−424

Wang MC, Bartnicki-Garcia S (1976) Synthesis of $\beta(1 \rightarrow 3)$ glucan microfibrils by a cell-free extract from *Phytophthora cinnamomi*. Arch Biochem Biophys 175:351−354

Wessels JGH (1984) Apical hyphal wall extension: do lytic enzymes play a role? In: Nombela C (ed) Microbial cell wall synthesis and autolysis. Elsevier, Amsterdam, pp 31−42

Wood PJ (1980) Specificity in the interaction of direct dyes with polysaccharides. Carbohydr Res 85:271−287

Chapter 8

Synthesis and Function of Glycosylated Proteins in *Saccharomyces cerevisiae*

W. TANNER[1]

1 Introduction

Proteins can be covalently modified in a number of ways. The most complex and evolved modification is glycosylation. Glycoproteins occur in all eukaryotic cells (Kornfeld and Kornfeld 1985; Tanner and Lehle 1987). They are also found frequently, but not always, as constituents of cell envelopes of archaebacteria (Sumper 1987). The possibility of rare occurrence in eubacteria (Messner and Sleytr 1988) is still under debate.

The carbohydrate moiety of glycoproteins generally amounts to about 20%, of the total molecular mass (Cohen and Ballou 1981) but in extreme cases can be 90%. The saccharide is either attached to the hydroxyl residue of serine, threonine, hydroxylysine and hydroxyproline (*O*-linked), or to the amide residue of asparagine (*N*-linked). In fungal cells, sugars have also been found *O*-linked to β-hydroxyphenylalanine and to β-hydroxytyrosine of proteins (Lin and Kolattukudy 1976).

Glycoproteins only occur in special cellular compartments:

(a) On the cell surface either associated with the plasma membrane or as a periplasmic and cell envelope (cell wall) component, respectively;
(b) In lysosomes and in organelles equivalent to lysosomes, like plant and fungal vacuoles;
(c) Associated with nuclei or nuclear envelopes (Holt and Hart 1986);
(d) In the endoplasmic reticulum, Golgi complex, and secretory vesicles, i.e. organelles involved in glycoprotein synthesis (see below).

This very specific localization of glycoproteins within and mainly outside cells led to the suggestion, that it may be the carbohydrate moiety that determines the destination of these molecules (Eylar 1965). Evidence for such a case has meanwhile been obtained for lysosomal proteins in certain mammalian cells (Kornfeld and Kornfeld 1985).

In the following article knowledge concerning synthesis, localization, and function of glycoproteins in fungal cells will be summarised. Similarities as well as differences to higher eukaryotes will be stressed. A more extensive review on the same subject has recently been published (Tanner and Lehle 1987).

---

[1] Lehrstuhl für Zellbiologie und Pflanzenphysiologie, Universität Regensburg, 8400 Regensburg, FRG

2 Biosynthetic Pathways

In the late sixties "lipid intermediates" were for the first time observed to be involved in the synthesis of complex carbohydrates in eukaryotic cells (Tanner 1969). Subsequently the lipophilic molecule transferring sugars to proteins had been identified in a number of laboratories as the phosphorylated polyprenol dolichol phosphate (Behrens and Leloir 1970; Tanner et al. 1971; Evans and Hemming 1973; Jung and Tanner 1973; Baynes et al. 1973). In fungal cells dolichol phosphate activated sugars were found to be involved in *O*- as well as in *N*-glycosylation (Babczinski and Tanner 1973; Tanner and Lehle 1987).

2.1 *O*-Glycosylation

Small unbranched oligosaccharides consisting of up to five 1,2- and 1,3-linked mannoses are attached to serine and threonine residues of *O*-glycosylated proteins in *S. cerevisiae* (Sentandreu and Northcote 1969; Nakajima and Ballou 1974). Such proteins are synthesized at the ER (Larriba et al. 1976; Haselbeck and Tanner 1983; Watzele et al. 1988) and the glycosylation proceeds according to the sequence shown in Fig. 1. The first mannose is transferred from Dol-P-Man to protein (Babczinski and Tanner 1973; Sharma et al. 1974) in a reaction of the ER (Haselbeck and Tanner 1983). The extension takes place by transfer of mannosyl residues directly from GDP-Man most likely in the Golgi complex.

   The reaction sequence of Fig. 1 has been observed in other fungal cells, such as in *Hansenula* (Bretthauer and Wu 1975), *Fusarium* (Soliday and Kolattukudy 1979), *Aspergillus* (Letoublon and Got 1974), and *Neurospora* (Gold and Hahn 1976).

   Thus, *O*-glycosylation of proteins in fungal cells differs in three ways from that of animal and plant cells:

(a) The sugar linking the saccharide to the protein is a mannosyl residue, whereas it is mainly Gal-NAc in mammalian cells (Beyer et al. 1981) and arabinose or galactose in those plant proteins analyzed so far (Selvendran and O'Neill 1982);

FORMATION OF O-MANNOSIDIC-BOND IN FUNGI

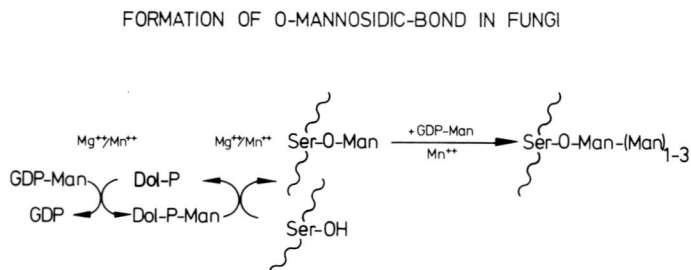

Fig. 1. Reaction sequence of *O*-glycosylation in *S. cerevisiae* and other fungal cells

(b) It is at the ER where the first mannose is transferred to protein in fungal cells, whereas, at least in mammalian cells, it is clear that Gal-NAc-linked saccharides are initiated in the Golgi complex (Roth 1984);

(c) The participation of dolichol phosphate activated sugars in *O*-glycosylation has been observed only in fungi so far.

2.2 *N*-Glycosylation

Highly branched saccharides, which may be more than one hundred mannosyl residues long, are typical for fungal *N*-linked chains (Cohen and Ballou 1981). Since these chains consist exclusively of mannoses except for the two GlcNAc residues forming the link to protein, yeast glycoproteins are generally called mannoproteins.

The biosynthetic pathway for *N*-glycosylation in *S. cerevisiae* has mainly been worked out in four laboratories (Nakajima and Ballou 1975; Lehle 1981; Parodi 1981; Byrd et al. 1982); it is shown in Fig. 2. The cellular localization has been established with the help of Schekman's sec mutants (Schekman 1985). The sequence of ER reactions is the same as that previously established for mammalian cells (Robbins et al. 1977; Li et al. 1978). Plant and animal cells are able to modify greatly the oligosaccharide structure attached to the protein; it may be trimmed to a $GlcNAc_2Man_3$ residue, which subsequently is extended again in Golgi reactions to a "complex type" oligosaccharide consisting of additional GlcNAc, Gal and Fuc residues, besides sialic acid in animals and xylose in plants (Kornfeld and Kornfeld 1985; Takahashi et al. 1986). In addition, glycoproteins of higher eukaryotes may also contain "high mannose"-type oligosaccharides, which are more or less identical to the saccharide structures synthesized at the ER (Fig. 2). A specific feature of fungal cells, on the other hand, is the lack of "complex type" oligosaccharides and the existence of partly very long mannosyl extensions, which are produced in the Golgi apparatus.

As shown in Fig. 2 the dolichol pyrophosphate-linked oligosaccharide is synthesized in two phases: first a heptasaccharide is put together using sugar nucleotides in each step as sugar donor, subsequently additional seven sugars are transferred, all from dolichol-phosphate intermediates, to yield $Dol-PP-GlcNAc_2Man_9Glc_3$. Evidence has been presented that up to the heptasaccharide synthesis proceeds at the cytoplasmic side of the ER, whereas the additional sugars are attached in the ER lumen (Snider and Rogers 1984), whereby the Dol-P intermediates act in some way as transmembrane translocators (Haselbeck and Tanner 1982, 1984).

Finally, the question which asparagines within a protein become glycosylated shall be discussed briefly. From sequence comparisons (Marshall 1974) and in vitro studies (Struck and Lennarz 1980; Lehle and Bause 1984) it is clear, that the sequence -AsN-X-Ser/Thr- is necessary for *N*-glycosylation, whereby the amino acid X is not allowed to be a proline. Since not all AsN-X-Ser/Thr sequences on proteins which are translocated into the ER lumen are glycosylated, additional requirements − not really understood − have to be met. Sequence or structural requirements for *O*-glycosylation are not known.

FORMATION OF N-GLYCOSYLATED PROTEINS IN YEAST

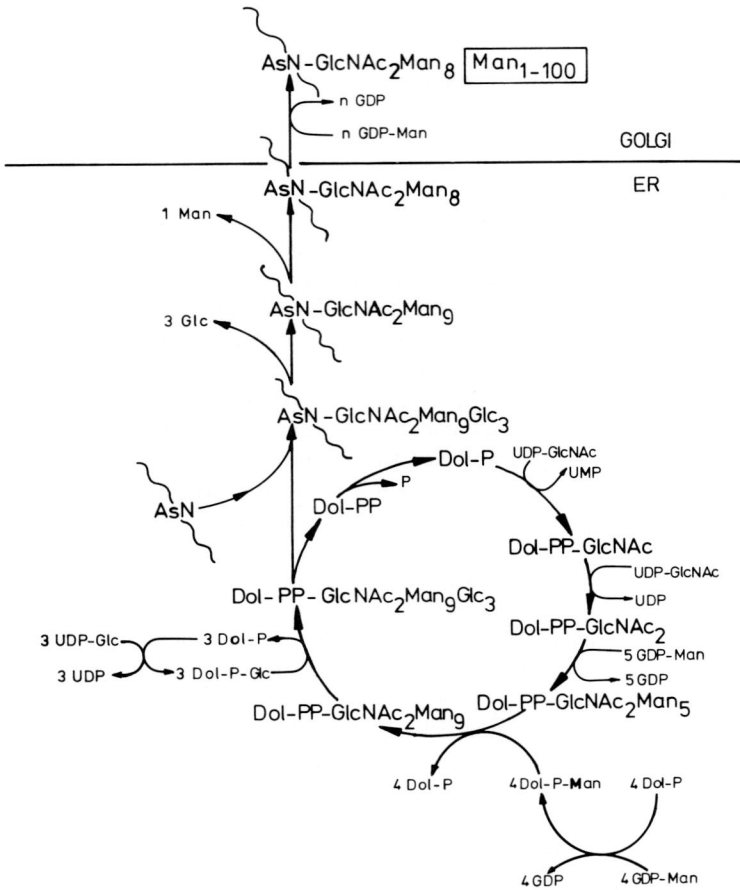

Fig. 2. The dolichol cycle: reaction sequence of *N*-glycosylation at the ER (ubiquitous). Subsequent modifications at the Golgi are partly specific for yeast. Abbreviations (here and in the text): Dol-P = dolichol phosphate, Dol-PP = dolichol pyrophosphate, Glc = glucose, Man = mannose, Gal = galactose, GlcNAc = *N*-acetylglucosamine, Fuc = fucose, GalNAc = *N*-acetylgalactosamine, UDP = uridine diphosphate, GDP = guanosine diphosphate

3 Functional Aspects

3.1 Possible Functions of Saccharide Moieties

As mentioned already, the reactions of *N*-glycosylation at the ER proceed in an identical manner in all eukaryotes. This evolutionary conservation from yeast to man certainly suggests an important function for this intricate and costly way of modifying proteins. Surprisingly, a number of glycoprotein enzymes are as active in their non-glycosylated form (Tanner and Lehle 1987). Among the possible functions of the car-bohydrate moiety most widely discussed are "sorting" during intracellular protein transport and cell-cell recognition related to processes like embryogenesis, cell differentiation, and mating. An example of each of these two possibilities will briefly be presented here in relation to *S. cerevisiae*.

3.1.1 Vacuolar Proteins

Experiments with human fibroblasts have shown that most lysosomal enzymes require N-linked saccharides and a specific mannose-6-phosphate residue thereon as a sorting signal to target these proteins into the lysosome (Hickman et al. 1974; Kaplan et al. 1977). Yeast vacuoles are functionally equivalent to lysosomes and although their proteins are glycosylated and contain mannose-6-phosphate residues, their carbohydrate moiety is not required for correct lysosomal targeting (Schwaiger et al. 1982). Indeed, recently evidence has been presented that the sorting signal is a specific amino acid sequence within the "pro" part of the immature carboxypeptidase Y (Johnson et al. 1987; Valls et al. 1987). Thus, at the moment it is certainly not possible to generalize the concept of saccharide moieties being sorting signals for intracellular protein transport.

3.1.2 Cell Surface Agglutinins

The haploid mating partners of a number of fungal species agglutinate in a mating type and species-specific way before fusion and zygote formation (Yanagishima 1984). Responsible for this phenomenon are specific cell surface glycoproteins, called ag-

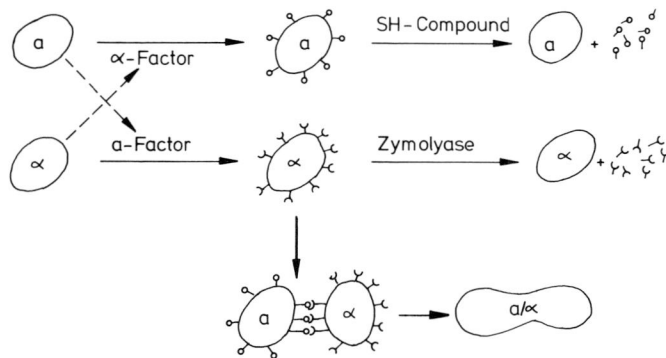

Mating Factor Induced Agglutination in S. cerevisiae

Fig. 3. The peptide pheromone *a* and *α* factor induce the formation of specific agglutinins on *S. cerevisiae*, which can be readily released from the cell surface by mild treatments

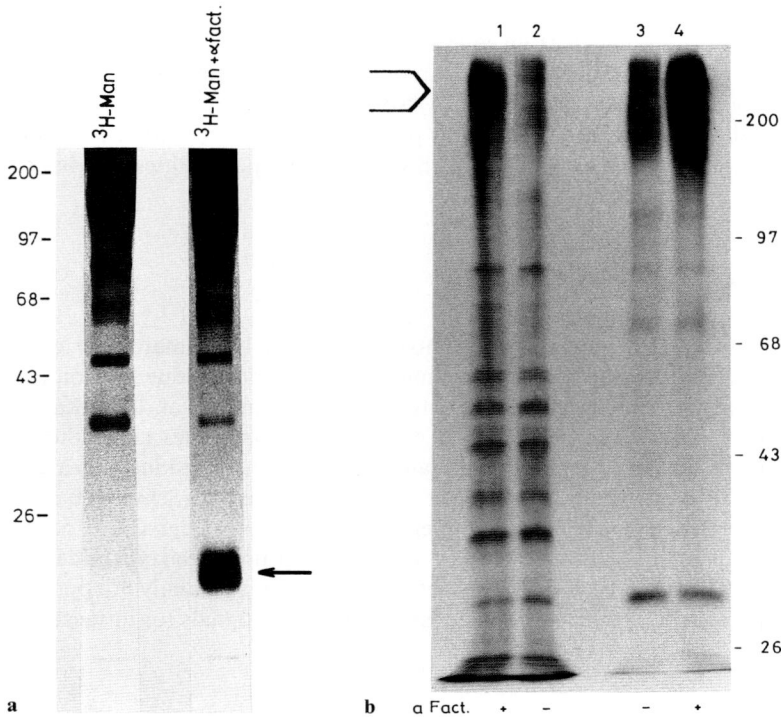

Fig. 4. a *S. cerevisiae a* cells were grown for 2 h with [2-³H]mannose in the presence and absence of
α factor. The cell surface components were released with mercaptoethanol (Orlean et al. 1986),
separated on SDS gel and made visible by fluorography. **b** *S. cerevisiae* α cells were grown for 90 min
with ³⁵SO₄²⁻ in the presence and absence of *a* factor. The cell surface components were released by
mild zymolyase treatment (Hauser 1988), separated on SDS gel and made visible by fluorography.
Tracks 1 and 2 = crude extract; tracks 3 and 4 biologically active fraction after separation by HPLC
gel filtration on a TSK 3000 column

glutinins (Cohen and Ballou 1981). In *S. cerevisiae* (Fig. 3) agglutination is strongly
increased by the mating pheromones *a* and α factor (Betz et al. 1978). The corre-
spondingly induced cell surface components can be released by mild treatment
without affecting the vitality of the cell (Fig. 3) and separated on SDS-gels. As shown
in Fig. 4 (a and b) *a* cells treated with α factor possess an additional glycosylated protein of 22 kD and extracts obtained from α cells treated with *a* factor show a diffuse
band of around and larger than 200 kD. The latter results in 5 defined bands of molecular weight 38 to 124 kD, when carbohydrate moieties are removed with Endo F
(K. Hauser 1988 and unpublished results). This agglutinin may be related to a con-
sititutive one from *S. cerevisiae* characterized by Terrance et al. (1987). The 22 kD protein of *a* cells contains about 20 O-linked saccharides attached to a protein of 13 kD
(Orlean et al. 1986; Watzele et al. 1988). After mild periodate oxidation the protein
shows identical behavior on HPLC-sizing columns but has lost its biological activity
(Watzele et al. 1988). Analogous observations have been made with agglutinins of *S.*
kluyveri and of *H. wingei* (Pierce and Ballou 1983; Yen and Ballou 1974) indicating

that the carbohydrate moieties of agglutinins play an essential role in the phenomenon of cell-cell interaction. A more solid proof may be available after expression of the cloned genes in *E. coli* and testing the biological activity of the carbohydrate-free proteins.

3.2 Protein Glycosylation Essential for Growth and/or Proliferation

The most direct way to find out how important protein modification by glycosylation is would be to prevent glycosylation by specific inhibitors or by gene disruption experiments and see whether such cells still grow and divide. A specific inhibitor for *N*-glycosylation is tunicamycin (Takatsuki et al. 1975; Tkacz and Lampen 1975; Lehle and Tanner 1976). In its presence yeast cells stop to initiate new buds, although buds already initiated are not halted in their growth (Arnold and Tanner 1982). After one generation all cells assume the appearance of cells typically arrested in G1 ($>86\%$ of the cells unbudded, $<15\%$ cells budded, but containing two nuclei (Ettenhuber 1985)). Since a temperature-sensitive alg mutant, defective in an early step of Dol-PP-GlcNAc$_2$Man$_9$Glc$_3$ formation (Huffacker and Robbins 1982) behaves in the same way at the non-permissive temperature (Klebl et al. 1984), the results with tunicamycin cannot be caused by a side effect of the antibiotic. Recently it has been shown that preventing *N*-glycosylation does not result in a true G1 arrest, since the nuclei of the cells not initiating a new bud end up as diploids (Vai et al. 1987), thus giving rise to a new phenotype: unbudded G2 cells. *N*-glycosylated proteins, therefore, seem to be obligatorily required for bud initiation and for nuclear division, not however for cell growth.

Whether protein *O*-glycosylation is essential for yeast cells or eukaryotic cells in general is not clear. A specific inhibitor for *O*-glycosylation is not available. An indication that *O*-glycosylation in *S. cerevisiae* is obligatorily required for growth and/or cell division has been obtained from a disruption of the gene for GDP-Man: Dol-P mannosyltransferase (P. Orlean and P. Robbins, personal communication). This disruption in haploid cells is lethal and although Dol-P-Man in fungal cells is involved in both *O*- and *N*-glycosylation, indirect arguments favour the interpretation that this disruption is lethal because the cells lack *O*-glycosylation.

4 Summary

S. cerevisiae contains a large number of mannose-rich glycoproteins. The saccharides are linked to the protein either *N*- or *O*-glycosidically, i.e. either via an asparagine or the hydroxy amino acids serine or threonine. The yeast cell lacks the enzymes for the synthesis of "complex type" saccharide chains. *N*-glycosylation up to the protein-bound GlcNAc$_2$Man$_9$Glc$_3$ proceeds at the ER and in an identical manner in all eukaryotes. *O*-glycosylation in fungal cells differs in a number of ways from higher eukaryotes: the first sugar (a mannose!) is transferred to the protein via dolichol phosphate in a reaction at the ER.

There is no evidence available from *S. cerevisiae* that *N*-glycosylation is involved in intracellular targeting of proteins, e.g. into the vacuole. Protein-linked saccharides, however, seem to be involved in agglutinin-mediated mating reactions.

Prevention of *N*- or *O*-glycosylation seems to be lethal to *S. cerevisiae*.

Acknowledgements. I am very grateful to numerous coworkers whose names are given in the publications cited from this laboratory. The pheromone *a* factor has been kindly supplied by Drs. Betz and Duntze, Bochum. The original work from this laboratory has been supported by the Deutsche Forschungsgemeinschaft (SFB 43) and by Fonds der Chemischen Industrie.

References

Arnold E, Tanner W (1982) An obligatory role of protein glycosylation in the life cycle of yeast cells. FEBS Lett 148:49–53

Babczinski P, Tanner W (1973) Involvement of dolichol monophosphate in the formation of specific mannosyl linkages in yeast glycoproteins. Biochem Biophys Res Commun 54:1119–1124

Baynes JW, Hsu AF, Heath EC (1973) The role of mannosyl-phosphoryl-dihydropolyisoprenol in the synthesis of mammalian glycoproteins. J Biol Chem 248:5693–5704

Behrens NH, Leloir LF (1970) Dolichol monophosphate glucose: an intermediate in glucose transfer in liver. Proc Natl Acad Sci USA 66:153–159

Betz R, Duntze W, Manney TR (1978) Mating-factor-mediated sexual agglutination in *Saccharomyces cerevisiae*. FEBS Lett 4:107–110

Beyer TA, Sadler JE, Rearick JI, Paulson JC, Hill RL (1981) Glycosyltransferases and their use in assessing oligosaccharide structure and structure-function relationships. Adv Enzymol 52:23–175

Bretthauer RK, Wu S (1975) Synthesis of the mannosyl-O-serine(threonine)-linkage of glycoproteins from polyisoprenylphosphate mannose in yeast (*Hansenula holstii*). Arch Biochem Biophys 167:151–160

Byrd JC, Tarentino AL, Maley F, Atkinson PH, Trimble RB (1982) Glycoproteins synthesis in yeast. J Biol Chem 257:14657–14666

Cohen RE, Ballou CE (1981) Mannoproteins: structure. In: Tanner W, Loewus FA (eds) Plant carbohydrates II. Encyclo Plant Physiol, N Ser, vol 13B. Springer, Berlin Heidelberg New York, pp 441–458

Ettenhuber C (1985) Untersuchungen phasenspezifischer Stoffwechselvorgänge in synchronisierten Hefekulturen. Diplomarb, Univ Regensburg

Evans PJ, Hemming FW (1973) The unambiguous characterization of dolichol phosphate mannose as a product of mannosyl transferase in pig liver endoplasmic reticulum. FEBS Lett 31:335–338

Eylar EH (1965) On the biological role of glycoproteins. J Theor Biol 10:89–113

Gold MH, Hahn HJ (1976) Role of mannosyl lipid intermediate in the synthesis of *Neurospora crassa* glycoproteins. Biochemistry 15:1808–1814

Haselbeck A, Tanner W (1982) Dolichyl phosphate-mediated mannosyl transfer through liposomal membranes. Proc Natl Acad Sci USA 79:1520–1524

Haselbeck A, Tanner W (1983) O-Glycosylation in *Saccharomyces cerevisiae* is initiated at the endoplasmic reticulum. FEBS Lett 158:335–338

Haselbeck A, Tanner W (1984) Further evidence for dolichol phosphate-mediated glycosyl translocation through membranes. FEMS Lett 21:305–308

Hauser K (1988) Hefe-Agglutinine: Optimierung eines Testsystems; Versuche zur Funktion der Kohlenhydratketten des *a*-Agglutinins; Anreicherung eines α-Agglutinins. Diplomarb, Univ Regensburg

Hickman S, Shapiro LJ, Neufeld EF (1974) A recognition marker required for uptake of a lysosomal enzyme by cultured fibroblasts. Biochem Biophys Res Commun 57:55–61

Holt GD, Hart GW (1986) The subcellular distribution of terminal N-acetylglucosamine moieties. J Biol Chem 261:8049–8057

Huffaker T, Robbins PW (1982) Temperature-sensitive yeast mutants deficient in asparagine-linked glycosylation. J Biol Chem 257:3203–3210

Johnson LM, Bankaitis VA, Emr SD (1987) Distinct sequence determinants direct intracellular sorting and modification of a yeast vacuolar protease. Cell 48:875–885

Jung P, Tanner W (1973) Identification of the lipid intermediate in yeast mannan biosynthesis. Eur J Biochem 37:1–6

Kaplan A, Archord DT, Sly WS (1977) Phosphohexosyl components of a lysosomal enzyme are recognized by pinocytosis receptors on human fibroplasts. Proc Natl Acad Sci USA 74:2026–2030

Klebl F, Huffaker TC, Tanner W (1984) A temperature-sensitive N-glycosylation mutant of *S. cerevisiae* that behaves like a cell-cycle mutant. Exp Cell Res 150:309–313

Kornfeld R, Kornfeld S (1985) Assembly of asparagine-linked oligosaccharides. Annu Rev Biochem 54:631–664

Larriba G, Elorza MV, Villanueva JR, Sentandreu R (1976) Participation of dolichol phosphomannose in the glycosylation of yeast wall mannoproteins at the polysomal level. FEBS Lett 71:316–320

Lehle L (1981) Biosynthesis of mannoproteins in fungi. In: Tanner W, Loewus FA (eds) Plant carbohydrates II. Encycl Plant Physiol, N Ser, vol 13 B. Springer, Berlin Heidelberg New York, pp 458–483

Lehle L, Bause E (1984) Primary structural requirements for N- and O-glycosylation of yeast mannoproteins. Biochim Biophys Acta 799:246–251

Lehle L, Tanner W (1976) The specific site of tunicamycin inhibition in the formation of dolichol-bound N-acetylglucosamine derivatives. FEBS Lett 71:167–170

Letoublon R, Got R (1974) Rôle d'un intermediaire lipique dans le transfert du mannose à des accepteurs glycoprotéique endogènes chez *Aspergillus niger*. FEBS Lett 46:214–217

Li E, Tabas I, Kornfeld S (1978) The synthesis of complex-type oligosaccharides. J Biol Chem 253:7762–7770

Lin TS, Kolattukudy PE (1976) Evidence for novel linkage in a glycoprotein involving β-hydroxyphenylalanine and β-hydroxytyrosine. Biochem Biophys Res Commun 72:243–250

Marshall RD (1974) The nature and metabolism of the carbohydrate-peptide linkages of glycoproteins. Biochem Soc Symp 40:17–26

Messner P, Sleytr UB (1988) Asparaginyl-rhamnose: a novel type of a protein-carbohydrate linkage in a eubacterial surface-layer glycoprotein. FEBS Lett 228:317–320

Nakajima T, Ballou CE (1974) Characterization of the carbohydrate fragments obtained from *Saccharomyces cerevisiae* mannan by alkaline degradation. J Biol Chem 249:7679–7684

Nakajima T, Ballou CE (1975) Yeast manno-protein biosynthesis: solubilization and selective assay of four mannosyltransferases. Proc Natl Acad Sci USA 72:3912–3916

Orlean P, Ammer H, Watzele M, Tanner W (1986) Synthesis of an O-glycosylated cell surface protein induced in yeast by α-factor. Proc Natl Acad Sci USA 83:6263–6266

Parodi AJ (1981) Biosynthesis mechanisms for cell envelope polysaccharides. In: Arnold WN (ed) Yeast cell envelopes: biochemistry, biophysics and ultrastructure, vol II. CRC Press, Boca Raton, pp 47–64

Pierce M, Ballou CE (1983) Cell-cell recognition in yeast. Characterization of the sexual agglutination factors from *Saccharomyces kluyveri*. J Biol Chem 258:3576–3582

Robbins PW, Hubbard SC, Turco SJ, Wirth DF (1977) Proposal for a common oligosaccharide intermediate in the synthesis of membrane glycoproteins. Cell 12:893–900

Roth J (1984) Cytochemical localization of terminal N-acetyl-D-galactosamine residues in cellular compartments of intestinal goblet cells: implications for the topology of O-glycosylation. J Cell Biol 98:399–406

Schekman R (1985) Protein localization and membrane traffic in yeast. Annu Rev Cell Biol 1:115–143

Schwaiger H, Hasilik A, Figura K von, Wiemken A, Tanner W (1982) Carbohydrate-free carboxypeptidase Y is transferred into the lysosome-like vacuole. Biochem Biophys Res Commun 104:950–956

Selvendran RR, O'Neill MA (1982) Plant glycoproteins. In: Loewus FA, Tanner W (eds) Plant carbohydrates I. Encycl Plant Physiol, N Ser, vol 13 A. Springer, Berlin Heidelberg New York, pp 515–583

Sentandreu R, Northcote DH (1969) The characterization of oligosaccharide attached to threonine and serine in mannan glycopeptides obtained from the cell wall of yeast. Carbohydr Res 10:584–585

Sharma CB, Babczinski P, Lehle L, Tanner W (1974) The role of dolichol monophosphate in glycoprotein biosynthesis in S. cerevisiae. Eur J Biochem 46:35–41

Snider MD, Rogers OC (1984) Transmembrane movement of oligosaccharide-lipids during glycoprotein synthesis. Cell 36:753–761

Soliday CL, Kolattukudy PE (1979) Introduction of O-glycosidically linked mannose into proteins via mannosyl phosphoryl dolichol by microsomes from Fusarium solani f. pisi. Arch Biochem Biophys 197:367–378

Struck DK, Lennarz WJ (1980) The function of saccharide-lipids in synthesis of glycoproteins. In: Lennarz WJ (ed) The biochemistry of glycoproteins and proteoglycans. Plenum, New York London, pp 35–83

Sumper M (1987) Halobacterial glycoprotein biosynthesis. Biochim Biophys Acta 906:69–79

Takahashi N, Hotta T, Ishihara H, Mori M, Tejima S, Bliguy R, Akazawa T, Endo S, Arata Y (1986) Xylose-containing common structural unit in N-linked oligosaccharides of laccase from Sycamore cells. Biochemistry 25:388–395

Takatsuki A, Kohno K, Tamura G (1975) Inhibition of biosynthesis of polyisoprenol sugars in chick embryo microsomes by tunicamycin. Agric Biol Chem 39:2089–2091

Tanner W (1969) A lipid intermediate in mannose biosynthesis in yeast. Biochem Biophys Res Commun 35:144–150

Tanner W, Lehle L (1987) Protein glycosylation in yeast. Biochim Biophys Acta 906:81–99

Tanner W, Jung P, Behrens NH (1971) Dolicholmonophosphates: Mannosyl acceptors in a particulate in vitro system of S. cerevisiae. FEBS Lett 16:245–248

Terrance K, Heller P, Wu Y-S, Lipke PN (1987) Identification of glycoprotein components of α agglutinin, a cell adhesion protein from Saccharomyces cerevisiae. J Bacteriol 169:475–482

Tkacz JS, Lampen JO (1975) Tunicamycin inhibition of polyisoprenyl N-acetylglucosaminyl pyrophosphate formation in calf-liver microsomes. Biochem Biophys Res Commun 65:248–257

Vai M, Popolo L, Alberghina L (1987) Effect of tunicamycin on cell cycle progression in budding yeast. Exp Cell Res 171:448–459

Valls LA, Hunter CP, Rothman JH, Stevens TH (1987) Protein sorting in yeast: the localization determinant of yeast vacuolar carboxypeptidase Y resides in the propeptide. Cell 48:887–897

Watzele M, Klis F, Tanner W (1988) Purification and characterization of the inducible a agglutinin of Saccharomyces cerevisiae. EMBO J 7:1483–1488

Yanagishima N (1984) Mating systems and sexual interactions in yeast. In: Linskens HF, Heslop-Harrion J (eds) Cellular interactions. Encycl Plant Physiol, N Ser, vol 17. Springer, Berlin Heidelberg New York, pp 403–423

Yen PH, Ballou CE (1974) Partial characterization of the sexual agglutination factor from Hansenula wingei Y-2340 type 5 cells. Biochemistry 13:2428–2437

Chapter 9

Lipids in the Structure and Function of Fungal Membranes

D. M. LÖSEL [1]

1 Introduction

Although the fungal thallus provides an excellent object for physiological and bio-chemical experimentation, being in immediate contact with its substrate and capable of rapid direct responses to manipulation of environmental factors, fungi have been relatively little exploited in fundamental studies on membrane composition and the roles of membrane lipids. Nevertheless, such investigations can often be carried out more directly and conveniently with these relatively simple eukaryotes than in tissues of more complex organisms.

2 The Membrane System

Dramatic changes in the amount and organization of membranes within fungal cells occur during the developmental cycle. From co-ordinated ultrastructural and bio-chemical studies, it is clear that membranes not only form a framework, which com-partmentalizes the different biochemical processes of the cell, controlling movement of substrates and products and providing an anchorage for enzymes and carriers, but that the membranes themselves form a continuous, dynamic system the composition of which differs in individual organelles and regions of the cell. These ideas have giv-en rise to membrane flow hypotheses (Morré and Mollenhauer 1974; Moore 1982; Robinson and Kristen 1982; Boller and Wiemken 1986; Harris 1986), which envisage a movement of membrane components from sites of synthesis associated with the en-doplasmic reticulum to Golgi bodies, at the margins of which membranous vesicles arise and migrate to the periphery of the cytoplasm or the hyphal tip (Fig. 1). Such vesicles fuse with the plasmalemma, adding to the membrane and releasing their con-tents of enzymes and/or wall components.

It is still unclear whether membrane material returns from the cell surface via an intracellular pool of membrane components or as vesicles passing back from the cell periphery to fuse with endoplasmic reticulum of Golgi membranes, as discussed by Robinson and Kristen (1982) with respect to higher plant cells. There is little evidence

[1] Department of Plant Sciences, University of Sheffield, Sheffield S10 2TN, UK

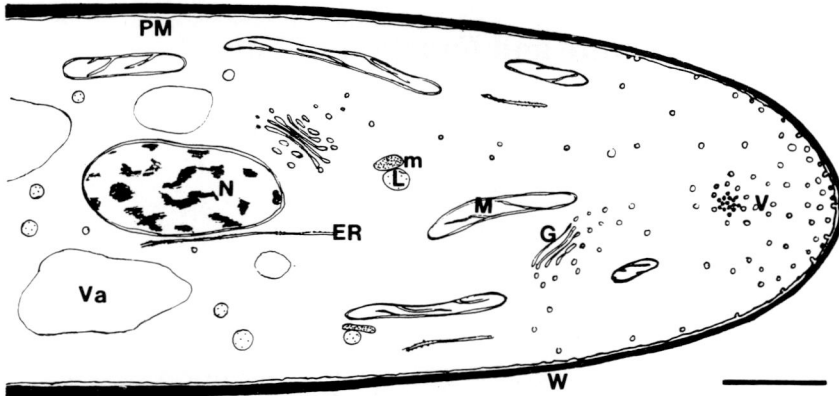

Fig. 1. Membranes and organelles in apical region of fungal hypha. *ER* = endoplasmic reticulum, *G* = Golgi body, *L* = lipid drop, *M* = mitochondrion, *m* = microbody, *N* = nucleus, *PM* = plasmalemma, *V* = vesicle, *Va* = Vacuole, *W* = wall, Scale bar = 1 µm

for endocytosis in walled cells of fungi (Wilson 1973) and plants, although endocytotic uptake of cationic ferritin by bean protoplasts has been demonstrated (Joachim and Robinson 1984) and there is some evidence that coated vesicles can give rise to partially coated reticulum (Harris 1986). Vacuoles, which are much more prominent features of fungi and higher plants than of animal cells and can occupy a substantial volume within mature cells, are bounded by the vacuolar membrane or tonoplast, which appears to originate by fusion of the numerous small vacuoles (also developing from vesicles?) present in younger cells. Many ultrastructural observations on the development of fungal cells are consistent with evidence, summarized by Boller and Wiemken (1986), for the growth of the plant cell tonoplast by incorporation of membranes from the Golgi body-endoplasmic reticulum-lysosome system and the linking of the tonoplast and plasmalemma by a pathway of membrane flow.

Since the original formulation of such hypotheses, much detail has been added. Chitosomes, membrane-bound organelles which convey chitin synthetase to sites of wall synthesis, have been isolated from several fungi and electron-micrographs have been published showing chitin fibrils extruded in vitro from chitosomes of *Mucor rouxii, Blastocladiella emersonii* and *Agaricus bisporus* (Bartnicki-Garcia et al. 1978; Mills and Cantino 1981; Hänseler et al. 1983). Membrane lipid, which accounted for 6% to 7% of the dry weight of chitosomes from *A. bisporus* and yeast cells of *M. rouxii* (Weete et al. 1985), comprised 2% and 4% sterols, respectively, 20% phospholipid, and a substantial proportion of other polar lipids which, in the case of *A. bisporus*, may have included glycolipid and sphingolipid.

Vesicles of similar size to chitosomes or somewhat larger are so abundant in the hyphal tip region that their accumulation was long recognized with the light microscope as a Spitzenkörper in Ascomycotina and Basidiomycotina, before being resolved by the electron microscope (Beckett et al. 1974) as an accumulation of vesicles, which derive from Golgi bodies and stream towards the hyphal tip at a rate which has been calculated by Trinci (1978) to be of the order of 40000 vesicles per minute. In

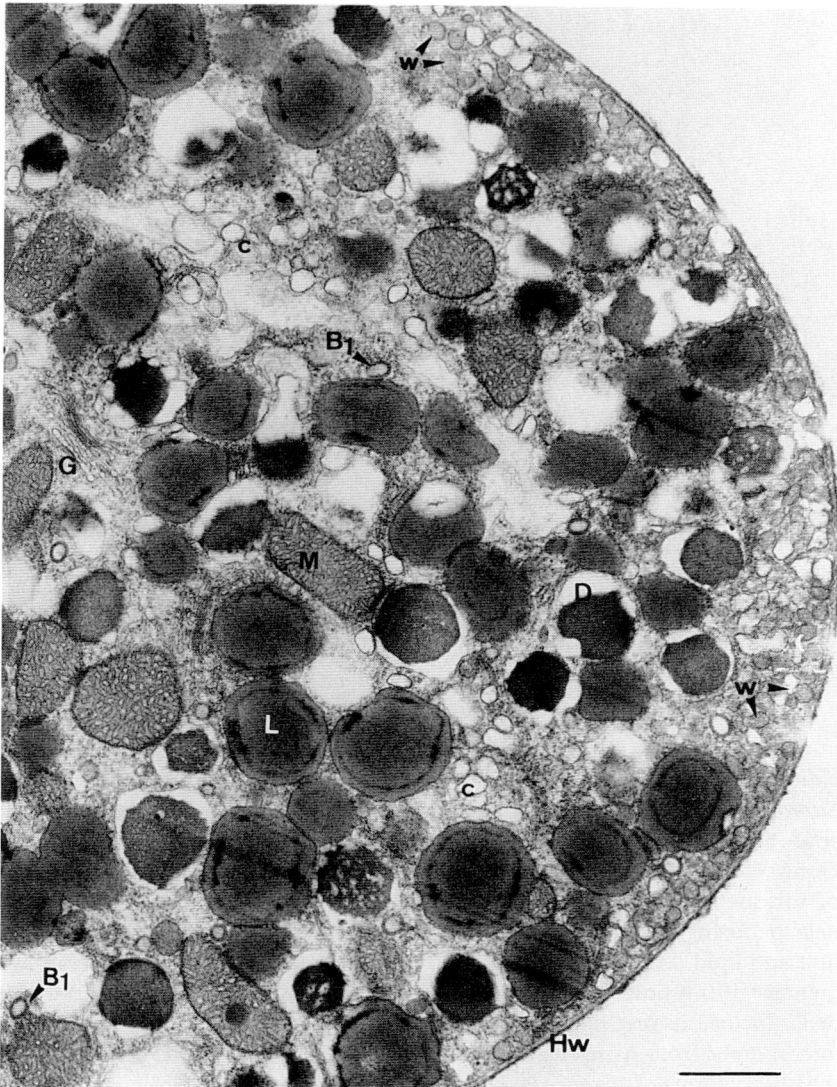

Fig. 2. Median section of hyphal tip in the Oomycete *Brevilegnia minutandra* (Saprolegniales), showing mitochondria (*M*), lipid bodies (*L*), Golgi body (*G*), hyphal wall (*HW*), wall vesicles (*w*), dense body vacuoles (*D*), vesicles with granular cortex (*B₁*). Scale bar = 1 μm. (From Armbruster 1982)

fungi lacking Golgi bodies or dictyosomes, e.g. in hyphal tips of *Allomyces arbuscula* (Class Chytridiomycetes), such vesicles are thought to arise from Golgi-like cisternae of the endoplasmic reticulum (Roos and Turian 1977). In the Oomycetes which, instead of chitin, have a wall of microfibrillar branched glucans and small amounts of cellulose, glucans are probably also transported in Golgi-derived vesicles to sites of wall synthesis (Gooday 1983).

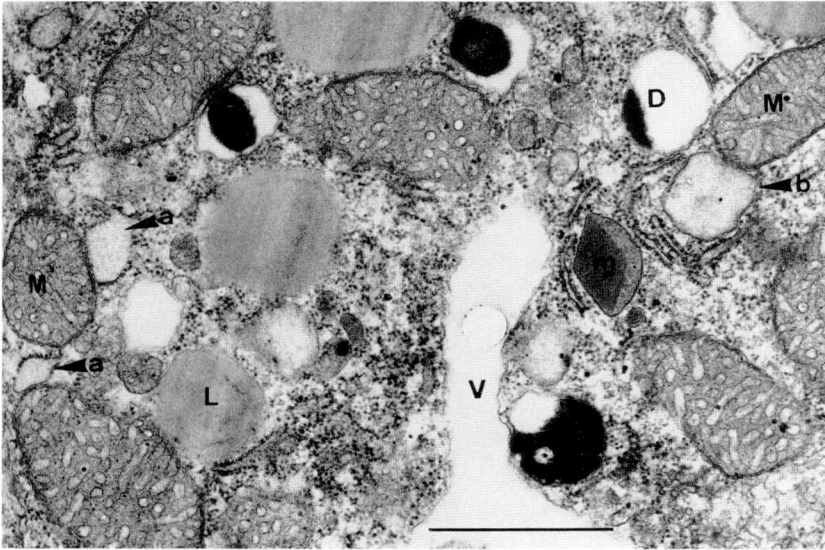

Fig. 3. Lipid drops (L), dense-body vesicles (D) and their postulated origin from the endoplasmic reticulum during presporangial development of an Oomycete, *Dictyuchus pseudodictyon* (Saprolegniaceae). a = expanded rough ER cisternae, b = intermediate stage of development, M = mitochondria, V = vacuole, m = microbody with crystalline inclusion. Other microbody-like organelles lie in close contact with lipid drops. Scale bar = 1 μm. (From Armbruster 1982)

Some similarity to the chitosome system is provided by electron microscope observations of the formation of microfibrils in vitro by membrane preparations from *Saprolegnia* (Class Oomycetes), when supplied with high concentrations of UDP-glucose (Fèvre and Rougier 1981; see also Chapter 7). Fig. 2 shows the hyphal tip in an Oomycete *Brevilegnia minutandra* (Order Saprolegniales), densely filled with vesicles of various types, emerging from the periphery of the Golgi body and accumulating near or fusing with the plasmalemma, as well as mitochondria, endoplasmic reticulum and lipid drops. Fig. 3 shows some of these features in greater detail in another member of this order and provides evidence that lipid drops and vesicles with electron dense inclusions originate from the endoplasmic reticulum. These and other electron micrographs from Armbruster's (1982) investigation of Oomycetes are consistent with the observations of Wanner et al. (1981) on the development of lipid bodies in *Neurospora* and various plant tissues from sites of lipid accumulation within the unit membrane bilayer of the endoplasmic reticulum. The resulting lipid bodies are considered to be bounded by a 'coat', derived from a half unit membrane and may also contain membrane-derived osmiophilic inclusions.

3 Membrane Structure

Fungal membranes, like those of other eukaryotes, are currently interpreted as lipid bilayers within which protein molecules, such as enzymes, are anchored at various

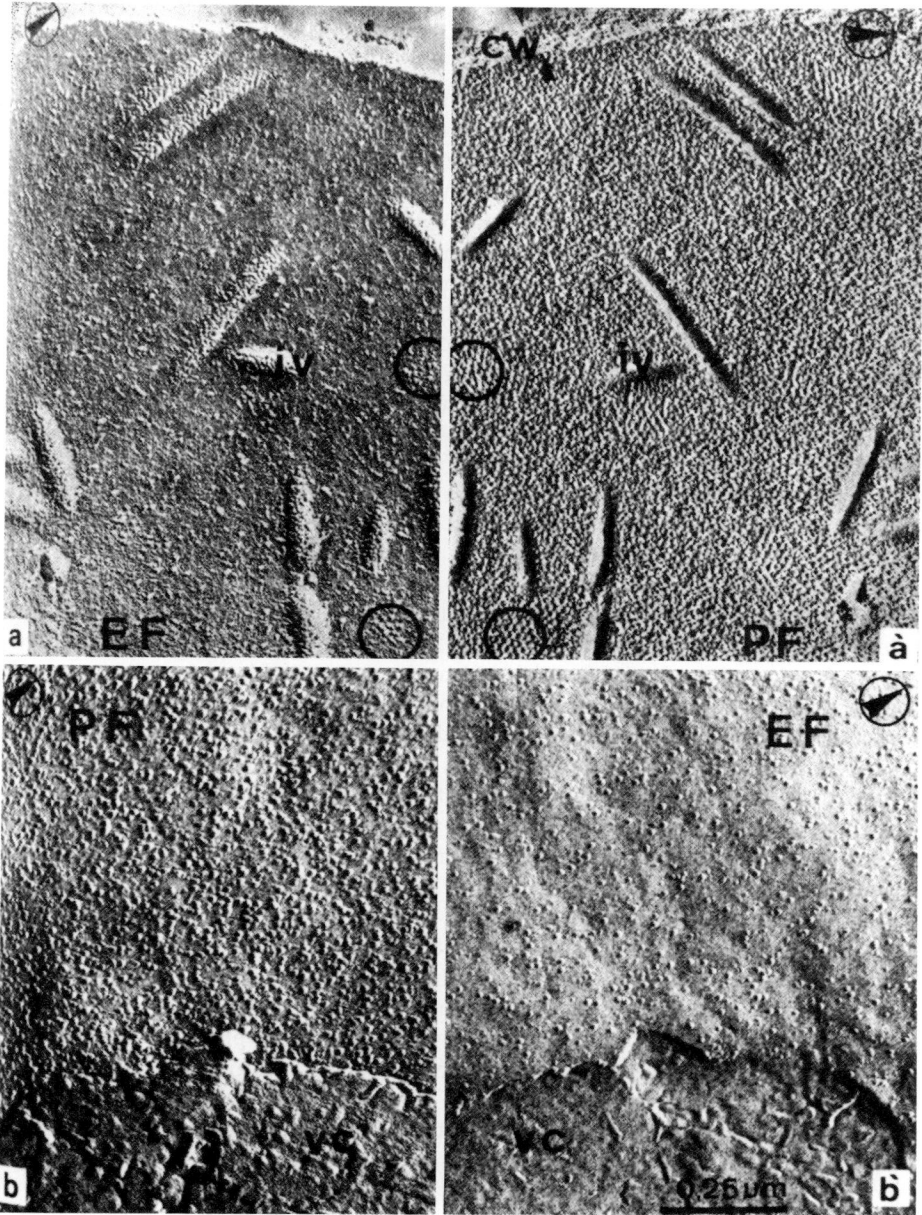

Fig. 4a–b. Complementary replicas of yeast cell plasmalemma and tonoplast. The encircled arrows indicate the direction of shadowing. **a** *EF* (exoplasmatic) and **à** *PF* (plasmatic-fracture face) of the plasmalemma at early stationary phase. *Encircled areas* show particles in hexagonal array. *cw*, Cell wall; *iv*, invagination. **b** *PF* and **b** *EF* of a tonoplast. *vc*, Vacuolar lumen. Except for the invaginations and the hexagonal arrays in **a**, particles in **a** and **b** appear to be distributed at random. (From Kramer et al. 1978)

depths. In a comparative study by Kramer et al. (1978), the plasmalemma and tonoplast of *Saccharomyces cerevisiae* differed markedly in composition, the ratio of protein to lipid being three times higher in the plasmalemma than in the tonoplast. Electron microscopic observations on freeze fracture preparations of these membranes (Fig. 4) showed particles which were thought to be protein, embedded in a smooth matrix of lipid, previously shown to be extractable with lipid solvents and detergents (Kopp 1972). The difference in particle density between plasmalemma and tonoplast corresponded well to the protein to lipid ratios, determined in the biochemical study. Other differences between the membranes included the presence of invaginations in the plasmalemma, corresponding to features on the inner surface of the cell wall (Kopp 1972), and patches of particles in paracrystalline arrangement in the plasmalemma, contrasting with the entirely random distribution of particles in the tonoplast. In keeping with its wall-building functions, the plasmalemma gave cytochemical reactions indicating the presence of carbohydrate. The electron micrographs published in this study corresponded well with earlier observations on *S. cerevisiae* by Takeo et al. (1976), who commented on the possible significance of the presence of small lipid granules, closely associated with the tonoplast and observed to fuse with it, in relation to the ability of the tonoplast bounding isolated vacuoles to swell and shrink.

The arrangement of the phospholipid molecules in membranes results from the polar head group being orientated towards the cytoplasm and the hydrophobic fatty acid tails directed inwards, towards the fatty acid tails of the opposite lipid layer. This picture is of course greatly simplified, ignoring other lipid components, such as sterols, glycolipids and sphingolipids, each of which comprises a diversity of molecular species. The several types of phospholipid present in membranes differ in their polar groups and the fatty acids within individual phospholipid classes vary both in chain-length and degree of unsaturation. Through the introduction of one or more double bonds, the configuration of the fatty acid chains is altered, affecting the fluidity of the membrane. It is likely that the orientation of glycolipid components of fungal membranes is similarly determined by their hydrophilic polar heads and hydrophobic fatty acid tails. Sterols are believed to contribute to the stability of fungal membranes through their molecules being of a suitable shape to lock across the fatty acid chains of the polar lipids (see also Chapter 10). Other neutral lipids, such as triacylglycerols have been reported from fungal membrane preparations but their location and functions are unclear. Some may lie between the phospholipid layers of membranes of the endoplasmic reticulum, possibly giving rise to lipid bodies (Wanner et al. 1981), or perhaps representing reserves of acyl lipid for further membrane development. The recurring reports of free fatty acids from membranes are widely regarded as indicating phospholipase activity during extraction.

4 Membrane Composition

Estimates of the lipid content of fungal membranes vary between 30% and 50% (Weete 1980). In the study of yeast membranes mentioned above (Kramer et al. 1978), the tonoplast lipid contained a higher proportion of phospholipid than in the plasma-lemma, whereas the plasmalemma showed a higher proportion of neutral lipid, 25% of which comprised sterol, compared with only 6% in this fraction of the tonoplast. A surprisingly high proportion of other neutral lipids was recorded from both membranes, particularly the plasmalemma. Phospholipids are generally the major class of membrane lipid, accompanied by sterols and in some reports other neutral lipids, glycolipids and smaller amounts of sphingolipids. A novel lipoprotein, representing a class of compounds not previously reported from fungi, has recently been studied in Oomycetes (Warner et al. 1986). The amounts of individual membrane lipids recorded from different fungi vary greatly and the problem of assessing these is complicated by variations in the stage of development investigated, the experimental conditions and the differing interests and methodologies of laboratories. Nevertheless, even with these limitations, significant chemotaxonomic correlations can be recognized and more extensive surveys reveal a diversity of lipid composition, in keeping with the diversity of the fungi (Weete 1980; Lösel 1988a). The basic similarity in ultrastructural organization, growth and physiology of cells of taxonomically diverse fungi, despite their characteristic differences in membrane lipid composition, suggests that appropriate combinations of differing components can provide functionally similar membrane systems.

Generalizations concerning fungal membrane lipid composition have been based mainly on yeast cells and relatively few filamentous fungi. The paucity of current knowledge of fungal lipids in general is demonstrated by counts on species included in recent reviews (Table 1). Most data derive from entire membrane systems or even total lipids, rather than separate analysis of individual membrane fractions. This more precise approach has been adopted for only a few fungi, e.g. *Neurospora* (Aaronson et al. 1982), *Aspergillus niger* (Letoublon et al. 1982) and, more recently, in an investigation of well-characterized membrane fractions from *Neurospora crassa* by Bowman et al. 1987 (Table 2). Their analysis of membrane fractions of high purity, as judged by the use of antibodies to marker enzymes, from mid-log mycelium of *N. crassa* showed that the plasma membranes, endoplasmic reticulum, vacuolar membranes and mitochondrial membranes differed markedly in their phospholipids, sterols and carbohydrate content, as well as in polypeptide components, but showed a generally similar fatty acid composition, apart from mitochondrial membranes which had a higher content of 18:2 and 18:3 unsaturated fatty acids. In general, vacuolar membranes, endoplasmic reticulum and plasmalemma were found to be more similar to each other in lipid composition than to the membranes of mitochondria.

In each of these membrane fractions from *N. crassa*, phosphatidyl choline and phosphatidyl ethanolamine, in differing proportions, together accounted for 80% or more of the total phospholipid. From the absence of phosphatidic acid and lysophospholipid, it appears that there was no enzymic degradation of phospholipids during extraction. Mitochondrial membranes differed from all other membrane fractions in their substantial amount of cardiolipin as well as their lack of sterol and carbohydrate. The sterol contents of plasmalemma and tonoplast in this study were in good agreement with those reported for yeast (Kramer et al. 1978). Significant amounts of phosphatidyl serine occurred only in the vacuolar membranes of *N. crassa*. In this connection, Bowman et al. (1987) mentioned preliminary evidence for the special role of phosphatidyl serine in the activation of vacuolar ATPase and point-

Table 1. Fungi in which lipids have been investigated

	Species recognized	% of Species analysed
Mastigomycotina	1 170	1.3
Zygomycotina	750	1.9
Ascomycotina	28 650	0.1
Basidiomycotina	16 000	0.3

(From data reviewed by Lösel 1988a)

Table 2. Lipid composition of *Neurospora* membranes (Bowman et al. 1987)

Phospholipid composition

Phospholipid	Vacuolar membranes	Mitochondrial membranes	Endoplasmic reticulum	Plasma membranes
Phosphatidylcholine	42.42%[a]	36.35%	50.51%	46.43%
Phosphatidylethanolamine	40.34%	43.42%	36.34%	39.38%
Phosphatidylinositol	12.16%	9.11%	13.15%	15.18%
Phosphatidylserine	6.9%	ND, ND[b]	ND, ND	ND, 1%
Phosphatidic acid	ND, ND	ND, ND	ND, ND	ND, ND
Cardiolipin	ND, ND	12.13%	ND, ND	ND, ND

	Carbohydrate content	Sterol composition
Fraction	Glucose equivalents (mg/mg protein)	Sterol (mol%)
Vacuolar membranes	2.84	8.8
Mitochondrial membranes	0.03	0.9
Endoplasmic reticulum	0.31	27.7
Plasma membranes	1.12	21.8

Fatty acid composition

Fatty acids	Vacuolar membranes	Mitochondrial membranes	Endoplasmic reticulum	Plasma membranes
16:0	23.30%[c]	15.16%	33.30%	26.25%
18:0	2.1%	1.1%	3.4%	2.2%
Saturated	25.31%	16.17%	36.34%	28.27%
18:1	5.4%	5.6%	8.6%	6.6%
18:2	40.40%	45.44%	31.32%	38.39%
18:3	30.25%	34.33%	25.27%	28.29%
Unsaturated	75.69%	84.83%	64.65%	72.74%

[a] For each membrane fraction the results from two independent experiments are shown. The numbers given represent the amount of phosphate measured for each type of lipid divided by the total phosphate in the identified lipids. Unidentified spots on the chromatograms contained less than 2% of the total phosphate.
[b] None detected.
[c] For each membrane fraction the results of two independent experiments are shown.

ed to a report by Xie et al. (1984) of specific activation by phosphatidyl serine of delipidated coated vesicle ATPase, which in some respects had similar properties to the fungal vacuolar membrane ATPase.

Apart from the above studies, and the lipid analyses of chitosomes, already mentioned, our knowledge of membrane lipids in fungi is largely gleaned from records of phospholipids, sterols, glycolipids and sphingolipids separated from total lipid extracts of cells in various conditions and stages of development. Some general information on the distribution and functions of these components has been assembled in recent reviews (Weete 1980; Lösel 1988 a, b).

5 Functions of Membrane Lipids

Our understanding of the roles of membrane lipids of fungi is derived from the distribution, both taxonomic and intracellular, of individual classes, as well as from compositional changes encountered during normal growth and morphogenesis, in response to experimental treatments or in biochemical mutants. Some of the resulting conclusions are summarized below but for more complete coverage, other reviews should be consulted.

5.1 Phospholipids

The occurrence of phosphatidyl choline and phosphatidyl ethanolamine as major phospholipids in all classes of fungi is in accord with their generally accepted structural role in membranes. Analogy with green plants, where some fatty acid desaturation reactions occur in the endoplasmic reticulum, on phosphatidyl choline as carrier (Murphy et al. 1984), suggests another possible function of this phospholipid in fungi. As in other eukaryotes, cardiolipin occurs specifically in mitochondria of *N. crassa* (Bowman et al. 1987), *A. niger* (Letoublon et al. 1982) and probably other fungi. Phosphatidyl serine, associated with vacuolar membranes of *N. crassa*, as discussed above, is generally low or absent in the Mastigomycotina but present in substantial amounts in Zygomycetes and higher fungi. Apart from a handful of records, phosphatidyl glycerol appears to be confined to the order Peronosporales of the Oomycetes. Phospholipid requirements have been demonstrated for chitin synthase activity in *Schizophyllum commune* and *Coprinus cinereus* (Vermeulen and Wessels 1983; Montgomery and Gooday 1985; see also Chapters 5 and 6) and for chitinase in *Mucor mucedo* (Humphreys and Gooday 1984; see also Chapter 5). Strong evidence has recently been provided for the role of phosphatidyl inositol in glucomannan synthesis by *Rhizoctonia cerealis* (Robson 1987). Phosphatidyl inositol, which is normally a substantial component of fungal membranes, is also implicated in various transfer and regulatory processes in other organisms, where its relationship with phosphoinositides is increasingly of interest, and has recently been recognized as a hydrophobic anchor for certain enzymic proteins (Low et al. 1986), which are released by phospholipase. The synthesis of phospholipids is described in Chapters 16 and 17.

5.2 Fatty Acids

Since the fatty acid moieties of phospholipid and glycolipid components are considered to have a key role in the structure and functioning of membranes, it is unfortunate that the fatty acid composition of individual classes of fungal membrane lipids is much less frequently recorded than total fatty acids, within which the proportion derived from membranes is highly variable. The broadest spectrum of fatty acid composition occurs in the lower fungi, with substantial amounts of polyunsaturated C_{20} fatty acids, particularly among the Oomycetes. Zygomycetes differ from other classes in synthesizing γ-linolenic acid instead of α-linolenic acid, but there are exceptions and a few species contain both forms. In Basidiomycotina, α-linolenic acid is the predominant fatty acid. Studies on the influence of temperature and other environmental factors on membrane lipid composition in fungi suggest that the ability to respond rapidly to stresses by alteration of fatty acid chain length or unsaturation is important in maintaining membrane function.

5.3 Sterols

The sterol composition of fungi shows many chemotaxonomic correlations (Lösel 1988a), including differences between higher and lower fungi in the relative abundance of cholesterol and ergosterol. Cholesterol forms a substantial proportion of the sterols of Chytridiomycetes and Oomycetes. Ergosterol, the characteristic sterol of Zygomycetes, is absent from most other lower fungi but occurs widely along with other sterols in the higher fungi. The pythiaceous Oomycetes, which are unable to synthesize sterols, can grow vegetatively on sterol-free medium, although they require sterols for reproduction. Lacking sterols in their membranes, they are insensitive to polyene antibiotics and sterol-inhibiting fungicides. Besides their structural role in membranes, already mentioned, certain sterols may have special regulatory or morphogenetic roles in fungi (Elliott 1977; see also Chapter 10).

The location of the sterol esters which are commonly quite abundant in cells is uncertain. Since the amounts of these vary at different times in the life cycle, often in inverse proportion to the sterol fraction, it appears likely that they represent a storage form of sterols and fatty acids.

5.4 Carotenoids

A wide range of carotenoids have been recorded from fungi (Goodwin 1980), mainly stored in lipid drops but also apparently located in membranes, e.g. the outer mitochondrial membrane of *Neurospora* (Ramadan-Talib and Prebble 1978). Membrane fractions containing plasma membranes and, in particular, endoplasmic reticulum possessed about 80% of the carotenogenic activity in cell-free preparations from *N. crassa* (Mitzka-Schnabel and Rau 1981). Carotenoid synthesis was stimulated eight-

Table 3. Fungal glycolipids

Nitrogen-free glycolipids
Acylated sugars and sugar alcohols
Hydroxy-acid glycosides
Glycosyl glycerides
Sterol glycosides
Polyprenol-containing glycosides

Sphingolipids
Free ceramides − acylated[a] sphingoid[b] base
Ceramide phosphate
Glycosyl ceramides
Complex sphingolipids
  inositol phosphoryl ceramides
  mannosylinositolphosphorylceramides
  glycosylmannosylinositolphosphorylceramides

---

[a] C_{12} to C_{26} saturated, unsaturated and hydroxy fatty acids.
[b] Normal, iso-, saturated or unsaturated sphingosine, sphinganine or phytosphingosine.

fold by previous illumination of the mycelium. In fungi, as in green plants, there are strong indications that carotenoids may have protective roles, not only in relation to photooxidation and light responses but also as scavengers of free oxygen radicals (Krinsky 1978).

5.5 Glycolipids

Glycolipids and sphingolipids are the least well studied of fungal lipids although a wide range of these complex lipids has been recorded (Table 3). They are often present only in small amounts in fungi and are studied by relatively few laboratories which often concentrate on different components, e.g. long chain bases, fatty acids, sugars, polyols or other moieties. Besides their probable structural role, glycolipids including polyprenol glycolipids are important in the synthesis of various cell wall polysaccharides (Brennan and Lösel 1978; Weete 1980) and may also be involved in recognition reactions. Since glycophosphosphingolipids have been found only in wall-bearing eukaryotes, it has been suggested that they may also have a function in wall synthesis, possibly a surface-anchoring role.

6 Influence of Environment, Nutrition, and Development on Membrane Composition and Function

Much of our understanding of the relationships between membrane composition and function has been derived from studies of the responses of fungi to nutritional and other environmental factors and of the requirements of mutants with deficiencies in membrane-lipid synthetic processes. There are numerous examples of growth at lower temperatures being characterized by higher degrees of unsaturation of fatty acids and vice-versa, but the relationship is by no means clear or constant and varies greatly with individual species (Weete 1980; Lösel 1988b), since other means of modifying membrane fluidity are available, such as intramolecular rearrangements in the fatty acids of acyl lipids and changes in sterol content. A study by Dexter and Cooke (1984) on the growth of psychrophilic and mesophilic species of *Mucor* at different temperatures revealed changes in sterol and carotene content as well as modification of fatty acid composition. Other environmental factors influencing membrane composition include oxygen, which is required for fatty acid desaturation reactions and for sterol synthesis. Thus growth in strictly anaerobic conditions is accompanied by special nutritional requirements, such as the availability of substrate sterol and unsaturated fatty acids (Emerson and Natvig 1981). The membrane composition of fungi in otherwise normal environments, can also be modified by the substrate, as was demonstrated by Ratledge (1982) in fungi grown on alkanes, where the fatty acid composition reflected that of the alkanes provided.

Among the clearest examples of cell membrane arrangements and functions changing in a clearly defined manner in relation to differing stages of development and activity are chytrid zoospores, a developmental phase in which most synthetic processes appear to be switched off and little endoplasmic reticulum is present. In the most thoroughly investigated case, the zoospore of *B. emersonii* (Fig. 5), a quarter of a century of study by Cantino and coworkers (references in Cantino and Mills 1976, 1983), using synchronous cultures, has resulted in a remarkably integrated picture of the inter-relationships between cell membrane organization and biosynthesis, biochemical activity, motility and morphogenesis. The zoospore swims at the expense of triacylglycerol stored in lipid drops, which are embedded in a large 'symphiomicrobody' (in which glyoxylate-cycle enzymes have been demonstrated), lying along the outer surface of a single giant microbody in the posterior region of the cell. At a rate of 0.19 pg lipid$\cdot$h$^{-1}\cdot$spore^{-1}, over 60% of the zoospore lipid was consumed in five hours (Suberkropp and Cantino 1973; Cantino and Mills 1976). During the motile period, most of the ribosomes of the cell are sequestered in the 'nuclear cap', delimited by a membrane, and the cytoplasm contains a number of membrane-enclosed, horse

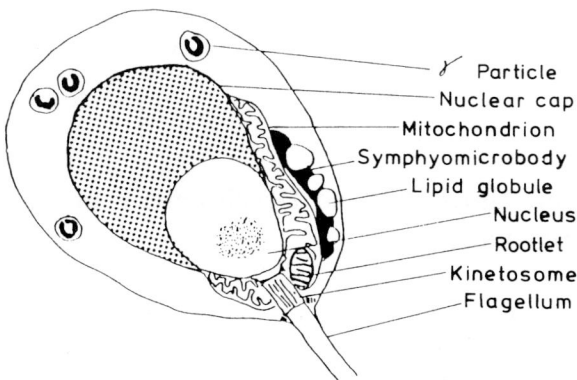

Fig. 5. The *B. emersonii* zoospore. The spore contains 12 ($\pm$7) γ particles, 8.2 ($\pm$1.5)$\times 10^5$ ribosomes in the nuclear cap, 9 ($\pm$2) lipid globules, and one each of the other labelled structures. The developmental origin of all of them has been established, and all but three of them (nucleus, rootlet, kinetosome) have been isolated and studied biochemically in vitro. (Cantino and Mills 1983)

shoe-shaped γ-particles, the lipid content of which includes mono- and diglycosyl-diacylglycerols and sterol (Suberkropp and Cantino 1973; Mills and Cantino 1974). After a period of swimming, the cell winds in its flagellum and within 15 minutes a wall is formed. During this time, the γ-particles begin to decay, giving rise to masses of 80 nm vesicles. These fuse with and extend the bounding membrane from which, in turn, other vesicles migrate to the periphery of the cell and fuse with the plasma-lemma (Truesdell and Cantino 1970). Like chitosomes of other fungi, these vesicles have been reported to form chitin fibrils in vitro, when supplied with acetyl glucos-amine (Mills and Cantino 1981). The encystment process is followed by normal spore germination, involving further membrane synthesis and all the other normal process-es of thallus development.

Particles of differing appearance, which appear to be similar in function to the γ-particles of *Blastocladiella*, occur in some other Chytridiomycetes (Cantino and Mills 1983). In *Allomyces macrogynus* Barstow and Pommerville (1980) observed calcofluor reactions, indicative of the presence of β-1,4-linked glucan around cells, 2–10 minutes after the induction of zoospore encystment. At the same time, vesicles of diameter 35–70 nm were seen near and fusing with the plasmalemma as electron-opaque fibrillar material began to accumulate. γ-Particles in the encysting cells had a crystalline inclusion, from which numerous vesicles arose. Although present in zoospores and female gametes of *Allomyces*, these particles were absent from the male gametes, which are unable to encyst.

7 Conclusions

Besides the highly specialized cells of the developing, swimming and encysting zoospores of *Blastocladiella*, other fungi represent interesting model systems for studying basic features of membrane biosynthesis and function. Spore germination and various synchronized developmental systems in fungi may also be expected to provide convenient tools for fundamental investigations of the roles of membrane lip-ids. Further comparative studies on well-characterized membrane fractions from a wider range of fungi are urgently required as well as critical evaluation of membrane responses to physiological and genetical manipulation. Such a background of infor-mation is essential if we are to understand not only the roles of lipids in maintaining membrane function in the face of imposed environmental or other stresses, in which membrane disruption may result from attacks on polyunsaturated fatty acids or other components, possibly mediated by free oxygen radicals or lipoxygenase, but also the significance of membrane lipids as targets for fungal pheromones, phytoalexins, fun-gicides and drugs.

References

Aaronson LR, Johnston AM, Martin CE (1982) The effects of temperature acclimation on membrane sterols and phospholipids of *Neurospora crassa*. Biochim Biophys Acta 713:456–462

Armbruster BL (1982) Sporangiogenesis in three genera of the Saprolegniaceae. I. Pre-sporangium hyphae to early primary spore initial stage. Mycologia 74:433–459

Barstow WE, Pommerville J (1980) The ultrastructure of cell wall formation and of gamma particles during encystment of *Allomyces macrogynus* zoospores. Arch Microbiol 179:179–189

Bartnicki-Garcia S, Bracker CE, Reyes E, Ruiz-Herrera J (1978) Isolation of chitosomes from taxonomically diverse fungi and synthesis of chitin fibrils in vitro. Exp Mycol 2:173–192

Beckett A, Heath IB, McLaughlin DJ (1974) An atlas of fungal ultrastructure. Longman, London

Boller T, Wiemken A (1986) Dynamics of vacuolar compartmentation. Annu Rev Plant Physiol 37:73–92

Bowman BJ, Borgeson CE, Bowman EJ (1987) Composition of *Neurospora crassa* vacuolar membranes and comparison to endoplasmic reticulum, plasmalemmas and mitochondrial membranes. Exp Mycol 11:197–205

Brennan PJ, Lösel DM (1978) Fungal lipid physiology. Adv Microb Physiol 17:47–179

Cantino EC, Mills GL (1976) Form and function in Chytridiomycete spores. In: Weber DJ, Hess WM (eds) The fungal spore. Wiley, New York, p 501

Cantino EC, Mills GL (1983) The Blastocladialean γ-particle: once viral endo-symbiont (?), now chitosome progenitor. In: Smith JE (ed) Fungal differentiation. Dekker, New York Basel, pp 176–204

Dexter Y, Cooke RC (1984) Fatty acids, sterols and carotenoids of the psychrophile *Mucor strictus* and some mesophile *Mucor* species. Trans Br Mycol Soc 83:455–461

Elliott CG (1977) Sterols in fungi. Their functions in growth and reproduction. Adv Microb Physiol 15:121–173

Emerson R, Natvig DO (1981) Adaptation of fungi to stagnant waters. In: Wicklow DT, Carroll GC (eds) The fungal community. Dekker, New York Basel, pp 109–128

Fèvre M, Rougier M (1981) β-1–3- and β-1–4-glucan synthesis by membrane fractions from the fungus *Saprolegnia*. Planta 151:232–241

Gooday GW (1983) The hyphal tip. In: Smith JE (ed) Fungal differentiation. Dekker, New York Basel, pp 315–356

Goodwin TW (1980) The biochemistry of the carotenoids, 2n edn. Chapman and Hall, London

Hänseler E, Nyhlén LE, Rast DM (1983) Dissociation and reconstitution of chitosomes. Biochim Biophys Acta 745:121–133

Harris N (1986) Organization of the endomembrane system. Annu Rev Plant Physiol 37:73–92

Humphreys AM, Gooday GW (1984) Phospholipid requirement of microsomal chitinase from *Mucor mucedo*. Curr Micrbiol 11:187–190

Joachim S, Robinson DG (1984) Endocytosis of cationic ferritin by bean leaf protoplasts. Eur J Cell Biol 34:212–216

Kopp F (1972) Zur Membranstruktur: Lokalisation von Membranlipiden im Hefeplasmalemma. Cytobiologie 6:287–317

Kramer R, Kopp F, Niedermeyer W, Fuhrmann GF (1978) Comparative studies of the structure and composition of the plasmalemma and the tonoplast in *Saccharomyces cerevisiae*. Biochim Biophys Acta 507:369–380

Krinsky NJ (1978) Non-photosynthetic functions of carotenoids. Philos Trans R Soc London Ser B 284:143–152

Letoublon R, Mayet B, Frot-Coutaz J, Nicolau C, Got R (1982) Subcellular distribution of phospholipids and of polyprenol phosphate in *Aspergillus niger* Van Tieghem. Biochim Biophys Acta 711:509–514

Lösel DM (1988a) Fungal lipids. In: Ratledge C, Wilkinson SG (eds) Microbial lipids, vol I. Academic Press, London New York, pp 699–806

Lösel DM (1988b) Functions of lipids. Specialized roles in fungi and algae. In: Ratledge C, Wilkinson SG (eds) Microbial lipids, vol II. Academic Press, London New York, pp 367–438

Low MG, Ferguson MAJ, Futerman AH, Silman I (1986) Covalently attached phosphatidylinositol as a hydrophobic anchor for membrane proteins. TIBS 11:212–215

Mills GL, Cantino EC (1974) Lipid composition of the zoospores of *Blastocladiella emersonii*. J Bacteriol 118:192–201

Mills GL, Cantino EC (1981) Chitosome-like vesicles from gamma particles of *Blastocladiella emersonii* synthesize chitin. Arch Microbiol 130:72–77

Mitzka-Schnabel U, Rau W (1981) Subcellular site of carotenoid biosynthesis in *Neurospora crassa*. Phytochemistry 20:63–69

Montgomery GWG, Gooday GW (1985) Phospholipid-enzyme interactions of chitin synthase of *Coprinus cinereus*. FEMS Microbiol Lett 27:29–33

Moore TS (1982) Phospholipid biosynthesis. Annu Rev Plant Physiol 33:235–259

Morré DJ, Mollenhauer HH (1974) The endomembrane concept: a functional integration of endoplasmic reticulum and Golgi apparatus. In: Robards AW (ed) Dynamics of plant ultrastructure. Mc-Graw-Hill, New York, pp 84–137

Murphy DJ, Latzko E, Woodrow IE, Mukherjee KD (1984) The oleate desaturase system of pea leaf microsomes. Elsevier, Amsterdam, pp 45–49

Ramadan-Talib Z, Prebble J (1978) Photosensitivity of respiration in *Neurospora* mitochondria. Biochem J 176:767–775

Ratledge C (1982) Microbial oils and fats: an assessment of their commercial potential. Prog Indust Microbiol 16:119–206

Robinson DG, Kristen U (1982) Membrane flow via the Golgi apparatus of higher plants. Int Rev Cytol 77:89–127

Robson GD (1987) The effect of Validomycin A on the inositol content and morphology of *Rhizoctonia cerealis* and other fungi. Ph D Thes, Univ Manchester

Roos U-P, Turian G (1977) Hyphal tip organization in *Allomyces arbuscula*. Protoplasma 93:231–247

Suberkropp KF, Cantino EC (1973) Utilization of endogenous reserves by swimming zoospores of *Blastocladiella emersonii*. Arch Mikrobiol 89:205–221

Takeo K, Shigeta M, Takagi Y (1976) Plasma membrane ultrastructural differences between the exponential and stationary phases of *Saccharomyces cereviseae* as revealed by freeze-etching. J Gen Microbiol 97:323–329

Trinci APJ (1978) Wall and hyphal growth. Sci Prog 65:75–99

Truesdell LC, Cantino EC (1970) Decay of γ-particles in germinating zoospores of *Blastocladiella emersonii*. Arch Mikrobiol 70:378–392

Vermeulen CA, Wessels JGH (1983) Evidence for a phospholipid requirement of chitin synthase in *Schizophyllum commune*. Curr Microbiol 8:67–71

Wanner J, Formanek H, Theimer RR (1981) The ontogeny of lipid bodies (spherosomes) in plant cells. Ultrastructural evidence. Planta 151:109–123

Warner SA, Eierman DF, Domnas AJ, Dean DD (1986) A novel lipoprotein from Oomycete fungi. Exp Mycol 10:315–322

Weete JD (1980) Lipid biochemistry of fungi and other organisms. Plenum, New York London

Weete JD, Furter R, Hänseler E, Rast DM (1985) Cellular and chitosomal lipids of *Agaricus bisporus* and *Mucor rouxii*. Can J Microbiol 31:1120–1126

Wilson CL (1973) A lysosomal concept for plant pathology. Annu Rev Phytopathol 11:247–272

Xie X-S, Stone DK, Racker E (1984) Activation and partial purification of the ATPase of clathrin-coated vesicles and reconstitution of the proton pump. J Biol Chem 259:11676–11678

Chapter 10

Importance and Role of Sterols in Fungal Membranes

H. VANDEN BOSSCHE [1]

1 Introduction

The basic role of biomembranes is to provide a barrier between a cell or organelle and its environment and at the same time to serve as a matrix for the association of proteins with lipids (Gibbons et al. 1982) or, as pointed out by Lewis Thomas (1974), "it takes a membrane to make sense out of disorder". In these membranes, sterols play a major role both architecturally and functionally. The most common membrane sterol in animals is cholesterol (Fig. 1).

Since the role of cholesterol in biological and artificial membranes has been the subject of many investigations (see for example Oldfield and Chapman 1972; Bloch 1983; Demel 1987; Dawidowicz 1987) we understand to some extent the function of this triterpenoid. It is known that the viscosity of the lipid core of the plasmamembrane is a function of the concentration of sterol. A decrease in the ratio of cholesterol to phospholipids indicates an increase in membrane fluidity, provided that compensatory changes in other membrane components do not occur (Demel and De Kruyff 1976). It has been shown that in sterol-deficient L-cells an increased fluidity of the membrane alters several of its functions. For example, an increase in the rubidium transport by the Na-K-activated ATPase, an increase in the passive efflux of rubidium from the cells (Chen et al. 1978) and an increased membrane fragility have been observed (Kandutsch et al. 1978). These sterol-depleted cells are also unable to take up soluble horseradish peroxidase by pinocytosis, indicating that vesicle formation and internalization of the plasmamembrane is dependent on the maintenance of a certain rigidity in the lipid core of the membrane (Heiniger et al. 1976).

In mixtures with phospholipids, cholesterol, by its condensing effect, increases the chain order in the liquid crystalline state and by its liquefying effect, decreases the chain order in the gel state (Demel 1987). Prerequisites for both the condensing and liquefying effect of cholesterol are the planar ring system (a consequence of its *trans*-antistereochemistry at the ring junctions), a 3β-hydroxyl group and a long flexible chain at C_{17} (Demel 1987). The protrusion of two angular methyl groups from the β plane (C_{18} and C_{19}) of the sterol nucleus accommodates with the asymmetry of the phospholipid acyl chain. Indeed in most membranes one of the acyl chains is saturated and the other has at least one *cis*-double bond.

---

[1] Deptartment of Comparative Biochemistry, Janssen Research Foundation, Turnhoutseweg 30, 2340 Beerse, Belgium

CHOLESTEROL

LANOSTEROL

1- (O - β - N - ACYLGLUCOSAMINYL) - 2, 3, 4 -
TETRAHYDROXYPENTANE - 29 -HOPANE

ERGOSTEROL

($\Delta^{5,7,22}$ - 24 β - methyl - cholesta - 3 β - ol)

CAMPESTEROL

(24 α - methylcholesterol)

STIGMASTEROL

($\Lambda^{5,22}$-24 α - ethylcholesta - 3 β - ol

SITOSTEROL

(24 α - ethylcholesterol)

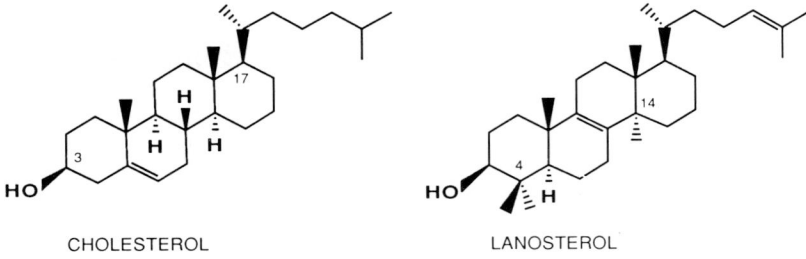

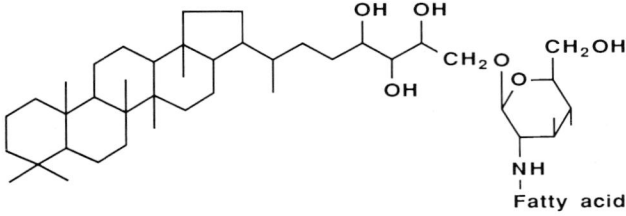

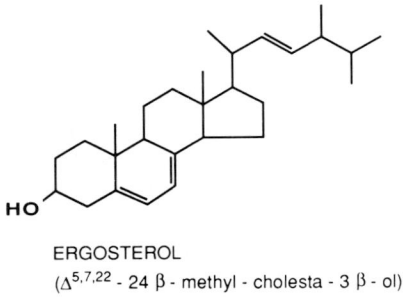

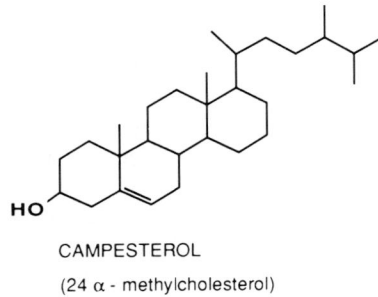

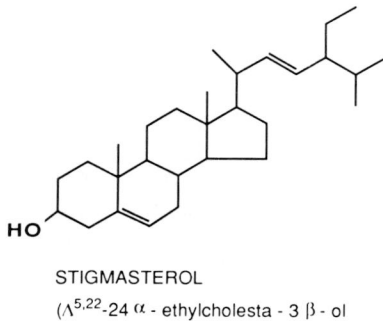

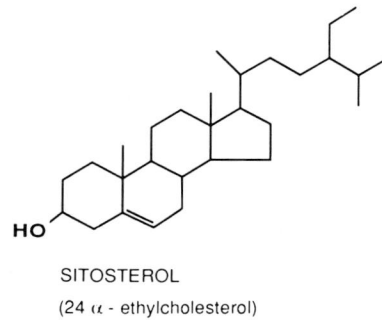

Fig. 1. Chemical structures of some triterpenoids

The molecular interactions which occur between sterols and phospholipids are not completely understood. The fact that neither the 3α-OH isomer nor 3β-thio-cholesterol interact with phospholipids as efficiently as cholesterol is consistent with the view that hydrogen bonding occurs between the carbonyl oxygen of the phospholipid-acyl side-chain and the 3β-OH of cholesterol (Cooper and Strauss 1984). However, according to Cooper and Strauss (1984) nonpolar interactions might be more important than hydrogen bonding.

The modulation of the membrane physical state is not the only function of cholesterol. Cholesterol serves as the precursor for bile acids and steroid hormones. Cornell et al. (1977) and Kandutsch et al. (1978) showed that in lipid-depleted medium, biosynthesis of cholesterol is also required for cellular proliferation and differentiation. Cornell et al. (1977) have shown that in the absence of exogenous lipid, inhibition of cholesterol biosynthesis by 25-hydroxycholesterol (a sterol derivative known to regulate cholesterol synthesis by reducing the levels of the HMG-CoA reductase) leads to the development of a reversible accumulation of cells (L_6, a rat myogenic cell line, and human fibroblasts from embryonic lung) in the G 1 phase of the cell cycle. Of great interest was the finding that a cycle of de novo sterol synthesis was a prerequisite for DNA synthesis in primary cultures of mouse or human lymphocytes, even when the cells were cultured in medium containing lipoprotein cholesterol dispersed with albumin (Chen et al. 1975). The requirement for de novo synthesis of cholesterol during a particular stage of cell division was further illustrated by studies with phytohemagglutinin (PHA)-stimulated lymphocytes (Pratt et al. 1977; Kandutsch et al. 1978). Soon after (within $4-7$ h) the addition of PHA to a lymphocyte culture, a cycle of sterol synthesis started and reached a peak after about 24 hours. Synthesis of DNA started after about 24 hours and reached its peak at 48 hours. If the synthesis of cholesterol was suppressed by an oxygenated sterol (25-hydroxycholesterol or 20α-hydroxycholesterol) during the first 24 hours of culture, the subsequent synthesis of DNA and cell division was also suppressed. However, after the sterol synthesis had reached its peak the addition of an oxygenated sterol did not affect DNA synthesis or cell division. That sterol biosynthesis is necessary for cell division of T-lymphocytes (T-lymphocytes are the major population triggered by mitogens to undergo DNA synthesis) has been proven by other investigators. For example, Cuthbert and Lipsky (1980) found a 20-fold increase in cholesterol synthesis by PHA-stimulated human peripheral blood mononuclear cells and showed that although sterol synthesis is not essential to initial lymphocyte activation, it is required for continuation of the response beyond the initial period. According to these authors this requirement might result from the necessity for the cells to synthesize adequate new membrane for blast transformation, cell division, and the multiple cycles of cell proliferation that normally occur thereafter.

Whether or not sterol synthesis is required to initiate cell activation or cell proliferation, it is certainly striking that cells proliferating in vivo synthesize cholesterol at high rates and are essentially independent of dietary cholesterol or of cholesterol blood levels. Examples of cells with high rates of sterol synthesis are found in intestinal mucosa, epidermis and developing brain. Nondividing muscle cells do not synthesize significant amounts of the sterol and mature brain also produces small amounts of cholesterol. As suggested by Kandutsch et al. (1978), the high rates of cholesterol synthesis in proliferating cells in vivo, largely independent from circulating

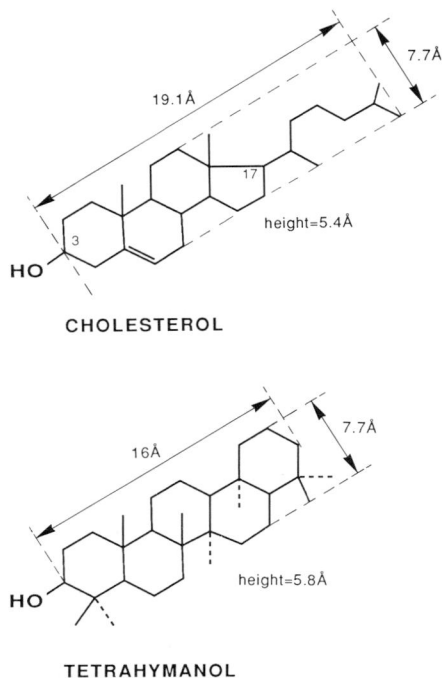

Fig. 2. Structures and dimensions of cholesterol and tetrahymanol. Dimensions were taken
from Rohmer et al. (1979)

cholesterol levels, seem to indicate that de novo synthesis of cholesterol is a requirement for these cells.

There are indications that in bacteria, triterpenes of the hopane family, the
hopaniods, may play the role of sterols (Taylor 1984; Prince 1987). For example the
hopanoid glycolipid (Fig. 1) and its aglycone from the gram-positive *Bacillus
acidocaldarius* [1-(*O*-β-*N*-acylglucosaminyl)-2,3,4-tetrahydroxypentane-29-hopane]
was found to condense lipid monolayers in a manner similar to cholesterol (Taylor
1984). When the structure of tetrahymanol, a hopanoid formed by *Tetrahymena
pyriformis*, is compared with that of cholesterol some striking similarities are seen
(Fig. 2). Tetrahymanol is similar in shape and dimensions to cholesterol with a flat
character of the same approximate length and with the hydroxyl group in the same
orientation. However, the molecule is somewhat thicker reaching the thickness of that
of lanosterol (Figs. 1, 4). As will be discussed later, *T. pyriformis* is able to compensate
for this difference by changing the fatty acyl groups of the phospholipids to get an
association similar to that of cholesterol and phospholipids in mammalian membranes (Gibbons et al. 1982).

It can be understood why anaerobic bacteria synthesize these pentacyclics instead
of the tetracyclic triterpenes. Indeed, although hopanoids are synthesized from the
same starting material as sterols, i.e. mevalonic acid (Fig. 3), in contrast with sterols
the synthesis of the pentacyclic sterol-like molecules does not need oxygen. Whereas
sterol synthesis proceeds from squalene via a 2,3-epoxide, to the sterol nucleus,
hopanoids are formed from squalene by a direct pentacyclisation, catalysed by a

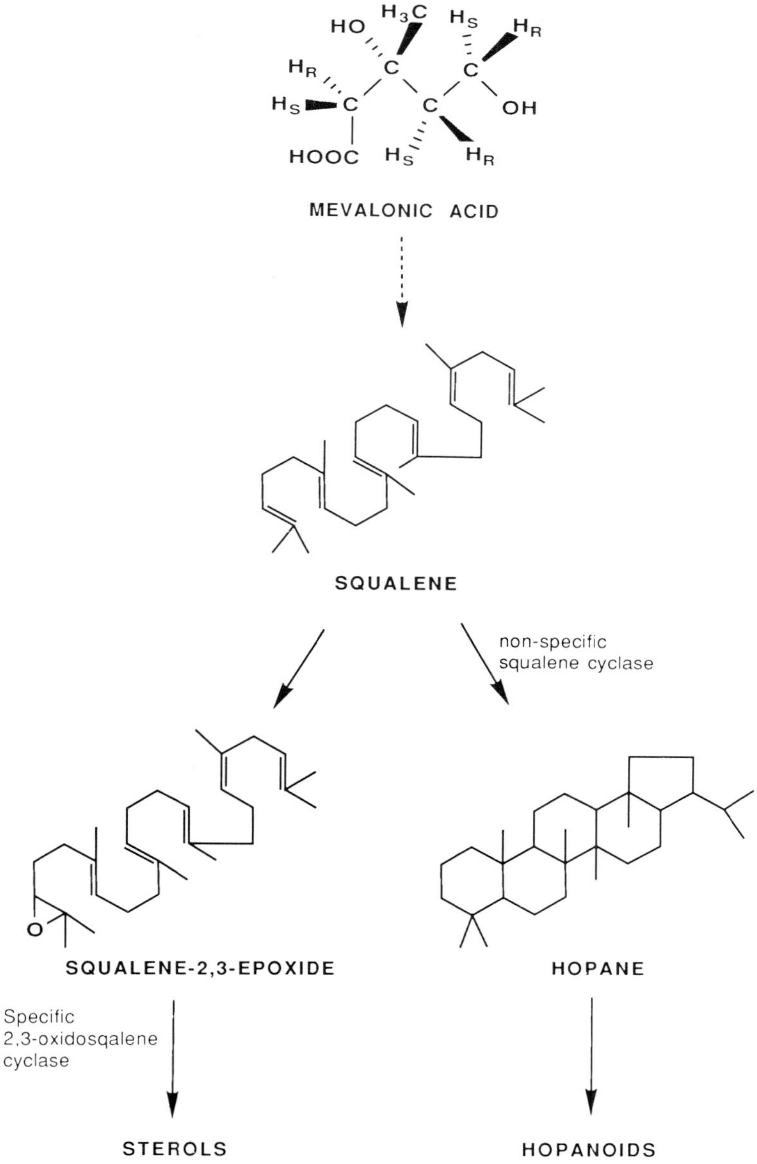

Fig. 3. Synthesis of sterols and hopanoids from mevalonic acid

primitive, non-specific squalene cyclase (Anding et al. 1976), to form the hopanoid nucleus (Fig. 3). The epoxidation requires oxygen and can therefore occur in aerobic organisms only. In some bacteria, e.g. the gram-negative, aerobic, methane oxidizer, *Methylococcus capsulatus*, two squalene cyclases have been reported, one cyclizing squalene or its epoxide to hopanoids (as found in the obligately aerobic *Acetobacter pasteurianum*) and the other cyclizing squalene epoxide to lanosterol or 3-epilano-sterol (Rohmer et al. 1979, 1980). The presence of this 3α-lanosterol together with the ability of a cell-free system of *M. capsulatus*, when provided with squalene-2,3-epoxide, to synthesize 3-hydroxyhopanoids, further illustrate the non-specificity of these bacterial cyclases (Taylor 1984). This contrasts with the squalene cyclase in eukaryotic cells which acts specifically on the 3(*S*) enantiomer of squalene epoxide and not on the 3(*R*) enantiomer (Barton et al. 1975; Bouvier et al. 1976; Ourisson et al. 1982). Thus, it is possible that an important factor in the evolution of the sterol molecule as a membrane component, is the appearance of a selective squalene-2,3-epoxide cyclase. In this context it is of interest to note that microorganisms able to synthesize sterols to some extent (e.g. *Azotobacter chroococcum, Streptomyces olivaceus* and *M. capsulatum*) have guanine plus cytosine (G+C) contents of >50 mol%. This and other studies indicate that the occurrence of sterols correlates with the G+C content (Taylor 1984).

Whereas in animals the major sterol is cholesterol, yeasts, fungi and plants pro-duce different sterols in which the molecule has been modified either by insertion of an extra double bond in the B ring of the nucleus and/or by the addition of 1 or 2 carbon atoms at C-24 of the side chain. The 24-alkylated sterol, ergosterol, was first isolated by Charles Tanret in 1889, from ergot of rye and Smedley-MacLean and Thomas (1920) found the sterol present in baker's and brewer's yeast identical to ergosterol isolated from ergot. Although ergosterol has been known for almost 100 years, we do not yet know enough about its physiological function. Indeed, as pointed out by Gibbons et al. (1982), one of the most intriguing questions in com-parative biochemistry is why animals generally contain cholesterol, whereas yeasts, fungi, some protozoa (e.g. *Leishmania*), and plants contain 24-alkylated sterols such as ergosterol, campesterol, stigmasterol and/or β-sitosterol (Fig. 1). As mentioned by Parks et al. (1984) these alkylated sterols are metabolically very expensive for the cell to synthesize. Therefore, it is hard to believe that the 24-alkylated sterols should not have a functional significance.

In this overview I will address the following questions:

1. Do yeasts and fungi really need ergosterol and what is the role or function of this alkylated sterol?
2. What are the consequences for the cell to use alkylated sterols instead of choles-terol?
3. Why ergosterol instead of cholesterol?

2 Ergosterol: Does It Meet the Requirements of Fungal Membranes?

2.1 Ergosterol and Anaerobic Growth

Since the work of Andreason and Stier (1953), it is known that yeast is capable of growing under anaerobic conditions only if the medium is supplemented with ergosterol and unsaturated fatty acid. Morpurgo et al. (1964) showed that under anaerobic conditions and in the absence of ergosterol, growth of *Saccharomyces cerevisiae* stopped after 5–7 generations and cells started to die 24 h later. In most of the cells cytoplasmic membrane, nuclear membrane, vacuole and mitochondria were cytologically undetectable. When a suspension of cells grown anaerobically in the presence of ergosterol were aerated for 180 min the respiration increased to the level of a similar suspension of aerobic cells. Cells grown in the absence of ergosterol on the contrary, were not able to increase their respiration rate even after many hours of aeration.

Yeast cells cultured anaerobically in the presence of ergosterol and oleic acid contain mitochondrial structures with an extensively folded inner membrane system and well developed cristae. Cells grown anaerobically on 10% glucose without ergosterol or oleic acid contain promitochondria only (Parks 1978).

The requirement of *S. cerevisiae*, cultured under anaerobic conditions, for sterols is partly met by cholesterol (about 23% of the growth observed in the presence of ergosterol). However, no detectable growth occurs when cholesterol is replaced by lanosterol (Vanden Bossche et al. 1980).

By operating under as strict anaerobic conditions as possible and in the presence of 2,3-iminosqualene (an 2,3-oxidosqualene-cyclase inhibitor), Pinto et al. (1983) were able to demonstrate that in the absence of sterols with a 24-methyl group in the β-confirmation no growth of *S. cerevisiae* occurred. 24β-Methylcholesterol, ergosterol and brassicasterol (a $\Delta^{5,22}$-24β-methyl sterol, ergosta-5,22-dienol) were active whereas 24α-methylcholesterol (campesterol) and cholesterol were totally inactive. Desmosterol (cholest-5,24-dien-3β-ol), was only active after it was metabolized to the active 24β-methylcholesterol (Nes 1984). The minimal ergosterol concentration needed to induce growth at 27 °C was only 600 ng/ml. Maximal growth occurred at 1.5 µg/ml. This was also achieved in the presence of 4.5 µg/ml of cholesterol and 500 ng/ml of ergosterol, whereas the presence of 5 µg/ml of cholesterol alone did not allow any significant growth to occur. The sparing activity of cholesterol indicates that there are at least two principle functions for sterols in yeasts. Both are satisfied by ergosterol but whereas only one can be played by 24-desmethylsterols such as cholesterol (Nes 1984).

2.2 Studies Using Lipid Vesicles

It is well known that lanosterol has detrimental effects on membrane properties (see for example Taylor and Parks 1980). Further evidence for the differential actions between ergosterol and lanosterol and phospholipids in membranes was obtained by using differential scanning calorimetry (Vanden Bossche et al. 1984a). The mid-transition temperature of multilamellar vesicles of dipalmitoylphosphatidylcholine is 41 °C. Addition of ergosterol lowered the transition temperature slightly; however, the enthalpy of melting decreased. When the ergosterol concentration reached 15 mol% almost no transition was observed. Similar results are described by Parks et al. (1984). When lanosterol instead of ergosterol was used a much smaller effect on the enthalpy of melting was found (Vanden Bossche et al. 1984a). However, by using glucose permeability as a parameter (Vanden Bossche et al. 1984b) it was found that the incorporation of ergosterol (phospholipid : ergosterol = 2 : 1) into unilamellar vesicles composed of phosphatidylcholine, phosphatidylethanolamine and diphosphatidyl-glycerol (5 : 3 : 1) reduced the release per hour of the entrapped glucose by 57%. When lanosterol replaced ergosterol in the vesicles the release of glucose became similar to that observed with vesicles composed of phospholipids alone. Electron spin resonance studies with egg lecithin liposomes containing different amounts of ergosterol or lanosterol demonstrated that lanosterol-containing liposomes are more fluid than ergosterol-containing ones (Nozawa and Morita 1986). Electron paramagnetic resonance studies on *S. cerevisiae* strains indicated an increased rigidity with increased ergosterol content (Lees et al. 1984). These studies indicate that ergosterol, as cholesterol, can, depending on the physical state of the membrane, increase (condensing effect) or decrease (liquefying effect) the chain order of phospholipid membranes. Bloch (1983) in his excellent review on sterol function and membrane function concludes from ^{13}C-NMR studies that lanosterol is considerably more mobile in the phospholipid bilayer than cholesterol. This is in agreement with microviscosity studies on phosphatidylcholine vesicles in which lanosterol and dihydrolanosterol gave microviscosity values ($\bar{\eta}$) only slightly higher than observed with sterol-free control liposomes. 4,4'-Dimethylcholesterol causes a substantial increase in microviscosity. Thus, demethylation at C-14 affects membrane fluidity more profoundly than dealkylation at C-4 (Bloch 1983). This may originate from the fact that the C-14 methyl group of lanosterol protrudes from the sterol α face in the lipid-lipid contact region and as such diminishes the van der Waals contacts allowing greater motional freedom to fatty acyl chains (Bloch 1983).

2.3 Studies Using Sterol Auxotrophs

Fungal sterol mutants are valuable tools in the study of the effects of altered sterol structures on physical properties of membranes and cell physiology. A number of mutants are discussed by Henry (1982) and Parks et al. (1985).

A mutant of *Ustilago maydis* that accumulates 14-methylfecosterol has been found to be viable. However, the doubling time of the mutant is about 6.5 h compared to about 2.25 h for the wild type. At this rate of doubling, projected cell yield of the mutant after 18 h is only about 5% of that of the wild strain (Sisler and Walsh 1981; Sisler et al. 1983).

Using sterol-auxotrophic mutant strains of *S. cerevisiae*, Servouse and Karst (1986) also noted that the generation time of mutant strains, even at optimal ergosterol supplementation, is 50% higher than in the wild strain. This might also be the case for a mutant (*erg* 6) which was blocked in C-24-sterol methyltransferase and able to synthesize zymosterol, cholesta-5,7,24-trien-3β-ol and cholesta-5,7,22,24-tetraen-3β-ol (Barton et al. 1974). Although for the latter mutant the growth yield after 48 h was similar to that of the wild strain (Woods 1971), it should be noted that a mutant of the *erg* 6 series (*nys* 3a) may contain small amounts of ergosterol (Woods 1971). Servouse and Karst (1986) also found that the wild strain is relatively impermeable to exogenous sterol in aerobic growth, whereas auxotrophic mutants are generally highly permeable to sterols. Mutant strains, with an ergosterol level of about

0.3%, presented optimal growth. But it was not possible to obtain a significant decrease in this value by sterol starvation since the generation time increased dramatically and growth stops before an appreciable decrease in ergosterol occurs. Hence, the authors assumed that the 0.3% ergosterol level is sufficient but probably minimal for aerobic yeast growth.

Using a nystatin-resistant mutant of *Rhodotorula gracilis*, that is defective in ergosterol biosynthesis, Künemund and Höfer (1983) found that the plasma membrane of the mutant displays a distinctly lower permeability for cations as compared to the wild strain. This mutant also did not show measurable cotransport of H^+ by the onset of monosaccharide transport and did not exchange external K^+ for protons. This indicates that ergosterol plays an important role in the cation (H^+ and K^+) permeability of the plasma membrane.

The excellent studies of Parks and his colleagues (Rodriguez and Parks 1983; Rodriguez et al. 1982, 1985) suggest that there are at least four different levels of function for sterols which they designated; sparking, critical domain, domain and bulk. Many sterols and stanols were found to satisfy bulk membrane functions. However, growth of yeast sterol auxotrophs on cholestanol is precluded unless minute amounts (1 – 10 ng/ml) of ergosterol are available. Rodriguez et al. (1985) designated this phenomenon the sparking of growth, in which cholestanol satisfies an overall membrane sterol requirement (bulk function) and ergosterol fulfills a high specificity sparking function. The sterol auxotroph, RD5-R (α, hem 1, erg 3, erg 6) was able to grow in media containing 5 µg/ml of lanosterol plus 100 ng/ml of ergosterol but not on lanosterol alone. The growth observed was similar to that in the presence of 0.5 µg/ml of ergosterol (Rodriguez et al. 1985). Since the additional 100 ng of ergosterol needed is insufficient to modulate overall plasma membrane fluidity, Rodriguez et al. (1985) suggest that the 100 ng ergosterol is only required for very restricted areas in a cellular membrane and have designated this role for ergosterol as critical domain function. JR4, a sterol mutant defective in C14-demethylation, is able to grow on 14α-methylfecosterol (Rodriguez et al. 1985). However, a critical analysis of the sterol pool from this mutant revealed the presence of small quantities of ergosterol approximately equal to the level required for growth of RD5-R on lanosterol (100 ng/ml). Two other strains, LS-60 and LS-61, which also synthesize 14α-methylfecosterol contain ergosterol levels comparable to the level found in JR4 (Rodriguez et al. 1985). These data further provide evidence for a critical domain function of ergosterol. These studies also revealed that when the RD5-R mutant was supplemented with less than 5 µg/ml of ergosterol, growth rates were unaffected but growth yields decreased significantly.

When the RD5-R mutant is grown in a medium supplemented with 5 µg/ml of ergosterol the plasma membranes will contain 100 nmol/ml of ergosterol, in contrast to 20 nmol/ml when grown on 1 µg/ml. At this latter concentration fluorescence anisotropy measurements indicate that the sterol is capable of preventing detectable changes in lipid properties and is, therefore, fulfilling a domain function. Below this membrane sterol level the cells are no longer able to undergo membrane expansion (Lewis et al. 1987).

Further support for the different roles of sterols in yeast comes from studies by Dahl and Dahl (1985) on *S. cerevisiae*, strain GL7 (a squalene-epoxide-cyclase deficient yeast mutant). The addition of 100 ng/ml ergosterol to cells, growing poorly on

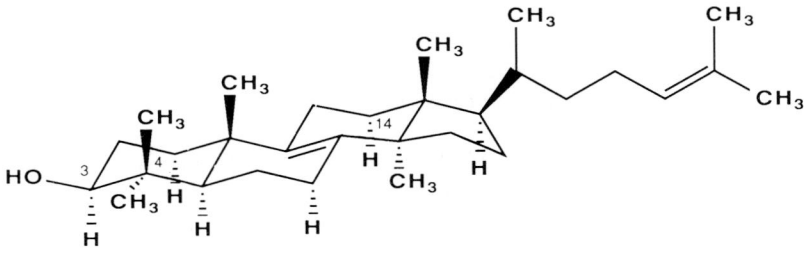

LANOSTEROL

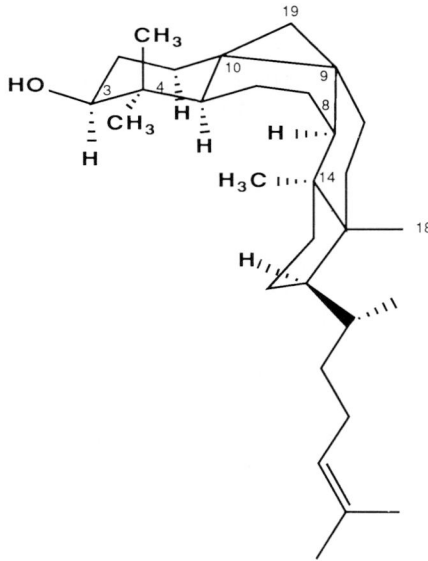

CYCLOARTENOL

Fig. 4. Chemical structures of lanosterol and cycloartenol. (After Bloch 1983)

cholesterol, elicited a sequential stimulation first of polyphosphoinositide metabolism, followed by phospholipid synthesis and cell proliferation. Within 10 min after addition of ergosterol to cells prelabelled with ^{32}Pi or [^{3}H]inositol, the isotope content of the polyphosphoinositides increased markedly followed by a rapid decrease. Subsequently, upon continuous labeling, ^{32}P incorporation into phosphatidylinositol, phosphatidylserine and phosphatidic acid (in decreasing order) increased. After a lag period of 3 hours, the growth rate increased. Stimulation of polyphosphoinositide metabolism was not inhibited by cycloheximide suggesting that the effect was not mediated by increased protein synthesis but might have originated from modulation by ergosterol of enzyme activities. Since cycloheximide inhibited the increased ^{32}P incorporation into phospholipids, protein synthesis might be involved

(Dahl and Dahl 1985). It is of interest that growth of this GL7 sterol auxotroph strain is supported under aerobic conditions by cycloartenol but not by lanosterol (Buttke and Bloch 1980). Cyclolaudenol [24(*S*)-24-methyl-9,19-cyclo-5 α, 9 β-lanost-25-enol], a 24 β-methyl compound with the same nucleus as cycloartenol is superior to cycloartenol in supporting growth of the GL7 strain. It has been suggested that cycloartenol, a molecule shaped like a butterfly (a non-planar conformation), is superior to lanosterol, a flat molecule with protruding methyl groups on the α-face, because of a more favourable spatial disposition of the nuclear methyl groups at C-4 and C-14 (Fig. 4). This will promote more effective interaction between the phospholipid acyl chains and the sterol α-face (Buttke and Bloch 1980).

Examining the transmethylation pathway for the synthesis of phosphatidylcholine, Kawasaki et al. (1985) observed an ergosterol-specific promotion of enzyme activity. They found that GL7 cells grown on ergosterol or ergosterol/cholesterol (1 : 3) mixtures, incorporate several times more ^{3}H from [methyl-^{3}H] methionine into phosphatidylcholine than cells that have been raised on cholesterol alone. Further, GL7-membrane fractions derived from ergosterol-grown cells (1 µg/ml) catalyse methyl transfer from *S*-adenosylmethionine to phosphatidylethanolamine more efficiently than enzyme preparations from cells raised on cholesterol. Incorporation of ^{32}P from [γ-^{32}P]ATP into the yeast membranes is rapid and greater when ergosterol-grown cells rather than cholesterol-grown cells were the source of membranes. This study clearly indicates that cholesterol meets the requirements of yeasts less effectively than ergosterol and furthermore suggests that the sterol molecule can function as a metabolic signal.

Parks et al. (1986) using that yeast auxotroph RD5-R grown on the saturated sterol, cholestanol, proved again the yeast has a specific growth requirement for a C-5,6-unsaturated sterol. Indeed, this yeast appeared to have a threshold concentration of 1.2 nM (476 ng/l) ergosterol. The investigators test aqueous extracts from a sterol wildtype, X2180-1A, and from commercial dried yeast, to determine if ergosterol was further metabolized to another form or if it caused the synthesis of another compound which was the active molecule in fulfilling the sparking requirement. In this study it was found that a sparking ergosterol replacement factor(s) (SERF) could fulfill the role played by sparking ergosterol. Using different analytical procedures, Parks et al. (1986) proved SERF to be different from ergosterol by solubility, thermostability, and thin-layer and liquid chromatography. They speculated that if SERF was derived directly from ergosterol, it would constitute a new end product of sterol metabolism that may have non-membrane-associated functions. Although the structure of this intriguing compound(s) is still not known, this study indicates that ergosterol is involved in critical functions in the cell and this at levels equivalent to those of vitamins, hormones and pheromones.

2.4 Studies with Azole Antifungals

Experiments with azole antifungals further suggested that changes in sterol structure might lead to an alteration in fatty acid composition to re-establish optimal lipid-lipid interaction. Azole antifungals are known to interact at nanomolar concentrations with the cytochrome P-450 (P450)-dependent 14α-demethylation of lanosterol or 24-methylenedihydrolanosterol in the endoplasmic reticulum of yeasts or fungi (for reviews see Vanden Bossche 1985; Vanden Bossche et al. 1987a, b; Yoshida 1988; see also Chapter 15). This results in ergosterol depletion and a concomitant accumulation of membrane disturbing 14α-methylsterols (for a review of the ultrastructural changes observed after azole treatment see Borgers 1988).

In *Candida albicans* grown for six hours in the presence of 0.1 µM ketoconazole or miconazole, a shift occurred from mono- to di-unsaturated fatty acids in the phospholipids. The linoleate content was found to be about 15% of the total amount of fatty acids in the phospholipid fraction of control cells and increased up to 22.5% and 25% in this fraction when *C. albicans* was incubated in the presence of ketoconazole or miconazole (Vanden Bossche et al. 1981, 1983). An increased desaturation of oleic acid leading to an accumulation of linoleic acid was also seen in sporidia of *Ustilago maydis* treated for 9.5 hours with 6 µM of the ergosterol biosynthesis inhibitor, triarimol. Linoleic acid was predominant and comprises 66% of the ^{18}C free fatty acids, whereas in controls the 18 : 2 content accounted for 35.5% only (Ragsdale 1975). Weete et al. (1983) found a higher degree of lipid unsaturation when *Taphrina deformans* was grown in a yeast-promoting medium supplemented with 0.21 µM of the 14α-demethylase inhibitor, propiconazole. Ragsdale (1975) speculates that the increased synthesis of unsaturated fatty acids, observed in cells treated with 14-demethylase inhibitors, might originate from the greater availability of NADPH which normally functions in reactions involved in ergosterol biosynthesis.

The observed shift to more unsaturated fatty acids may reflect an attempt by the yeast cell to change membrane fluidity, compensating for the azole-induced alteration of ergosterol synthesis. An adaptation of the phospholipids to changes in the structure of triterpenoids is seen in *Tetrahymena pyriformis*. This protozoan changes the fatty acyl groups of phospholipids on transition from a sterol-containing to sterol-free medium by introducing fatty acyl groups with *cis*-double bonds thus providing the membranes with phospholipids with a larger cros-section (Ferguson et al. 1975).

A shift to more unsaturated fatty acids (increased oleate and di-unsaturated fatty acid contents) was observed after cholesterol depletion of mouse LM cells defective in sterol synthesis (Freter et al. 1979). According to Freter et al. these auxotrophs compensate in this way for lowered sterol levels by providing lower melting membrane phospholipids whose "fluidizing " effect would be normally provided by sterols. Low et al. (1985) studying a sterol auxotroph yeast mutant found that the fatty acids of phospholipids in cells grown on ergosterol showed a shift toward a higher degree of saturation when compared to cells grown in media supplemented with cholesterol. This finding is in support of a co-ordinately regulated membrane sterol and fatty acid composition.

C. albicans incubated in the presence of 14-demethylase inhibitors is not able to maintain the synthesis of unsaturated fatty acids. Indeed, in *C. albicans*, grown for 16 hours in the presence of 10^{-8} M miconazole (Vanden Bossche et al. 1981) or ketoconazole (unpublished results), the shift from oleic to linoleic acid was replaced by a shift to palmitate (Vanden Bossche et al. 1981, 1983). An increase in palmitate was also found by Georgopapadakou et al. (1987) in *C. albicans* grown in the presence of 10^{-6} M clotrimazole, econazole, miconazole or ketoconazole. The increased synthesis of saturated fatty acids suggests an effect on the Δ^9 desaturase, a microsomal enzyme which, in *S. cerevisiae* (Tamura et al. 1976), contains similar components as the liver system, i.e. NADH, NADH-cytochrome b_5 reductase, cytochrome b_5, a cyanide-sensitive factor and phospholipids (Strittmatter et al. 1974). The requirement for phospholipids indicates that this enzyme, as other membrane-bound enzymes, is only active at a defined fluidity of the environment. It is thus possible that the azole-induced ergosterol depletion and accompanying accumulation of 14-methylsterols alter the fluidity in such a way that the desaturase is inhibited. From antagonistic effects of unsaturated fatty acids it can be deduced that the decreased availability of unsaturated fatty acids may contribute to the antifungal activity of azole antifungals. For example, studies of Yamaguchi (1977) revealed that unsaturated fatty acids, such as palmitoleic acid and oleic acid were effective in antagonizing the effects of clotrimazole and miconazole. The effects of 10^{-6} M of clotrimazole, econazole, miconazole and ketoconazole on the growth of *C. albicans* were found to be reversed by 1 mg/ml of oleic acid or by 1% of the oleic acid derivative, Tween 80 (Georgopapadakou et al. 1987). The vapour phase activity of imazalil on *Penicillium italicum* was partly antagonized by high concentrations ($10^{-5}-10^{-4}$ M) of oleic acid (Van Gestel 1986) and the effects of high concentrations of miconazole (10^{-5} M -3×10^{-4} M) on the viability of *C. albicans* was also partly antagonized by 10^{-5} to 10^{-4} M of oleic acid (Vanden Bossche et al. 1982). However, the growth inhibition seen in the presence of 5×10^{-8} M and 10^{-7} M miconazole was not antagonized nor was the growth inhibition induced by ketoconazole concentrations up to 3×10^{-4} (Vanden Bossche et al. 1982).

From the azole antifungal-induced ergosterol depletion, accumulation of 14α-methylsterols and consequent changes in the fatty acids, important alterations in the properties of fungal membranes can be expected. Indeed studies of Sancholle et al. (1984) on the effects of propiconazole on *Taphrina deformans* indicate that the observed inhibition of the 14α-demethylase results in a steady loss of radioactive substances from cells that had been incubated with ^{32}P. The 14α-demethylase inhibitor, miconazole, inhibits, at concentrations lower than those affecting growth, the uptake of purines by *C. albicans*. An effect already observed 1 h after the addition of this imidazole derivative (Vanden Bossche 1974).

Measuring the effects of sterol structures on viscosity changes in the lipid layers of yeast mitochondria, by using the hydrophobic fluorescent probe, 1,6-diphenyl-1,3,5-hexatriene (DHT), Parks et al. (1984) found that replacement of a sterol other than the ergosterol normally present in the mitochondrial membrane, affected the mobiblity of the fluorescent probe. These investigators also used the kynurenine-3-hydroxylase, an enzyme located in the outer membrane of yeast mitochondria, to test the effects of sterol structure modifications on the membrane activity. Arrhenius plots of enzymatic activity from a sterol mutant were found to have discontinuities, whereas kynurenine hydroxylase in mitochondria isolated from ergosterol accumulators exhibited no changes in activation energy. The discontinuities observed with the sterol mutant occurred at temperatures similar to those found in the experiments with DHT. These studies show that structural changes in the sterols impart substantial effects not only on the physical but also on enzymatic properties of membranes. Therefore, treatment of cells with 14α-demethylase inhibitors should also affect the activities of at least some of the membrane-bound enzymes. An example is the chitin synthase. This enzyme catalyses the synthesis of the β-1,4-polysaccharide, chitin, a major component of the primary septum in, for example, *S. cerevisiae* and the yeast form of *C. albicans* and of the septa and primary wall of the mycelium form of *C. albicans* and hyphae of fungi (Cabib et al. 1984; Soll 1985; Ruiz-Herrera 1985). The chitin synthase present in chitosomes isolated from different fungi is mostly zymogenic (Ruiz-Herrera 1985). This is as expected. Indeed, it has been shown that chitosomes from *Mucor rouxii* are particularly rich in ergosterol (Lopez-Romero et al. 1985) and Chiew et al. (1982) showed that high concentrations of ergosterol inhibit chitin synthesis. Furthermore, mutants of *C. albicans*, with a low ergosterol content, showed increased activity of this synthase (Pesti et al. 1981). From these studies it could be deduced that ergosterol biosynthesis inhibitors should disturb chitin synthesis. An itraconazole-induced increase in the ratio chitin:total carbohydrate was found in *C. albicans* grown for 24 h in a yeast form-promoting medium. The highest ratio was found at an itraconazole concentration of 35 ng/ml (Vanden Bossche 1985). This enhanced chitin synthesis corresponds with the increased and irregular distribution of chitin observed in ketoconazole- (Vanden Bossche et al. 1984b) or itraconazole-treated (Vanden Bossche 1985) *C. albicans*. The distribution of chitin is also altered in *Aspergillus fumigatus* treated for 4 h with 35 ng/ml of itraconazole (Marichal et al. 1985; Vanden Bossche et al. 1988). Similar effects were observed with the 14-demethylase inhibitor, imazalil, or the Δ^8-Δ^7 isomerase inhibitor, fenpropimorph (Kerkenaar and Barug 1984).

In *A. fumigatus*, itraconazole was found to induce an enhanced branching and swelling of the apex (Vanden Bossche et al. 1988). Sisler and Ragsdale (1984) found that treatment of *U. maydis* sporidia with 14-demethylase inhibitors leads to the production of large sporidia that are multicelled and frequently branched. Sporidia failed to separate, which suggests a possible interference with septum formation. *C. albicans* treated with 1–4 μg/ml bifonazole (Barug et al. 1983) or 50–500 ng/ml miconazole (De Nollin and Borgers 1975) formed chains or clusters of interconnected cells. Single cells were rarely seen and several buds and bud scars were present on one cell. As chitin is important in the primary wall synthesis and septum formation, the morphological alterations could be partly explained in terms of irregular deposition of this polysaccharide.

An enhanced branching was also observed in *Achlya bisexualis* treated with the cytochalasins A and E, compounds that exert their effects by blocking the elongation of actin filaments or disrupting microfilament networks (Harold and Harold 1986). In fungi, actin has been implicated in morphogenesis and tip growth (Heath 1987), in the polarization of growth and in the localized deposition of new cell wall in *S. cerevisiae* (Adams and Pringle 1984; Kilmartin and Adams 1984; Novick 1985). In stationary phase cells of *C. albicans*, actin granules were found to be distributed throughout the cytoplasmic cortex (Anderson and Soll 1986). Just before evagination, actin granules clustered at the site of evagination, then filled the early evagination in both budding and hypha-forming cells. During bud growth the actin granules redistributed throughout the cytoplasmic cortex whereas during hyphal growth, the majority of actin granules clustered at the hyphal apex (Anderson and Soll 1986). These studies suggest that actin is involved in the zones of wall expansion. In yeast, actin fibres appear to be similar to those seen in many other cells. According to Adams and Pringle (1984) the fibres are presumably bundles of individual actin filaments and next to the fibres cortical spots are observed. They speculate that these actin spots correspond to sites of anchorage of the cytoplasmic network of actin filaments to the cell membrane. The actin localization, as viewed by using the fluorochrome-labelled phalloidin (a specific probe for actin), indicates that the ring of actin patches is apparently coincident with the ring of chitin formed in the region of the neck. Based on these observations, Kilmartin and Adams (1984) proposed a role for the actin patches in localizing the synthesis of chitin to the neck. In the Oomycete, *Saprolegnia*, apical fibrillar caps of actin appear to be intimately associated with the plasmalemma and are postulated to be primarily functioning in tip morphogenesis or cytoplasmic migration (Heath 1987).

Novick and Botstein (1985) studying two temperature-sensitive lethal actin mutations found evidence for the direct role of actin in the organization, assembly and function of the yeast cell surface. For example, they found that following a shift from 23° to 37°C the actin-staining pattern changed in the mutants. Cables were no longer visible, leaving only randomly distributed surface patches. The actin bars seen at the permissive temperature in *act1-2* are lost gradually over a period of 1 h. These cells exhibit disorganized fine actin cables and patches are seen near the surface of both the mother cell and bud. They also tested the effect of actin mutations on the localization of chitin deposition. Following a shift to the restrictive temperature, chitin depositions become delocalized and are similar to the depositions found after treatment with ergosterol-biosynthesis inhibitors (see above). Novick and Botstein (1985) interpret their results as follows: actin may be a member of a set of proteins that serve to define the site at which budding will occur, i.e. the place where the chitin will be formed. Chitin deposition may become delocalized in the actin mutants because the active enzyme is no longer constrained in the plane of the plasma membrane. Temperature-sensitive cell division cycle mutants, defective in genes *cdc 3, cdc 10, cdc 11*, or *cdc 12* that are unable to complete cytokinesis at restrictive temperature (36°C), undergo multiple cycles of budding, DNA synthesis and nuclear division. The buds are abnormally elongated and the cells develop into multinucleate, quasi-mycelial forms (Adams and Pringle 1984). Other temperature-sensitive mutants, defective in gene *cdc 4*, produce multiple abnormally shaped buds (as described above for *C. albicans* treated with azoles) in the absence of a continuing nuclear cycle, and thus arrest development as multibudded, uninucleate cells (Adams and Pringle 1984).

Based on these studies, it can be speculated that the effects of azole antifungals on ergosterol synthesis might affect the fluidity of critical domains in the membrane so that the actin assembly is disrupted. This will result in delocalized deposition of chitin and, since actin is also implicated in the transport of vesicles, an accumulation of secretory vesicles can be expected as was described by Novick and Botstein (1985) in their actin mutants kept at the restrictive condition. An accumulation of vesicles was also observed by De Nollin et al. (1977) in *C. albicans* exposed to miconazole.

In summary, there is little doubt that the alkylated sterol, ergosterol, is a prerequisite for the yeast und fungal membrane, that ergosterol is implicated in a variety of cell functions and that interference with its synthesis may lead to growth inhibition, stasis and finally to cell death.

3 What Are the Consequences for the Cell to Synthesize Alkylated Sterols?

The conversion of lanosterol into cholesterol is a complex multistep process, catalysed by enzymes in the endoplasmic reticulum. The methyl groups on C-4 and C-14 must be removed, nuclear double-bond transformations ($\Delta^8 \to \Delta^7 \to \Delta^{5,7} \to \Delta^5$) must occur and the side chain should be saturated. The synthesis of ergosterol is even more complex (Fig. 5). In order for ergosterol to be synthesized the following requirements have to be met: a methyl group must be introduced at C-24, a Δ^8-Δ^7 isomerase has to be present, insertion of the 5,6- and 22,23-double bonds and saturation of the 24(28)- and of the 25,26-double bonds are needed (for a review see Mercer 1984). The 24β-methylation involves S-adenosylmethionine. In *Saccharomyces*, the Δ^{24}-sterol methyltransferase shows a preference for zymosterol. To achieve this, zymosterol must be transported from the endoplasmic reticulum to the mitochondria where the transferase is localized (Thompson et al. 1974). 24-Methylenezymosterol (fecosterol) should then be transported to the microsomes to be converted into ergosterol. In most fungi, lanosterol is the preferred substrate. The product of this methylation, 24-methylene-24,25-dihydrolanosterol, has been identified in a wide range of yeast and fungi such as *C. albicans, A. fumigatus, P. expansum, Mucor pusillus, Monilinia fructigena, Neurospora crassa,* and *U. maydis* (see Mercer 1984; Weete 1980). That in these fungi the 14-demethylation is preceeded by a methylation at C-24 means that they need to transport lanosterol to the mitochondria and 24-methylene-24,25-dihydrolanosterol back to the endoplasmic reticulum.

In liver and adrenals, a noncatalytic carrier protein (s) (sterol carrier protein) is present (Scallen et al. 1971, 1985; Scallen and Vahouny 1985; Vahouny et al. 1984)

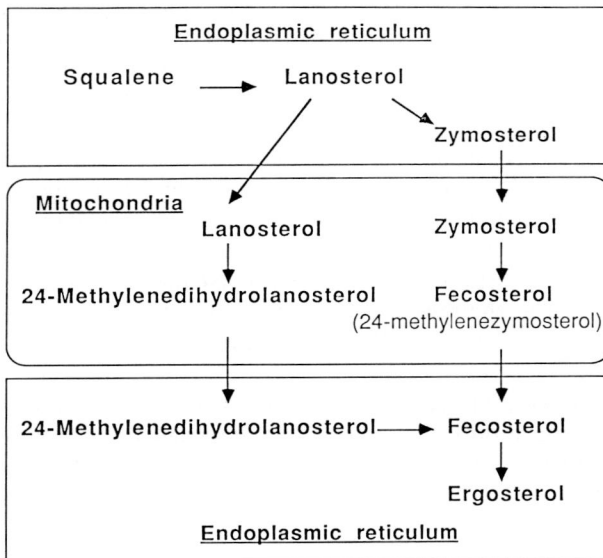

Fig. 5. Compartmentation of ergosterol synthesis in *S. cerevisiae* (24-methylation of zymosterol to form fecosterol) and most other fungal cells (24-methylation of lanosterol to form 24-methylenedihydrolanosterol)

that in, for example, adrenals seems to modulate transfer of cholesterol from adrenal lipid inclusion droplets to mitochondria (Chanderbhan et al. 1982). This carrier protein or a similar peptide may also modulate the transfer of cholesterol to the inner membrane site of cholesterol side-chain cleavage. In mammalian cells evidence exists that the cytoskeleton also plays an important role in the intracellular sterol distribution. The studies of Osawa et al. (1984) and of Hall (1985) indicate that the cytoskeleton and in particular actin is involved in the transport of cholesterol to the inner membrane of the mitochondria where the sterol binds to the cytochrome P-450 that catalyses the side-chain cleavage to form pregnenolone. However, studies on the transfer of cholesterol from the endoplasmic reticulum to the plasmamembrane of Chinese hamster ovary cells indicate that although a vesicular traffic is possible the cytoskeletal network seems not to be involved (for a review see Dawidowicz 1987).

Our preliminary studies indicate the presence of a soluble protein in *S. cerevisiae* that stimulates (in vitro) ergosterol synthesis. Whether or not this protein is involved in the transfer of sterols is not known, its exact role should be further investigated. We have also tested the effects of cytochalasin A on a possible accumulation of lanosterol or 24-methylene-24,25-dihydrolanosterol in exponentially growing *A. fumigatus*. Although the cytochalasin induced enhanced branching, even at 10^{-5} M, no inhibition of ergosterol synthesis was observed. These preliminary results do not favour a role for actin in the transfer of these ergosterol precursors to or from the microsomal membranes. Microtubules may play a role in the intracellular movement of very low density lipoproteins (VLDL) in liver (Le Marchand et al. 1975) and the results of Rajan and Menon (1985) suggest that microtubules are involved in the transport and utilization of low density lipoprotein (LDL) in cultured rat luteal cells, known to utilize cholesterol from LDL. To investigate the possibility that microtubules may be involved in sterol transport in the *S. cerevisiae* sterol auxotroph, FY3, Lorenz et al. (1986) tested the effects of the anti-microtubule agents, nocodazole and methyl benzimidazol-2-yl carbamate (MBC) on growth and sterol transport. Their results seem to indicate that microtubules are not involved in the intracellular transport of sterol for esterification. As far as the author is aware no studies are available on the role the components of the cytoskeleton in yeast and fungi might play in intracellular transport of ergosterol precursors. Since it is obvious that yeast and fungi need a system to obtain the proper temporal and spatial transfer of the ergosterol precursors, more investigations in this important area of intracellular sterol transport are needed.

4 Why Ergosterol Instead of Cholesterol?

The studies summarised here point to ergosterol as being the sterol that best meets the requirements of yeast and fungal membranes. This is understandable for yeast cells, e.g. most *S. cerevisiae* strains, that are unable to synthesize multienoic fatty acids. In most cells the latter fatty acids are available for phospholipid biosynthesis and to maintain the preferred degree of disorder in their membranes. Side-chain methylation might have been selected in yeast to maintain the required fluidity. Indeed, as compared with cholesterol, ergosterol has a greater disordering effect (attributed to the bulky methyl substituent at C-24, that might weaken the van der Waals interactions) when introduced (concentration >8 mol%) in lecithin vesicles (Bloch 1983). Weakening of the van der Waals interactions gives the acyl chains of the phospholipids greater mobility and permits cells to grow at relatively low temperatures. Indeed, a number of *S. cerevisiae* strains are able to grow between 0° and 42°C (Stokes 1971). It is of interest to note that cold acclimation of rye seedlings resulted in substantinal changes in the lipid composition of the plasma membranes; free sterols increased from 34 to 45 mol%, mainly as a result of an increase (from 21 to 32 mol%) in the 24α-ethylsterol, sitosterol, content (Lynch and Steponkus 1987).

Most of the fungi discussed in this paper, such as *Candida* spp., *Aspergillus* spp., *Penicillium* spp. and *Ustilago* spp. produce dienoic fatty acids (linoleic acid) and in for example *C. albicans*, 9.6% of the fatty acid content is represented by the trienoic acid, linolenic acid (for a review see Weete 1980). These organisms can also grow over a broad range of temperatures but the presence of these di- and trienoic acids gives them a greater flexibility to overcome temperature changes so that they do not need ergosterol to maintain a high fluidity in their membranes. Therefore, although the maintenance of a certain fluidity is an important function of ergosterol, it is probably not the only one.

In his review on metabolism of sterols in yeast, Parks (1978) makes the observation that all cells that have ergosterol as their major sterol, have a rigid cell wall next to the plasma membrane. In cholesterol-containing membranes the condensing property of this sterol can impart to the membrane-mechanical strength needed to maintain cell integrity. In ergosterol-containing membranes this requirement might be eliminated by the protective cell wall (Parks 1978). However, the major sterols synthesized by *Leishmania* spp. are 24-methylene-7-dehydrocholesterol, 24-methylenelathosterol, 24(18)-dehydroergosterol and ergosterol, all 24-alkylated sterols (Holz 1985). That the latter sterols play a role in these protozoa can be deduced from the fact that the 14α-demethylase inhibitors ketoconazole (Berman et al. 1986) and itraconazole (Hart et al. 1989) inhibit ergosterol synthesis in *Leishmania* amastigotes and proved to have clinically useful antileishmanial activities (Weinrauch et al. 1987; Borelli 1987). Although the *Leishmania* surface membrane must present "refractory" properties to withstand the lysosomal environment in the host macrophage, its chemical composition does not indicate a rigidity similar to that of the fungal cell wall.

In a study, using *Gibberella fujikuroi*, Nes et al. (1986) found that the addition of 24-epiiminolanosterol, an inhibitor of the introduction of a methyl group into the sterol side chain, lead to aberrant mycelial membranes resulting in leakage of cytoplasmic contents, growth inhibition and induction of asexual sporulation. The latter is normally operative at growth arrest. This again indicates that the 24-alkyl substitutent is obligatory for at least some of the sterol-controlled functions. One of these functions might be to block the signal that is needed to switch from hyphal growth to sporulation. But again, why ergosterol is able to function in this way and not lanosterol is still open to speculation. Or should we speculate here that the "sparking ergosterol replacement factor (SERF)" described by Parks et al. (1986), in which the synthesis of ergosterol might be involved, is the real signal molecule for the initiation of growth and that for this function other sterols cannot substitute for this C-5,6-unsaturated sterol?

In conclusion ergosterol seems to be the "better" sterol than cholesterol and other 24-desmethyl sterols for a number of important functions in yeast and fungi, but we do not yet know enough to understand really why evolution has chosen to maintain ergosterol as the key sterol in fungi.

References

Adams AEM, Pringle JR (1984) Relationship of actin and tubulin distribution to bud growth in wildtype and morphogenetic-mutant *Saccharomyces cerevisiae*. J Cell Biol 98:934–945

Anderson JM, Soll DR (1986) Differences in actin localization during bud and hypha formation in the yeast *Candida albicans*. J Gen Microbiol 132:2035–2047

Anding C, Rohmer M, Ourison G (1976) Nonspecific biosynthesis of hopane triterpenes in a cell-free system from *Acetobacter rancens*. J Am Chem Soc 98:1274–1275

Andreason AA, Stier SJB (1953) Anaerobic nutrition of *Saccharomyces cerevisiae*. I. Ergosterol requirement for growth in defined medium. J Cell Comp Physiol 41:23–26

Barton DHR, Corrie JET, Widdowson DA, Bard M, Woods RA (1974) Biosynthetic implications of the sterol content of ergosterol-deficient mutants of yeast. JCS Chem Commun 30–31

Barton DHR, Jarman TR, Watson KC, Widdowson DA, Boar RB, Damps K (1975) Investigations on the biosynthesis of steroids and terpenoids. XII. Biosynthesis of 3β-hydroxy-triterpenoids and 3-hydroxy-steroids from 3S-2,3-epoxide-2,3-dihydrosqualene. J Chem Soc Perkin Trans (I):1134–1138

Barug D, Samson RA, Kerkenaar A (1983) Microscopic studies of *Candida albicans* and *Torulopsis glabrata* after in vitro treatment with bifonazole. Arzneim Forsch 33:528–537

Berman JD, Goad LJ, Black DH, Holz Jr GG (1986) Effects of ketoconazole on sterol synthesis by *Leishmania mexicana mexicana* amastigotes in murine macrophage tumor cells. Mol Biochem Parasit 20:85–92

Bloch KE (1983) Sterol structure and membrane function. Crit Rev Biochem 14:47–92

Borgers M (1988) Ultrastructural correlates of antimycotic treatment. In: McGinnis MR (ed) Current topics in medical mycology, vol 2. Springer, Berlin Heidelberg New York Tokyo, pp 1–39

Borelli D (1987) A clinical trial of itraconazole in the treatment of deep mycoses and leishmaniasis. Rev Inf Dis 9:(Suppl 1):S 57–S 63

Bouvier P, Rohmer M, Benveniste P, Ourisson G (1976) $\Delta^{8(14)}$-steroids in the bacterium *Methylococcus capsulatus*. Biochem J 159:267–271

Buttke TM, Bloch K (1980) Comparative responses of the yeast mutant strain GL7 to lanosterol, cycloartenol, and cyclolaudenol. Biochem Biophys Res Commun 92:229–236

Cabib E, Kang MS, Bowers B, Elango N, Mattia E, Slater MI, Au-Young J (1984) Chitin synthesis in yeast, a vectorial process in the plasma membrane. In: Nombela C (ed) Microbial cell wall synthesis and autolysis. Elsevier, Amsterdam, pp 91–100

Chanderbhan R, Noland BJ, Scallen TJ, Vahouny GV (1982) Sterol carrier protein$_2$. Delivery of cholesterol from adrenal lipid droplets to mitochondria for pregnenolone synthesis. J Biol Chem 257:8928–8934

Chen HW, Heiniger H-J, Kandutsch AA (1975) Relationship between sterol synthesis and DNA synthesis in phytohemagglutinin-stimulated mouse lymphocytes. Proc Natl Acad Sci USA 72:1950–1954

Chen HW, Heiniger H-J, Kandutsch AA (1978) Alteration of ^{86}Rb$^+$ influx and efflux following depletion of membrane sterol in L-cells. J Biol Chem 253:3180–3185

Chen HW, Leonard DA, Fischer RT, Trzaskos JM (1988) A mammalian mutant cell lacking detectable lanosterol 14α-methyldemethylase activity. J Biol Chem 263:1248–1254

Chiew YY, Sullivan PA, Shepherd MG (1982) The effects of ergosterol and alcohols on germ-tube formation and chitin synthase in *Candida albicans*. Can J Biochem 60:15–20

Cooper RA, Strauss JF III (1984) Regulation of cell membrane cholesterol. In: Shinitzky M (ed) Physiology of membrane fluidity, vol 1. CRC Press, Boca Raton, pp 73–97

Cornell R, Grove GL, Rothblat GH, Horwitz AF (1977) Lipid requirement for cell cycling. The effect of selective inhibition of lipid synthesis. Exp Cell Res 109:299–307

Cuthbert JA, Lipsky PE (1980) Sterol metabolism and lymphocyte function: Inhibition of endogenous sterol biosynthesis does not prevent mitogen-induced human T-lymphocyte activation. J Immunol 124:2240–2246

Dahl JS, Dahl CE (1985) Stimulation of cell proliferation and polyphosphoinositide metabolism in Saccharomyces cerevisiae GL7 by ergosterol. Biochem Biophys Res Commun 133:844–850

Dawidowicz EA (1987) Dynamics of membrane lipid metabolism and turnover. Annu Rev Biochem 56:43–61

Demel RA (1987) Structural and dynamic aspects of membrane lipids. In: Stumpf PK, Mudd JB, Nes WD (eds) The metabolism, structure, and function of plant lipids. Plenum, New York London, pp 145–152

Demel RA, De Kruyff B (1976) The function of sterols in membranes. Biochim Biophys Acta 457:109–132

De Nollin S, Borgers M (1975) Scanning electron microscopy of Candida albicans after in vitro treatment with miconazole. Antimicrob Agents Chemother 7:704–711

De Nollin S, Van Belle H, Goosens F, Thoné F, Borgers M (1977) Cytochemical and biochemical studies of yeasts after in vitro exposure to miconazole. Antimicrob Agents Chemother 11:500–513

Ferguson KA, Davis FM, Conner RL, Landrey JR, Mallory FB (1975) Effect of sterol replacement in vivo on the fatty acid composition of Tetrahymena. J Biol Chem 250:6998–7005

Freter CE, Landenson RC, Silbert DF (1979) Membrane phospholipids alterations in response to sterol depletion of LM cells. J Biol Chem 254:6909–6916

Georgopapadakou NH, Dix BA, Smith SA, Freudenberger J, !Funke PT (1987) Effect of antifungal agents on lipid biosynthesis and membrane integrity in Candida albicans. Antimicrob Agents Chemother 31:46–51

Gibbons GF, Mitropoulos KA, Myant NB (1982) Biochemistry of cholesterol. Elsevier Biomed Press, Amsterdam

Hall PF (1985) The role of the cytoskeleton in the supply of cholesterol for steroidogenesis. In: Strauss JF, Menon KMJ (eds) Lipoprotein and cholesterol metabolism in steroidogenic tissues. Stickley, Washington, pp 207–217

Harold RL, Harold FM (1986) Ionophores and cytochalasin modulate branching in Achlya bisexualis. J Gen Microbiol 132:213–219

Hart DT, Lauwers WJ, Willemsens G, Vanden Bossche H, Opperdoes FR (1989) Perturbation of sterol biosynthesis by itraconazole and retoconazole in Leishmania mexicana mexicana infected macrophages. Mol Biochem Parasit 33:123–134

Heath IB (1987) Preservation of a labile cortical array of actin filaments in growing hyphal tips of the fungus Saprolegnia ferax. Eur J Cell Biol 44:10–16

Heiniger H-J, Kandutsch AA, Chen HW (1976) Depletion of L-cell sterol depresses endocytosis. Nature (London) 263:515–517

Henry SA (1982) Membrane lipids of yeast: Biochemical and genetic studies. In: Strathern JN, Jones EW, Broach JR (eds) The molecular biology of the yeast Saccharomyces. Metabolism and gene expression. Cold Spring Harbor Laboratory, Cold Spring Harbor, pp 101–158

Höfer M, Huh H, Künemund A (1983) Membrane potential and cation permeability. A study with a nystatin-resistant mutant of Rhodotorula gracilis (Rhodosporidum toruloides). Biochim Biophys Acta 735:211–214

Holz GG (1985) Lipids of leishmanias. In: Chang K-P, Bray RS (eds) Leishmaniasis. Elsevier, Amsterdam, pp 79–92

Kandutsch AA, Chen HW, Heiniger H-J (1978) Biological activity of some oxygenated sterols. Science 201:498–501

Kawasaki S, Ramgopal M, Chin J, Bloch K (1985) Sterol control of the phosphatidylethanolamine-phosphatidylcholine conversion in yeast mutant GL7. Proc Natl Acad Sci USA 82:5715–5719

Kerkenaar A, Barug D (1984) Fluorescence microscope studies of Ustilago maydis and Penicilliun italicum after treatment with imazalil or fenpropimorph. Pestic Sci 16:199–205

Kilmartin JM, Adams AEM (1984) Structural rearrangements of tubulin and actin during the cell cycle of the yeast Saccharomyces. J Cell Biol 98:922–933

Künemund A, Höfer M (1983) Passive fluxes of K^+ and H^+ in wild strain and nystatin-resistant mutant of Rhodotorula gracilis (ATCC 26194). Biochim Biophys Acta 735:203–210

Lees ND, Kemple MD, Barbuch RJ, Smith MA, Bard M (1984) Differences in membrane order parameter and antibiotic sensitivity in ergosterol-producing strains of *Saccharomyces cerevisiae*. Biochim Biophys Acta 776:105–112

Le Marchand Y, Singh A, Patzelt C, Orci L, Jeanrenaud B (1975) In vivo and in vitro evidences for a role of microtubules in the secretory processes of liver. In: Borgers M, De Brabander M (eds) Microtubules and microtubule inhibitors. North-Holland Publishing Company, Amsterdam, pp 153–164

Lewis TA, Rodriguez RJ, Parks LW (1987) Relationship between intracellular sterol content and sterol esterification and hydrolysis in *Saccharomyces cerevisiae*. Biochim Biophys Acta 921:205–212

Lopez-Romero E, Monzon E, Ruiz-Herrera J (1985) Sterol composition of chitosomes from yeast cells of *Mucor rouxii*: comparison with whole cells. FEMS Microbiol Lett 30:369–372

Lorenz RT, Rodriguez RJ, Lewis TA, Parks LW (1986) Characteristics of sterol uptake in *Saccharomyces cerevisiae*. J Bacteriol 167:981–985

Low C, Rodriguez RJ, Parks LW (1985) Modulation of yeast plasma membrane composition of a yeast sterol auxotroph as a function of exogenous sterol. Arch Biochem Biophys 240:530–538

Lynch DV, Steponkus PL (1987) Plasma membrane alterations following cold acclimation: Possible relevance to freeze tolerance. In: Stumpf PK, Mudd JB, Nes WD (eds) The metabolism, structure, and function of plant lipids. Plenum, New York London, pp 213–215

Marichal P, Gorrens J, Vanden Bossche H (1985) The action of itraconazole and ketoconazole on growth and sterol synthesis in *Aspergillus fumigatus* and *Aspergillus niger*. Sabouraudia: J Med Vet Mycol 23:13–21

Mercer EI (1984) The biosynthesis of ergosterol. Pestic Sci 15:133–155

Morpurgo G, Serlupi-Crescenzi G, Tecce G, Valente F, Venettacci D (1964) Influence of ergosterol on the physiology and the ultra-structure of *Saccharomyces cerevisiae*. Nature (London) 201: 897–899

Nes WR (1984) Uniformity vs. diversity in the structure, biosynthesis, and function of sterols. In: Nes WR, Fuller G, Tsai L-S (eds) Isopentenoids in plants biochemistry and function. Dekker, New York Basel, pp 325–347

Nes WR, Hanners PK, Parish EJ (1986) Control of fungal sterol C-24 transalkylation: Importance to developmental regulation. Biochem Biophys Res Commun 139:410–415

Novick P (1985) Intracellular transport mutants of yeast. TIBS (November):432–434

Novick P, Botstein D (1985) Phenotypic analysis of temperature-sensitive yeast actin mutants. Cell 40:405–416

Nozawa Y, Morita T (1986) Molecular mechanisms of antifungal agents associated with membrane ergosterol. Dysfunction of membrane ergosterol and inhibition of ergosterol biosynthesis. In: Iwata K, Vanden Bossche H (eds) In vitro and in vivo evaluation of antifungal agents. Elsevier, Amsterdam, pp 111–122

Oldfield E, Chapman D (1972) Dynamics of lipids in membranes: Heterogeneity and the role of cholesterol. FEBS Lett 23:285–297

Osawa S, Betz G, Hall PF (1984) Role of actin in the responses of adrenal cells to ACTH and cyclic AMP: inhibition by DNase. J Cell Biol 99:1335–1342

Ourisson G, Albrecht P, Rohmer M (1982) Predictive microbial biochemistry – from molecular fossils to procaryotic membranes. TIBS (July):236–239

Parks LW (1978) Metabolism of sterols in yeast. CRC Crit Rev Microb 6:301–341

Parks LW, Bottema DK, Rodriguez RJ (1984) Physical and enzymatic function of ergosterol in fungal membranes. In: Nes WD, Fuller G, Tsai K-S (eds) Isopentenoids in plants. Biochemistry and function. Dekker, New York Basel, pp 433–452

Parks LW, Bottema DK, Rodriguez RJ, Lewis TA (1985) Yeast mutants as tools for the study of sterol metabolism. Methods Enzymol 111:333–345

Parks LW, Rodriguez RJ, Low C (1986) An essential fungal growth factor derived from ergosterol: a new end product of sterol biosynthesis in fungi? Lipids 21:89–91

Pesti M, Campbell JM, Peberdy JF (1981) Alteration of ergosterol content and chitin synthase activity in *Candida albicans*. Curr Microbiol 5:187–190

Pinto WJ, Lozano R, Sekula BC, Nes WR (1983) Stereochemically distinct roles for sterol in *Saccharomyces cerevisiae*. Biochem Biophys Res Commun 122:47–54

Pratt HP, Fitzgerald PA, Saxon A (1977) Synthesis of sterol and phospholipid induced by the interaction of phytohemagglutinin and other mitogens with human lymphocytes and their relation to blastogenesis and DNA synthesis. Cell Immun 32:160–170

Prince RC (1987) Hopanoids: the world's most abundant biomolecules? TIBS (December):455 – 456

Ragsdale NN (1975) Specific effects of triarimol on sterol biosynthesis in *Ustilago maydis*. Biochim Biophys Acta 380:81 – 96

Ragsdale NN, Sisler HD (1972) Mode of action of triarimol on sterol biosynthesis in *Ustilago maydis*. Biochim Biophys Acta 380:81 – 96

Rajan VP, Menon KMJ (1985) Role of microtubules in lipoprotein transport in cultured rat luteal cells. In: Strauss JF, Menon KMJ (eds) Lipoprotein and cholesterol metabolism in steroidogenic tissues. Stickley, Washington, pp 197 – 200

Rodriguez RJ, Parks LW (1983) Structural and physiological features of sterols necessary to satisfy bulk membrane and sparking requirements in yeast auxotrophs. Arch Biochem Biophys 225:861 – 871

Rodriguez RJ, Taylor FR, Parks LW (1982) A requirement for ergosterol to permit growth of yeast sterol auxotrophs on cholestanol. Biochem Biophys Res Commun 106:435 – 441

Rodriguez RJ, Low C, Bottema CDK, Parks LW (1985) Multiple functions for sterols in *Saccharomyces cerevisiae*. Biochim Biophys Acta 837:336 – 343

Rohmer M, Bouvier P, Ourisson G (1979) Molecular evolution of biomembranes: structural equivalents and phylogenic precursors of sterols. Proc Natl Acad Sci USA 76:847 – 851

Rohmer M, Bouvier P, Ourisson G (1980) Non-specific lanosterol and hopanoid biosynthesis from the bacterium *Methylococcus capsulatus*. Eur J Biochem 112:557 – 560

Ruiz-Herrera J (1985) Dimorphism in *Mucor* species with emphasis on *M. rouxii* and *M. bacilliformis*. In: Szaniszlo PJ, Harris JL (eds) Fungal dimorphism. Plenum, New York London, pp 361 – 384

Sancholle M, Weete JD, Montant C (1984) Effects of triazoles on fungi: I. Growth and cellular permeability. Pestic Biochem Physiol 21:31 – 44

Scallen TJ, Vahouny GV (1985) The participation of sterol carrier proteins in cholesterol biosynthesis, utilization and intracellular transfer. In: Strauss JF, Menon KMJ (eds) Lipoprotein and cholesterol metabolism in steroidogenic tissues. Stickley, Washington, pp 219 – 236

Scallen TJ, Schuster MW, Dhar AK (1971) Evidence for a noncatalytic carrier protein in cholesterol biosynthesis. J Biol Chem 246:224 – 230

Scallen TJ, Noland BJ, Gavey KL, Bass NM, Ockner RK, Chandebhan R, Vahouny GV (1985) Sterol carrier protein 2 and fatty acid-binding protein. J Biol Chem 260:4733 – 4739

Servouse M, Karst F (1986) Regulation of early enzymes of ergosterol biosynthesis in *Saccharomyces cerevisiae*. Biochem J 240:541 – 547

Sisler HD, Ragsdale NN (1984) Biochemical and cellular aspects of the antifungal action of ergosterol biosynthesis inhibitors. In: Trinci APJ, Ryley JF (eds) Mode of action of antifungal agents. Cambridge Univ Press, Cambridge, pp 257 – 282

Sisler HD, Walsh R (1981) Mutant of *Ustilago maydis* genetically blocked in sterol C-14 demethylation. Neth J Plant Pathol 87:235 – 236

Sisler HD, Walsh RC, Ziogas BN (1983) Ergosterol biosynthesis: a target of fungitoxic action. In: Matsunaka S, Hutson DH, Murphy SD (eds) Pesticide chemistry: Human welfare and the environment, vol 3. Mode of action, metabolism and toxicology. Pergamon, New York, pp 129 – 134

Smedley-MacLean I, Thomas EM (1920) XL. The nature of yeast fat. Biochem J 14:483 – 493

Soll DR (1985) *Candida albicans*. In: Szaniszlo PJ, Harris JL (eds) Fungal dimorphism. Plenum, New York London, pp 167 – 195

Stokes JL (1971) Influence of temperature on the growth and metabolism of yeasts. In: Rose AH, Harrison JS (eds) The yeasts, vol 2. Academic Press, London New York, pp 119 – 134

Strittmatter P, Spatz L, Corcoran D, Rogers MJ, Setlow B, Redline R (1974) Purification and properties of rat liver microsomal coenzyme A desaturase. Proc Natl Acad Sci USA 71:4565 – 4569

Tamura Y, Yoshida Y, Sato R, Kumaoka H (1976) Fatty acid desaturase system of yeast microsomes. Involvement of cytochrome b 5-containing electron-transport chain. Arch Biochem Biophys 175:284 – 294

Tanret C (1889) Sur un nouveau principe immédiat de l'ergot de seigle, l'ergostérine. CR Acad Sci 108:98 – 100

Taylor FR, Parks LW (1980) Adaptation of *Saccharomyces cerevisiae* to growth on cholesterol: selection of mutants defective in the formation of lanosterol. Biochem Biophys Res Commun 95:1437 – 1445

Taylor RF (1984) Bacterial triterpenoids. Microbiol Rev 48:181 – 198

Thomas L (1974) The lives of a cell – Notes of a biology watcher. Viking Press, New York, p 170

Thompson ED, Bailey RB, Parks LW (1974) Subcellular location of S-adenosylmethionine: Δ^{24}-sterol methyltransferase in *Saccharomyces cerevisiae*. Biochim Biophys Acta 334:116–126

Vahouny GV, Dennis P, Chanderbhan R, Fiskum G, Noland BJ, Scallen TJ (1984) Sterol carrier protein$_2$ (SCP$_2$)-mediated transfer of cholesterol to mitochondrial inner membranes. Biochem Biophys Res Commun 122:509–515

Vanden Bossche H (1974) Biochemical effects of miconazole on fungi: I. Effects on the uptake and/or utilization of purines, pyrimidines, nucleosides, amino acids and glucose by *Candida albicans*. Biochem Pharmacol 23:887–899

Vanden Bossche H (1985) Biochemical targets for antifungal azole derivatives: Hypothesis on the mode of action. In: McGinnis MR (ed) Current topics in medical mycology, vol 1. Springer, Berlin Heidelberg New York Tokyo, pp 313–351

Vanden Bossche H, Willemsens G, Cools W, Cornelissen F, Lauwers WF, Van Cutsem JM (1980) In vitro and in vivo effects of the antimycotic drug ketoconazole on sterol synthesis. Antimicrob Agents Chemother 17:922–928

Vanden Bossche H, Willemsens G, Cools W, Lauwers WF (1981) Effects of miconazole on the fatty-acid pattern in *Candida albicans*. Archiv Int Physiol Biochem 89:B134

Vanden Bossche H, Ruysschaert JM, Defriese-Quertain F, Willemsens G, Cornelissen F, Marichal P, Cools W, Van Cutsem J (1982) The interaction of miconazole and ketoconazole with lipids. Biochem Pharmacol 31:2609–2617

Vanden Bossche H, Willemsens G, Cools W, Marichal P, Lauwers W (1983) Hypothesis on the molecular basis of the antifungal activity of N-substituted imidazoles and triazoles. Biochem Soc Trans 11:665–667

Vanden Bossche H, Lauwers W, Willemsens G, Marichal P, Cornelissen F, Cools W (1984a) Molecular basis for the antimycotic and antibacterial activity of N-substituted imidazoles and triazoles: the inhibition of isoprenoid biosynthesis. Pestic Sci 15:188–198

Vanden Bossche H, Willemsens G, Marichal P (1984b) Cytochrome P-450 inhibitors at the origin of deteriorated fungal membranes. A summary. In: Nombela C (ed) Microbial cell wall synthesis and autolysis. Elsevier, Amsterdam, pp 307–312

Vanden Bossche H, Willemsens G, Marichal P (1987a) Anti-Candida drugs – The biochemical basis for their activity. CRC Crit Rev Microb 15:57–72

Vanden Bossche H, Marichal P, Gorrens J, Bellens D, Verhoeven H, Coene M-C, Lauwers W, Janssen PAJ (1987b) Interaction of azole derivatives with cytochrome P-450 isozymes in yeast, fungi, plant and mammalian cells. Pestic Sci 21:289–306

Vanden Bossche H, Marichal P, Geerts H, Janssen PAJ (1988) The molecular basis for itraconazole's activity against *Aspergillus fumigatus*. In: Vanden Bossche H, Mackenzie DWR, Cauwenbergh G (eds) *Aspergillus* and aspergillosis. Plenum, New York London, pp 171–197

Van Gestel J (1986) The vapour phase activity of antifungal compounds: a neglected or a negligible phenomenon? In: Iwata K, Vanden Bossche H (eds) In vitro and in vivo evaluation of antifungal agents. Elsevier, Amsterdam, pp 207–218

Weete JD (1980) Lipid biochemistry. Plenum, New York London, pp 49–95

Weete JD, Sancholle MS, Montant C (1983) Effects of triazoles on fungi: II. Lipid composition of *Taphrina deformans*. Biochim Biophys Acta 752:19–29

Weinrauch I, Livshin R, El-On J (1987) Ketoconazole in cutaneous leishmaniasis. Br J Med 117(5):666–668

Woods RA (1971) Nystatin-resistant mutants of yeast: Alterations in sterol content. J Bacteriol 108:69–73

Yamaguchi H (1977) Antagonistic action of lipid components of membranes from *C. albicans* and various other lipids on two imidazole antimycotics. Antimicrob Agents Chemother 12:16–25

Yoshida Y (1988) Cytochrome P450 of fungi: Primary target for azole antifungals. In: McGinnis MR (ed) Current topics in medical mycology, vol 2. Springer, Berlin Heidelberg New York Tokyo, pp 389–418

Chapter 11

HMG-CoA to Isopentenyl Pyrophosphate –
Enzymology and Inhibition

R. H. Abeles [1]

1 Hydroxy Methylglutaryl-Coenzyme A (HMG-CoA) Reductase

Compactin (*1*, Fig. 1) and related compounds (Endo et al. 1976; Alberts et al. 1980) are powerful inhibitors ($K_i \approx 10^{-10}$M) of HMG-CoA reductase (Fig. 2). Examination of the structure of compactin does not suggest any obvious reasons why it should be an effective inhibitor of this enzyme.

The hydroxy acid-side chain of compactin and mevinolin (*2*, Fig. 1) resembles mevalonic acid, and it is likely that the hydroxy acid moiety of these inhibitors occupies the hydroxymethylglutaryl binding site on the enzyme; however, mevalonic acid is not a potent inhibitor of HMG-CoA reductase. It is not apparent how the "lower portion" (decalin moiety or chlorophenol (*3*, Fig. 1)) facilitates binding of the inhibitor to the enzyme. Experiments were therefore done to define the binding site of compactin and compound *3* in order to understand the basis for strong binding (Nakamura and Abeles 1985). For the purpose of this analysis it is convenient to subdivide the active site of HMG-CoA reductase into three domains:

(1) A domain which binds the hydroxymethylglutarate portion of HMG-CoA;
(2) A domain which binds CoA;
(3) A domain which binds NADP or NADPH.

We then determined whether substances that specifically interact with each of these domains can affect the rate of binding of compactin or related inhibitors. We also determined how strongly fragments of compactin interact with the enzyme. From these binding constants of components of compactin, we can evaluate the "binding advantage" gained from covalently linking the two components.

Table 1 shows the effect of substrate molecules on the binding of compactin (*1*), compound (*3*), and compound (*6*) to HMG-CoA reductase.

Consider the result obtained with compound (*3*). HMG-CoA but not CoASH or NADP prevents binding of (*3*). We conclude the following: (1) Inhibitor (*3*) interacts with hydroxymethylglutaryl binding site, but not with the CoASH binding site. (2) Compound (*3*) does not interact with the NADP binding site. The behavior of compound (*6*) is analogous to that of (*3*). Compactin differs from (*3*) and (*6*) in that the rate of binding of compactin is reduced by CoASH.

[1] Graduate Department of Biochemistry, Brandeis University, Waltham, Massachusetts 02254, USA

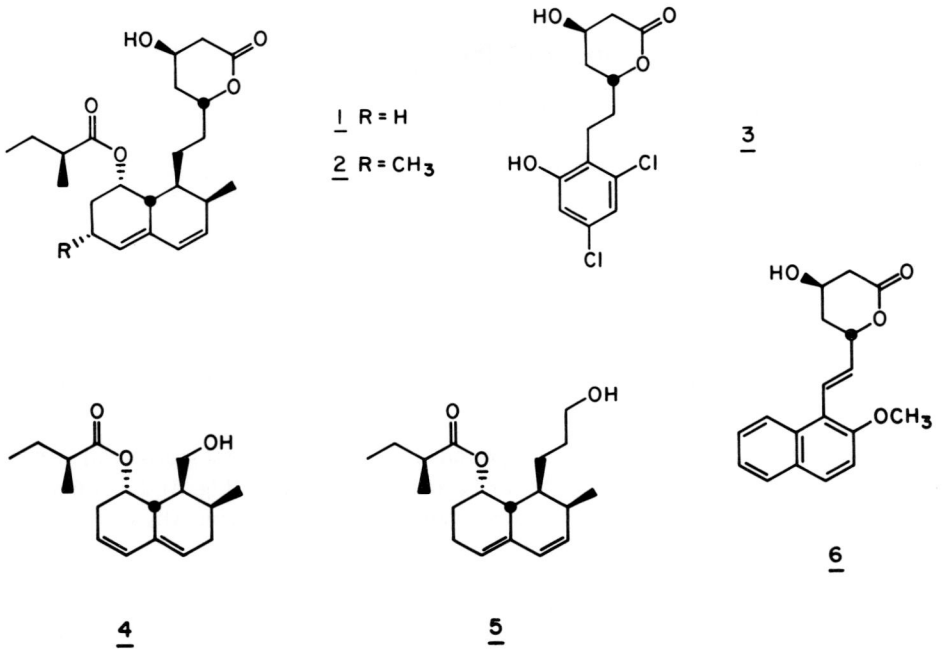

Fig. 1. Inhibitors of HMG-CoA reductase

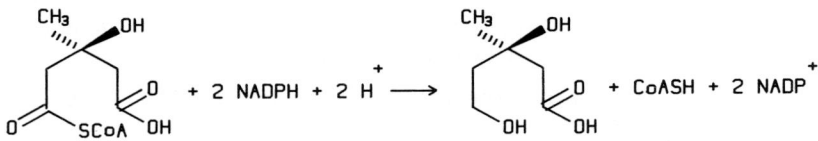

Fig. 2. Reaction catalyzed by HMG-CoA reductase

Table 1. Effect of substrates on rate of binding of inhibitors

	1	3	6
HMG-CoA	+	+	+
NADP	−	−	−
CoA-SH	+	−	−
K_D	0.2×10^{-9} M	0.2×10^{-9} M	0.8×10^{-6} M

+ Prevents binding of inhibitor.
− Does not prevent binding of inhibitor.

We assume that the major binding interactions of (3) and compactin are very similar, but compactin is larger than (3) and its binding area extends into the CoASH site. This interaction probably does not make a major contribution to the K_i of compactin. We propose that the binding of (3) and compactin, as well as other related compounds, involves interaction of the upper portion (hydroxy acid moiety) at the hydroxymethylglutaryl binding domain of the active site and the lower portion (decalin or aromatic moiety) at a hydrophobic area near, but not part of, the active site. We refer to this hydrophobic area as the "hydrophobic anchor". Tight binding then is due to the *simultaneous* interaction at these two binding sites. Hydrophobic anchors are probably involved in the binding of inhibitors for other enzymes.

In order to define the effect of connecting the upper and lower portion of compactin, we measured the interaction of the upper portion of compactin and the lower portion of compactin with HMG-CoA reductase. The inhibition constants of DL-mevalonate and DL-3,5-dihydroxyvalerate (determined from inhibition of HMG-CoA reductase vs HMG-CoA in initial velocity experiments) were found to be 11 ± 1 and 33 ± 6 mM, respectively. These values are consistent with dissociation constants obtained by protection of HMG-CoA reductase from inactivation by iodoacetic acid, 4 ± 2 and 21 ± 9 mM, respectively. These compounds resemble the upper portion of compactin and compound (3). No inhibition was detected with compounds (4) and (5) which resemble the lower portion of compactin. Simultaneous addition of these compounds with DL-3,5-dihydroxyvalerate did not enhance the inhibition over that observed in the presence of DL-3,5-dihydroxyvalerate alone.

We believe that the high affinity of compactin and (3) for HMG-CoA reductase is due to simultaneous interaction at two separate binding areas: the hydroxymethylglutarate domain and the hydrophobic region. It has been pointed out that a compound which interacts simultaneously with two binding sites of an enzyme can show a very high binding constant (Jencks 1981). If one represents the compound as A−B and the separate components as A and B, then the binding "advantage" in connecting A and B is equivalent to the ratio of dissociation constants $[K_{D(A)} \times K_{D(B)}/(K_{D(A-B)})]$. This advantage can be as high as 10^8 M. No inhibition by compounds (4) and (5) could be detected, possibly, due to their limited solubility. A lower limit for their dissociation constants may be conservatively estimated to be 0.5 mM. The advantage of connecting the lactone $(K_i \approx 4-20$ mM) and decalin portions of compactin together is then $\geq 5 \times 10^4$ M. The number 10^8 M represents a maximum value that

Table 2. Rate and equilibrium constants for the reaction of enzyme with compactin and HMG-CoA

Reaction	k_{on} (M^{-1} s^{-1})	k_{off} (S^{-1})	K_D (M) $= k_{off}/k_{on}$
E + HMG-CoA $\underset{k_{off}}{\overset{k_{on}}{\rightleftharpoons}}$ ES	1.9×10^5	0.11 ± 0.09	$(0.59 \pm 0.17) \times 10^{-6}$
E + compactin $\underset{k_{off}}{\overset{k_{on}}{\rightleftharpoons}}$ EC	$(2.7 \pm 0.3) \times 10^7$	6.5×10^{-3}	0.24×10^{-9}

would be observed if binding portion A (or B) of the compound so restricted portion B (or A) that portion B "fit" on the enzyme would occur without further loss of entropy. This is an improbable situation that would not be expected to be observed.

For compactin k_{on} and k_{off} were also determined. The values of these constants are given in Table 2 and are compared with those for HMG-CoA. Note that k_{on} for compactin is essentially diffusion controlled and is approximately 10^2 faster than k_{on} for HMG-CoA. Presumably the enzyme undergoes a conformational change before forming a productive complex with HMG-CoA, but interacts directly with compactin.

2 Isopentenyl Pyrophosphate Isomerase (IPPI)

IPPI catalyzes the reaction shown in Fig. 3. The reaction could proceed through a carbonium ion or a carbanion mechanism. The substrate analogue (Z)-3-(trifluoromethyl)-2-butenyl pyrophosphate reacts at $<1.8\times10^{-6}$ times the rate of dimethylallyl pyrophosphate (Reardon and Abeles 1986). This slow reaction rate is consistent with a carbonium ion mechanism. Electron withdrawal by the trifluoromethyl group is expected to greatly destabilize carbonium ion formation. In a similar study on prenyl transferase, substitution of a methyl group in dimethylallyl pyrophosphate with a trifluoromethyl group resulted in a 10^7-fold decrease in V_{max} (Poulter and Satterwhite 1977). This decrease was attributed to decreased rate of formation of the allylic carbonium ion intermediate. In nonenzymatic model reactions there was a 10^7-fold decrease in the rate of S_N1 solvolysis for a methanesulfonate derivative of the trifluoromethyl analogue relative to dimethylallyl methanesulfphonate itself, a result in agreement with the enzymatic studies (Poulter and Satterwhite 1977).

To obtain additional evidence for the carbonium ion mechanism, we synthesized the putative transition state analogue, 2-(dimethylamino)ethyl pyrophosphate and examined its effect on the reaction catalyzed by IPPI. The compound is an excellent inhibitor of the enzyme. It appeared to be an irreversible inhibitor. However, a more detailed examination of the inhibition mechanism revealed that the inhibition involved no covalent interaction between inhibitor and enzyme, but k_{off} was extremely slow. The kinetic data for 2-(dimethylamino)ethyl pyrophosphate and related compounds are summarized in Table 3. It is apparent that for maximal inhibition, both the positively charged nitrogen and the pyrophosphate moiety are required. Note, that the K_i for the phosphate analogue is $\sim 10^8$-fold higher.

We believe that inhibition of IPPI by 2-(dimethylamino)ethyl pyrophosphate provides strong evidence for a carbonium ion mechanism. There are now several ex-

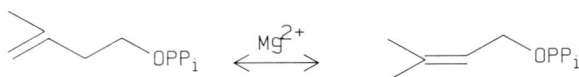

Fig. 3. Reaction catalyzed by IPPI

Table 3. Kinetic constants for inhibitors of IPPI

Compound	k_{on} (M^{-1} min^{-1})	k_{off} (min^{-1})	K_i (M)
2-Aminoethyl pyrophosphate	5.2×10^6	nd	nd
2-(Methylamino)ethyl pyrophosphate	1.9×10^6	nd	nd
2-(Dimethylamino)ethyl pyrophosphate	2.1×10^6	3.0×10^{-5}	1.4×10^{-11}
2-(Trimethylammonio)ethyl pyrophosphate	4.4×10^4	0.030	6.8×10^{-7}
Isopentenyl pyrophosphate	nd	nd	3.5×10^{-5} (k_m)
3-Bromo-3-butenyl pyrophosphate	nd	nd	4.5×10^{-5}
Methyl pyrophosphate	nd	nd	7×10^{-5}
Isoamyl pyrophosphate	nd	nd	4×10^{-4}
2-(Dimethylamino) ethyl phosphate	nd	nd	2.0×10^{-3}

amples in which substitution of a carbon atom, which acquires a carbonium ion character in the transition state, by a positively charged nitrogen leads to compounds that are very good inhibitors, presumably transition-state analogues (Sandifer 1982; Narula 1981; Rahier 1985; see also Chapters 12 and 14). It is surprising that an ammonium ion can take the place of a carbonium ion, since the geometry of the two structures is quite different. The fact that these compounds are inhibitors indicates that electrostatic interactions and not geometry are of overwhelming importance. It is likely that this interaction occurs with a negatively charged group on the enzyme, which in the catalytic process stabilizes the carbonium ion. Some evidence exists that in glycosidases, where carbonium ion mechanisms are involved, a carboxylate group stabilizes the carbonium ion. The nature of the negatively charged group at the active site of IPPI is not known. Possibly, it is a sulfhydryl group, since it has been established that the enzyme is sensitive to sulfhydryl reagents.

3 Inhibition of Mevalonate-5-Pyrophosphate Decarboxylase

We have previously shown that (Z)-3-(trifluoromethyl)-2-butenyl pyrophosphate is a very poor substrate for IPPI and is also an inhibitor of the isomerization of isopentenyl pyrophosphate (Reardon and Abeles 1986). 3-(Fluoromethyl)-2-butenyl pyrophosphate and 3-(fluoromethyl)-3-butenyl pyrophosphate are irreversible inactivators of IPPI (Muehlbacher and Poulter 1985; see also Chapter 12). Both (E)-3-(trifluoromethyl)-2-butenyl pyrophosphate and (Z)-3-(trifluoromethyl)-2-butenyl pyrophosphate are very poor substrates for prenyltransferase (Poulter and Satterwhite 1977). These fluorinated analogues of isopentenyl pyrophosphate and dimethylallyl pyrophosphate could be effective inhibitors of cholesterol biosynthesis. However, the pyrophosphate group would most likely prevent their uptake by intact cells where inhibition of cholesterol synthesis is of primary interest. We, therefore, tried to devise an approach that would allow introduction of these compounds into intact cells. One way of achieving this could be by exposing the cell to mevalonate fluorinated at C-6. If these fluorinated mevalonate analogues undergo the same biochemical transforma-

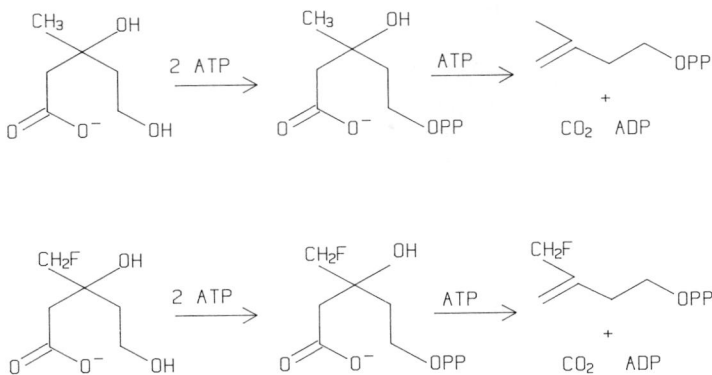

Fig. 4. Is the metabolism of mevalonate applicable to 6-fluoro-mevalonate (I)?

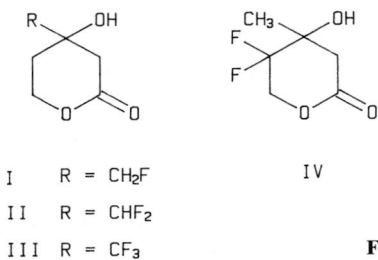

I   R = CH_2F IV

II   R = CHF_2

III R = CF_3 **Fig. 5.** Fluorinated mevalonate analogues

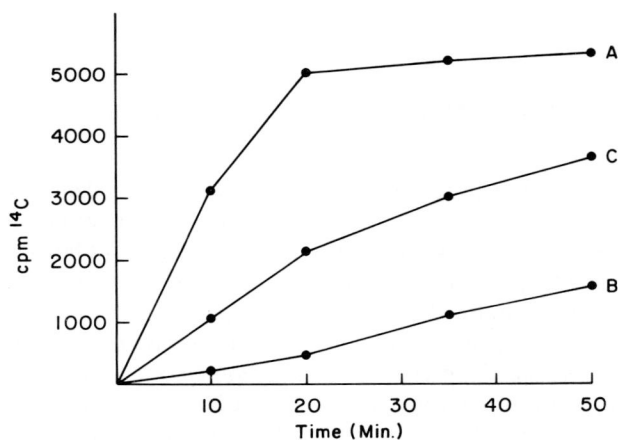

Fig. 6. Inhibition of non-saponfiable lipid biosynthesis by fluorinated mevalonate analogues. Enzyme: S_{10} rat liver homogenate; substrate: sodium mevalonate; inhibitors: *A* none, *B* 10 µM 6-fluoromevalonate (*I*), *C* 300 µM trifluoro-mevalonate (4)

Table 4. Inhibition of mevalonate incorporation into nonsaponifiable lipids

Compound	I_{50} (μM)
6-Fluoromevalonate	6.5
6,6-Difluoromevalonate	10.0
6,6,6-Trifluoromevalonate	300
4,4-Difluoromevalonate	1000

tions as mevalonate normally does, then the desired fluorinated inhibitors would be generated by the cell (Fig. 4). To test this hypothesis, we synthesized the fluorinated mevalonate analogues (*I–IV*) as shown in Fig. 5. While this work was in progress it was reported (Nave et al. 1985) that (*I*) inhibits cholesterol biosynthesis in liver homogenates. Their data showed that (*I*) was not transformed to 3-(fluoromethyl)-3-butenyl pyrophosphate. They proposed that inhibition of cholesterol biosynthesis most probably results from inhibition of mevalonate-5-pyrophosphate decarboxylase by the pyrophosphorylated form of (*I*).

The S_{10} supernatant of a rat liver homogenate contains all of the enzymes required for biosynthesis of cholesterol from mevalonate (Popjak 1969). Addition of compounds *I–IV* to the incubation results in a decreased rate of steroid biosynthesis. The time course for incorporation of mevalonate into nonsaponifiable lipids as well as the effect on this incorporation by (*I*) and (*III*) is shown in Fig. 6. The I_{50} values for compounds (*I–IV*), determined during the linear portion of the assay, are shown in Table 4. Although compounds (*I–IV*) inhibit the incorporation of [2-^{14}C]mevalonate into nonsaponifiable lipids, they do not affect the rate of incorporation of [4-^{14}C]IPP. This result suggests that the site of action of compounds (*I–IV*) is prior to IPPI.

To further pinpoint the site of action of these compounds, the distribution of radioactivity among the water-soluble metabolites of mevalonate was examined. In the control incubation the only water-soluble metabolites detected were isopentenyl pyrophosphate (IPP) and farnesyl pyrophosphate (FPP). Addition of compounds (*I*), (*III*) and (*IV*) results in a dramatic reduction in the levels of FPP, and two new peaks corresponding to mevalonate 5-phosphate and mevalonate 5-pyrophosphate are observed. The fluorinated mevalonate analogues appear to inhibit the conversion of mevalonate 5-pyrophosphate to IPP. This inhibition could be due to the fluoromevalonate analogues, or more likely, to metabolic derivatives of these analogues, possibly pyrophosphorylated compounds.

To determine whether the analogues were metabolically modified, the water-soluble products formed upon incubation of [5-^{3}H]-*I* and [5-^{3}H]-*III* with the S_{10} supernatant were examined. It was established that mevalonate analogues (*I* and *III*) were phosphorylated and pyrophosphorylated in rat liver homogenates. These results suggested that phosphorylated or pyrophosphorylated fluoromevalonate analogues (*I–IV*) inhibit the conversion of mevalonate to FPP, possibly by inhibition of mevalonate decarboxylase. It has already been shown that pyrophosphorylated (*I*) inhibits the decarboxylase (Nave et al. 1985). We, therefore, examined the effect of phosphorylated and pyrophosphorylated (*I*) and (*III*) on the decarboxylase. The results are summarized in Table 5. The pyrophosphorylated forms of the inhibitors

Table 5. Inhibitors of mevalonate-5-pyrophosphate decarboxylase

Inhibitor	K_i (μM)
6-Fluoromevalonate-5-pyrophosphate	0.01
6,6,6-Trifluoromevalonate-5-pyrophosphate	0.5
6-Fluoromevalonate-5-phosphate	0.9
Mevalonate-5-phosphate	105
Mevalonate-5-pyrophosphate	9.9 (K_m)

are effective inhibitors of the decarboxylase. The phosphorylated species are less effective, and essentially no inhibition is seen with unphosphorylated compounds.

Particularly noteworthy is the observation that pyrophosphorylated (*I*) has a remarkably low K_i (10 nM). The basis for this low K_i is not clear. Presumably the decarboxylation involves the phosphorylation of the 3-hydroxy group of mevalonate pyrophosphate, followed by decarboxylation with elimination of P_i.

Possibly the presence of fluorine on the carbon α to the carbon bearing the OH group that becomes phosphorylated reduces the rate of phosphorylation. Under these conditions, an enzyme-mevalonate pyrophosphate-ATP complex may accumulate, and K_{diss} for that complex may be relatively low. Some evidence in support of the retarding effect of adjacent fluorine in phosphorylation is provided by the results with (*IV*). Our data suggest that (*IV*) was phosphorylated more slowly than (*I*) or (*III*). If fluorine adjacent to the carbinol indeed retards the rate of phosphorylation of the carbinol, two effects could be responsible:

(1) The larger size of the fluorine compared to that of the hydrogen could prevent the proper alignment of the OH group. If that were the case, this reaction shows unusually high steric requirements.
(2) The reaction does not involve general-base catalysis i.e. the OH group does not lose a proton in the transition state. Under these conditions positive charge develops in the transition state and the electron-withdrawing effect of the fluorine would destabilize the transition state. However, until the reaction is further characterized, it is premature to speculate further concerning the basis of the low K_i for (*I*).

References

Alberts AW, Chen J, Kuron G, Hunt V, Huff J, Hoffman C, Rothrock J, Lopez M, Joshua H, Harris E, Patchett A, Monaghan R, Currie S, Stapley E, Albers-Schonberg G, Hensens O, Hirshfield J, Hoogsteen K, Liesch J, Springer J (1980) Mevilonin: A highly potent competitive inhibitor of HMG-CoA reductase and a cholesterol-lowering agent. Proc Natl Acad Sci USA 77:3957–3961

Endo A, Kuroda M, Tanzawa K (1976) Competitive inhibition of 3-hydroxy-3-methyl glutaryl coenzyme A reductase by ML 236A and ML 236B fungal metabolites having hypocholesterolemic activity. FEBS Lett 72:323–326

Jencks WP (1981) On the attribution and additivity of binding energies. Proc Natl Acad Sci USA 78:4046–4050

Muehlbacher M, Poulter CD (1985) Isopentenyl Diphosphate Dimethylallyl diphosphate isomerase. Irreversible inhibition of the enzyme by active-site directed covalent attachment. J Am Chem Soc 107:8307–8308

Nakamura CE, Abeles RH (1985) Mode of interaction of β-Hydroxy-β-methylglutaryl Coenzyme A reductase with strong binding inhibitors: Compactin and related compounds. Biochemistry 24:1364

Narula AS, Rahier A, Benveniste P, Schuber F (1981) 24-Methyl-25-azacycloartenol, an analogue of a carbonium ion high-energy intermediate, is a potent inhibitor of (S)-Adenosyl-L-methionine sterol C-24-methyl transferase in higher plant cells. J Am Chem Soc 103:2408–2410

Nave JF, d'Orchymont H, Ducep JB, Piriou F, Jung MJ (1985) Mechanism of the inhibition of cholesterol biosynthesis by 6-fluoro mevalonate. Biochem J 227:247–252

Popjak G (1969) Enzymes of sterol biosynthesis in liver and intermediates of sterol biosynthesis. Methods Enzymol 15:393–454

Poulter CD, Satterwhite DM (1977) Mechanism of the prenyl transfer reaction. Studies with (E)- and (Z)-3-trifluormethyl-2-buten-1-yl pyrophosphate. Biochemistry 16:5470–5478

Rahier A, Taton M, Schmitt P, Benveniste P, Place P, Anding C (1985) Inhibition of Δ^8-Δ^7-sterol isomerase and of cycloeucalenol-obtusifoliol isomerase by N-benzyl-8aza-4,10-dimethyl-trans-decal-3-ol, an analogue of a carbocation high energy intermediate. Phytochemistry 24:1223–1232

Reardon JE, Abeles RH (1986) Mechanism of action of isopentenyl pyrophosphate isomerase: Evidence for a carbonium ion intermediate. Biochemistry 25:5609–5616

Sandifer RM, Thompson MD, Gaughan RG, Poulter CD (1982) Squalene synthetase. Inhibition by an ammonium analogue of a carbocationic intermediate in the conversion of presqualene pyrophosphate to squalene. J Am Chem Soc 104:7376–7378

Chapter 12

Isopentenyl Diphosphate to Squalene — Enzymology and Inhibition

C. D. POULTER[1]

1 Introduction

The conversion of isopentenyl diphosphate to squalene requires five distinct steps mediated by three enzymes — isopentenyl diphosphate: isomerase, farnesyl diphosphate
synthetase, and squalene synthetase. These transformations constitute a fundamental
building phase in the sterol pathway, during which carbons for the basic steroidal skeleton are assembled from isopentenyl diphosphate. Isomerase mediates the conversion

Isomerase

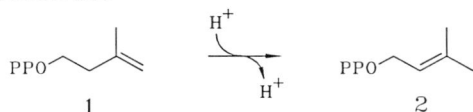

Farnesyl Diphosphate Synthetase

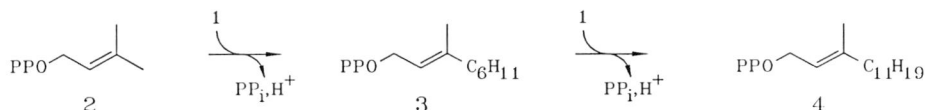

Squalene Synthetase

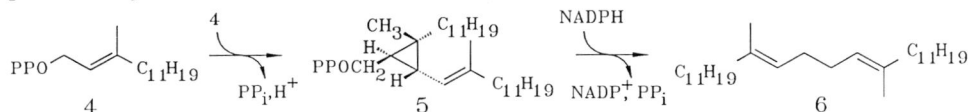

of isopentenyl diphosphate (1) to its allylic isomer dimethylallyl diphosphate (2), a
highly reactive, powerful alkylating agent (Poulter and Rilling 1981 a). Farnesyl
diphosphate synthetase catalyzes the first of two important isoprenoid polymerization reactions by joining 1 and 2 by a 1′−4-linkage[2] to give geranyl diphosphate (3),
and then using 3 as a substrate for a second condensation with 1 to generate farnesyl

---

[1] Department of Chemistry, University of Utah, Salt Lake City, Utah 84112, USA
[2] The numbering system is described in Poulter et al. 1977.

diphosphate (4) (Poulter and Rilling 1981 b). Squalene synthetase then links two molecules of 4 by a cl'−2−3-condensation to produce pre-squalene diphosphate (5) and subsequently catalyzes the rearrangement and reduction of 5 to generate squalene (6) (Poulter and Rilling 1981 c). In this manner, the hydrocarbon moieties of six molecules of isopentenyl diphosphate are incorporated into squalene and thereafter into steroid.

The majority of bond-forming reactions in the central portion of the isoprenoid pathway are catalyzed by prenyltransferases, a generic name given to the family of enzymes that mediate alkylation of electron-rich substrates by the hydrocarbon moieties of isoprenoid allylic diphosphates (Poulter and Rilling 1978). By combining different allylic substrates and acceptors, it is possible to generate a vast array of metabolites,

and it is not surprising that prenyltransferases are often the first pathway-specific enzymes in branches of the isoprenoid pathway. For example, in most organisms two prenyltransferases commonly compete for 2. The majority of the substrate is consumed by farnesyl diphosphate synthetase in chain elongation (Poulter and Rilling 1978). However, a proportion of the allylic substrate is diverted into tRNA by tRNA: dimethylallyl diphosphate dimethylallyl transferase, an enzyme that alkylates the amino group of adenine-37 in tRNAs which read codons beginning in uracil with the dimethylallyl moiety (Dihanich et al. 1987). The modification is thought to promote fidelity during protein biosynthesis.

Several prenyltransferases can compete for 4. One group is involved in a continuation of 1'−4 chain elongation and provides geranylgeranyl diphosphate for carotenoid biosynthesis, all-trans polyprenyl diphosphates for unbiquinone biosynthesis, and dolichols needed during sugar transport and glycoprotein biosynthesis (Poulter and Rilling 1981 b). Some hemes contain a farnesyl side chain, presumably attached to the vinyl group in unmodified heme by a prenyltransferase. However, in many of the organisms that require sterols, the major diversion of 4 is into that pathway. It is interesting to note that the first step catalyzed by squalene synthetase is a prenyltransfer reaction. In this instance, 4 serves as both the prenyl donor and acceptor. The hydrocarbon moiety of one molecule is inserted into the C2−C3 double bond of the second with concomitant loss of a proton to generate the cyclopropane ring in 5. This is an essential step in the biosynthesis of 1'−1-fused triterpenes derived from 6. An identical cl'−2−3-condensation between two molecules of geranylgeranyl diphosphate is the first pathway-specific step in carotenoid biosynthesis (Gregonis and Rilling 1974). The cyclopropanation is also thought to be a required step (Poulter and Epstein 1973) in the biosynthesis of 1'−3-fused isoprenoids, a less common group represented by the botryococcenes (Huang et al. 1988) and the artemesia family of monoterpenes, and the 2−1'−3-fused santolinyl monoterpenoids.

2 Isopentenyl Diphosphate: Dimethylallyl Diphosphate Isomerase

2.1 General Considerations

Isopentenyl diphosphate: dimethylallyl diphosphate isomerase (EC 5.3.3.2) catalyzes the interconversion of isopentenyl diphosphate and dimethylallyl diphosphate. Although the isomerization is formally a [1, 3] sigmatropic rearrangement of a hydrogen atom, it is obvious from labelling experiments and stereochemical studies that the raction is more complex. During the course of the interconversion, the pro-R hydrogen at C2 is removed (Cornforth 1968) and a hydrogen is added to C4 from the re-face

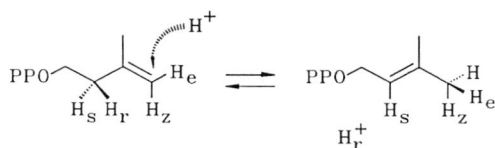

of the double bond (Cornforth et al. 1972). The newly formed methyl group is exclusively in the E position, and the rearrangement is, therefore, antarafacial (Poulter and Rilling 1981 a). The hydrogens readily exchange with protons from water, and upon incubation in D_2O, extensive incorporation of deuterium is seen in the E-methyl group and the pro-R locus at C2. Lack of incorporation of deuterium into the Z-methyl of 1 or the pro-S locus at C2 in 2, even upon prolonged incubation, demonstrates that the isomerization is stereospecific (Satterwhite 1985).

Dimethylallyl diphosphate is slightly favored in equilibrium ($K_{eq} = 3$) (Muehlbacher and Poulter 1988), although it is unlikely that equilibrium is normally achieved in vivo. The chain elongation reactions typically consume several molecules of 1 for each 2. In crude extracts from sources such as yeast and vertebrate liver, where there is a high commitment to sterol metabolism and two isopentenyl diphosphates are consumed per dimethylallyl diphosphate, activities for farnesyl diphosphate synthetase are typically slightly higher than isomerase. However, in E. coli, where most of the pathway is directed toward biosynthesis of ubiquinones and bacterial dolichols, the activity of isomerase is considerably below that of the chain elongation enzymes (M. M. Sherman, C. D. Poulter unpublished results). Thus, it is possible that the relative activities of isomerase and prenyltransferases are adjusted to provide sufficient amounts of 2 needed to prime chain elongation while maintaining a large pool of 1 for the subsequent polymerization steps.

2.2 Enzymology

2.2.1 Purification

Extensive purifications of isomerase from several different sources have been reported. Although an earlier report of a successful purification of the enzyme from porcine liver (Banthorpe et al. 1977) was not reproducible in other laboratories, recent proto-

Table 1. Purification of isomerase from *Claviceps purpurea* 26245[a]

Purification step	Protein (mg)	S.A. (μmol min^{-1}mg^{-1})	Purification (fold)	Yield (%)
pH 5.5 supernatant	7500	0.019	–	–
45% – 75% (NH$_4$)$_2$SO$_4$	2000	0.03	3	81
n-Butylsepharose	290	0.16	16	61
Fractogel TSK 650S	28	0.57	57	21
Protein Pak DEAE 5 PW	0.8	3.6	330	4

[a] From 50 g of freeze-dried mycelia.

cols for isomerase from *Claviceps sp.* (Bruenger et al. 1986) and baker's yeast (Reardon and Abeles 1986) gave enzyme preparations estimated to be 90% pure.

We also developed a procedure for the Claviceps enzyme (Muehlbacher and Poulter 1988). Total isomerase activity, as measured per gram of dry cell weight, in *Claviceps purpurea* (ATCC strain 26245) was determined as a function of growth time. It was found to increase sharply between days 4 and 6 and remained relatively constant thereafter at approximately 250 μmol min^{-1} g^{-1} for up to 16 days. However, the specific activity based on total protein in the cell free extracts decreased continually past day 6 as other proteins were expressed. The mycelia were typically harvested before day 7 and disrupted in a ball mill as described by Bruenger et al. (1986). The purification summarized in Table 1 gave enzyme judged to be >90% pure from the intensity of bands on SDS gels stained with Coomassie blue. Solutions of the purified enzyme (0.1–1.0 mg/mL) in buffer containing 20% glycerol could be stored at −20 °C for more than 6 months without loss of activity.

Isomerase from *Claviceps* (mol wt 35000) (Bruenger et al. 1986; Muehlbacher and Poulter 1988) and yeast (mol wt 39000) (Reardon and Abeles 1986) are monomers of similar size, and both are substantially larger than the porcine liver enzyme (mol wt 22000) (Bruenger et al. 1986). In general, attempts to purify the enzyme from vertebrate sources were frustrated because of problems associated with instability. The specific activity of isomerase from *Claviceps purpurea* was 3.6 μmol min^{-1} mg^{-1}, a value similar to those reported for the enzyme from *Claviceps sp.* (3.0 μmol min^{-1} mg^{-1}) (Bruenger et al. 1986) and yeast (2.7 μmol min^{-1} mg^{-1}) (Reardon and Abeles 1986). The only requirement for activity is a divalent cation (Mg^{2+}).

2.2.2 Steady State Kinetics

The reversible interconversion of isopentenyl diphosphate and dimethylallyl diphosphate is a simple UniUni process. The Michaelis constant for 1 was determined in the standard manner from double reciprocal plots of initial velocities and substrate concentrations. The value of $K_M = 2.2$ μM for the *C. purpurea* enzyme was similar to $K_M = 5.4$ μM reported for isomerase from *Claviceps sp.* (Bruenger et al. 1986) and substantially below $K_M = 35$ μM found for the yeast enzyme (Reardon and Abeles 1986). The Michaelis constant for 2 cannot be determined in a straightforward manner because the acid lability assay is specific for the conversion of 1 to 2. A value can, however, be calculated from product inhibition studies of initial velocities in the forward direction according to Eq (1) (Segal 1975).

$$V = \frac{V_{max}[1]}{K_M^1[1+([2]/K_M^2)]+[1]} \tag{1}$$

Double reciprocal plots of velocity versus the concentration of 1 at different concentrations of 2 gave a family of straight lines with a common intercept on the v^{-1} axis ($V_{max} = 3.6 \, \mu mol \, min^{-1} \, mg^{-1}$). Replots of $K_{M_{App}}$ for 1 versus [2] also gave a straight line with a y-intercept at $K_M^1 = 1.1 \, \mu M$ and an x-intercept at $K_M^2 = 0.9 \, \mu M$. The maximal velocity in the reverse direction ($V_{max}^{reverse} \simeq 2.5 \, \mu mol \, min^{-1} \, mg^{-1}$) was calculated from the Haldane relationship, where $K_{eq} = 3$.

2.3 Mechanistic Studies

2.3.1 General

Studies on isomerase-catalyzed exchange of deuterium and tritium from water into 1 and 2 are consistent with either carbocationic or carbanionic mechanisms. Recently support for carbocations was provided by structure/reactivity correlations. Reardon and Abeles (1986) (see also Chapter 11) discovered that (Z)-3-trifluoromethyl-2-butenyl diphosphate was not a substrate for isomerase and estimated that the fluorinated analog was at least 10^6 times less reactive than 2. Electron withdrawal by the trifluoromethyl moiety is predicted to drastically decrease the stability of a carbocationic species, hence decreasing the rate of isomerization, and to exert a moderate stabilizing effect on a carbanion. A related study with farnesyl diphosphate synthetase, where an excellent linear free energy correlation was observed between the rates of enzymes catalyzed 1'–4-condensations and model solvolytic reactions upon progressive replacement of hydrogens by fluorine, is discussed in section 3.3.2.

Reardon and Abeles (1986) and Muehlbacher and Poulter (1985) observed that ammonium analogs of carbocations proposed during rearrangement are potent inhibitors of isomerase. The absence of side reactions between substrates and water and the stereoselectivity found for the isomerization indicate that solvent is excluded from the active site during catalysis and that the reversible protonation-deprotonation steps

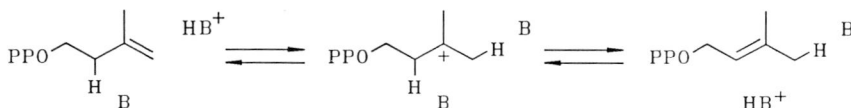

are orchestrated by active site bases and their conjugated acids. Furthermore, the antarafacial stereochemistry of the reaction implicates two bases which function in tandem as a free base-conjugate acid pair to promote isomerization (Cornforth et al. 1972).

2.3.2 Irreversible Inhibitors

2.3.2.1 2-(Dimethylamino)ethyl Diphosphate

2-(Dimethylamino)ethyl diphosphate (7) is a potent inhibitor of isomerase at pH 7. The ammonium analog inactivates the *Claviceps* enzyme in a time-dependent manner ($k_{on} \approx 900\,M^{-1}\,s^{-1}$). Under normal conditions in standard assay buffers, release of

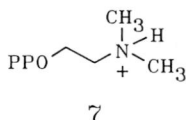

7

radiolabeled ammonium analog is barely detectable ($k_{off} < 1.1 \times 10^{-7}\,s^{-1}$) and $K_D < 10^{-10}\,M$ for the *Claviceps* enzyme. An even lower value of K_D was estimated for 7 with yeast isomerase by Reardon and Abeles (1986) (see also Chapter 11).

The isomerase-7 complex is highly stable. Treatment with 6 M urea at 37 °C for 1 hour fails to release significant amounts of the inhibitor ($\sim 0.3\%$) and under more stringent denaturing conditions (0.015 M SDS, 37 °C, 1 h) only 35 – 50% of 7 is released. When radiolabel is positioned in the diphosphate (^{32}P), ethylene (^{3}H), and methyls (^{14}C), some degradation of the inhibitor was noticed upon denaturation of the enzyme-inhibitor complex. However, the bulk of the material was released intact; thus, 7 was not attached covalently. This conclusion was dramatically reinforced when the enzyme-7 complex was treated with a basic buffer. In the presence of 6 M urea at 37 °C, pH 12, for 30 min, 83% of the analog was released, and when the pH was raised above 13, release was complete within 15 min, even in the absence of urea. Under these conditions the polypeptide chain was not hydrolyzed. Thus, release is most likely caused by deprotonation of 7 within the active site of the enzyme followed by expulsion of the free amine. The marked difference in the affinities of 7 and the free amine for isomerase strongly supports a carbocationic mechanism.

2.3.2.2 Allylic Fluorides

A second type of irreversible inhibition is seen with allylic fluorine-containing analogs of 1 and 2. In an attempt to conduct linear free energy studies similar to those that proved so successful for farnesyl diphosphate synthetase, we prepared diphosphates 8 and 9. Both inhibited isomerase irreversibly with pseudo first order kinetics which

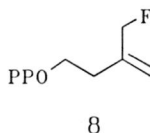

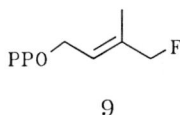

8   **9**

showed saturation at high concentrations of inhibitor. Kinetic constants are listed in Table 2. The inactivation process was retarded in the presence of substrate, and gave a 1 : 1 enzyme-inhibitor complex stable to prolonged dialysis, to treatment with urea or SDS, and during electrophoresis on SDS polyacrylamide gels. Colorimetric analy-

Table 2. Kinetic constants for irreversible inhibition of isomerase

Inhibitor (I)	k_2 (min^{-1})	K_I (μm)
8	0.22 ± 0.07	0.085 ± 0.04
9	0.97 ± 0.52	0.97 ± 0.55
10	0.13 ± 0.01	0.011 ± 0.004
11	0.24 ± 0.06	8.74 ± 1.83

sis of the buffer using eriochrome cyanine R (Megregian 1978) indicated release of a stoichiometric amount of fluoride during inactivation. It was discoverd that some enzymatic activity was restored upon treatment with DTT. This was accomplished by removing excess 8 or 9 by microconcentration. The inactive complex was treated with 100 mM DTT for 20 min at 37 °C along with an inhibited control without DTT. After treatment, the control had no detectable isomerase activity, while samples incubated with DTT regained 10% of their original activity. These experiments demonstrate that the covalent enzyme-inhibitor linkage is reactive and can be ruptured by a potent nucleophile.

The mechanism of inhibition was established by an NMR experiment. A sample of 8 containing ^{13}C ($>98\%$) at C4 was synthesized. ^{13}C and ^{19}F spectra of the inhibitor and isomerase were recorded independently (Fig. 1). The solutions were chilled to 5 °C and combined. Spectra were recorded at 5 °C, and the sample was then warmed to 37 °C. Upon warming the sharp ^{13}C resonance at 117.2 ppm for the inhibitor decreased in intensity, while a new broad peak typical of a carbon attached to the enzyme formed at 119.6 ppm. Concomitantly, the ^{19}F resonance at -213.5 ppm decreased in intensity, and a new peak due to fluoride was seen at -116 ppm. After 12 h, excess inhibitor was removed by microconcentration, and spectra were recorded for retained and passed solutions. The retained solution contained isomerase with the ^{13}C NMR signal at 119.6 ppm and no resonances in the ^{19}F region. The passed solution showed resonances for unreacted inhibitor (^{13}C at 117.2 ppm and ^{19}F at -213.5 ppm) and the fluoride peak at -116 ppm. These experiments demonstrate the formation of a covalent bond between 8 and a residue in isomerase by displacement of fluoride. In addition, the chemical shift of the ^{13}C resonance in isomerase treated with 8 demonstrates that the labeled carbon is olefinic. Thus, the reaction is an Sn2 displacement rather than an Sn2' process.

2.3.2.3 Epoxides

Based on our observations with allylic fluorides 8 and 9 and the requirement for protonation of the double bonds in 1 and 2, it was reasonable to assume that epoxide derivatives might be potent suicide inhibitors. As shown in Table 2, this hypothesis

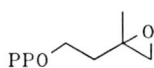

10

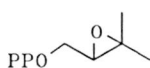

11

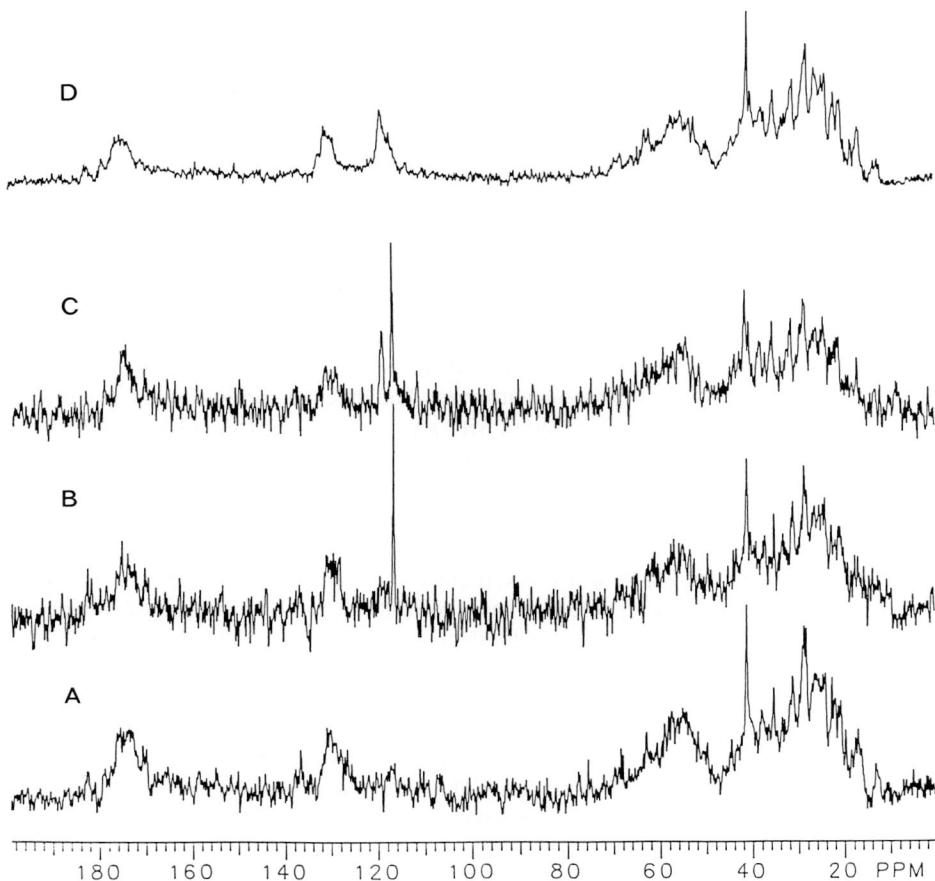

Fig. 1A–D. ^{13}C NMR spectra of isomerase and isomerase treated with fluoromethylisopentenyl diphosphate (8). **A** 7.5 mg isomerase, 4 °C. **B** 7.5 mg isomerase, 1 mM 8, 4 °C. **C** 7.5 mg isomerase, 1 mM 8, 37 °C for 3-$\frac{1}{2}$ hours. **D** 7.5 mg isomerase after filtration

proved to be correct. Epoxides 10 and 11 were potent irreversible inhibitors. As discussed for 8 and 9, inactivation was inhibited by substrate, the stoichiometry of inhibition was 1:1, and the complex did not dissociate upon being subjected to stringent treatment with SDS.

A plausible mechanism for the inhibition involves activation of the expoxide analogs by protonation of the oxirane oxygen. Certainly, residues in the catalytic site capable of protonating the double bonds in 1 and 2 are sufficiently acidic to protonate the epoxides. Once activated, 10 and 11 are susceptible to direct displacement by active-site nucleophiles to covalently link the inhibitors to isomerase.

3 Farnesyl Diphosphate Synthetase

3.1 General Considerations

Farnesyl diphosphate synthetase (EC 2.5.1.1) catalyzes the sequential addition of isopentenyl diphosphate (1) to dimethylallyl diphosphate (2) and geranyl diphosphate (3) to yield farnesyl diphosphate (4). Both steps in the chain elongation process are irreversible, and both reactions follow the same stereochemical course. The hydrocarbon moiety of the allylic substrate is added to the si-face of the double bond in 1, and the pro-R proton is removed from C2 with concomitant movement of the double bond to the C2−C3 location to generate the next higher homolog in the 1′−4-linked series (Cornforth 1968). Thus, the net stereochemistry for 1′−4-condensation catalyzed by farnesyl diphosphate synthetase is a syn addition-elimination.

Farnesyl diphosphate synthetase is but one of several commonly occurring prenyltransferase that catalyze 1′−4-condensation. Typically it coexists with enzymes that produce higher polyprenyl diphosphates needed for the biosynthesis of carotenoids, ubiquinones, and dolichols. All of these enzymes exhibit high levels of specificity for the stereochemistry of the double bonds and the number of isoprene units in the allylic diphosphates accepted as substrates (Poulter and Rilling 1981 b). They also generate only one double bond isomer regardless of how many individual isopentenyl units are added. Although farnesyl diphosphate synthetase releases and rebinds 3 before the second 1′−4-condensation, the binding mechanism of higher prenyltransferases is not clear.

3.2 Enzymology

3.2.1 Purification

Extensive purifications of farnesyl diphosphate synthetase have been reported from fungal and liver sources. Rilling and coworkers developed procedures for farnesyl diphosphate synthetase from baker's yeast (Eberhardt and Rilling 1975), *Phycomyces blakesleeanus* (H. C. Rilling unpublished results), avian liver (Yeh and Rilling 1977), and porcine liver (Reed and Rilling 1975). The porcine and human liver enzymes were purified in Popjak's laboratory (Barnard et al. 1978; Barnard 1985). Bartlett et al. (1985) developed an affinity column using a phosphonate-phosphate analog of geranyl diphosphate as the affinity ligand and reported a 100-fold purification from crude ammonium sulfate fractions, although the amount of enzyme that could be processed was limited.

More recently we modified the original Rilling procedures to obtain farnesyl diphosphate synthetase (>95%) from avian liver. The initial stages of the purification were identical to the early Rilling protocol. However, the steps after hydroxylapatite were replaced with a chromatofocusing column. This is a high resolution step which, when coupled with a gel filtration to remove the ampholites used to generate the pH gradient, provides homogeneous protein. The protocol is summarized in Table 3.

Table 3. Purification of farnesyl diphosphate synthetase from avian liver[a]

Purification step	Protein (mg)	S.A. (μmol min^{-1}mg^{-1})	Purification (fold)	Yield (%)
45% – 75% $(NH_4)_2SO_4$	21 000	0.01	–	–
DE52	430	0.15	16	33
Hydroxylapatide	20	2.5	250	24
Electrofocusing	8	3.8	538	21
Sephacryl	5	4.5	550	13

[a] From 2 kg of frozen liver.

A similar procedure was developed for the yeast enzyme. Commonly two peaks of activity are seen during chromatography, a poorly resolved pattern on DE-52 and two well-resolved peaks on hydroxylapatite. Inclusion of phenylmethylsulfonyl fluoride (PMSF) in the original extraction buffer alters the ratio of the peaks, thus, presumably the yeast enzyme is susceptible to proteolytic degradation. The major band in PMSF-treated solutions elutes from hydroxylapatite first and can be purified by chromatofocussing followed by gel filtration.

Once purified, farnesyl diphosphate synthetase is stable. The avian liver and yeast enzymes can be stored at $-20\,°C$ in buffer containing 20% glycerol for over one year with minimal loss of activity.

The enzymes from avian liver (Reed and Rilling 1975) and yeast (Eberhardt and Rilling 1975) have similar properties. Both are α_2 dimers with mol wts of 43 000 and 42 000, respectively. Specific activities are approximately 4 μmol min^{-1} mg^{-1}. The monomagnesium salts of 1 and 2 or 3 are the preferred substrates, and catalysis does not occur in the absence of divalent metal ions. Dissociation constants (K_D) for 1 (1.4 μM), 2 (1.8 μM), and 3 (0.17 μM) are similar (Reed and Rilling 1976), and only increase approximately 5-fold in metal-free buffers.

3.2.2 Steady State Kinetics

A steady state kinetic analysis of avian liver farnesyl diphosphate synthetase was conducted by Laskovics et al. (1979). The Michaelis constants, $K_M^1 = 0.1\ \mu$M and $K_M^3 = 0.2\ \mu$M, were similar to previously reported values. In addition, they found non-linear Linewaver-Burke plots when 1 was the varied substrate characteristic of substrate inhibition. This observation agreed with a report by Reed and Rilling (1976) that the enzyme binds two molecules of 1 per catalytic site in a non-cooperative manner, but only one molecule of 2 or 3.

Laskovics also found that 4 was a competitive inhibitor of 3; whereas, PP_i gave non-competitive patterns when 1 or 3 were varied. These results point to the tendency of 4 to bind with its diphosphate moiety in the allylic site, a hypothesis strengthened by the discovery that 4 is also a substrate for the slow addition of a third molecule of 1 to give geranylgeranyl diphosphate. Farnesyl diphosphate is also a substrate for abnormal hydrolysis and cyclization reactions catalyzed by the enzyme. The product inhibition studies suggest that PP_i binds to both the homoallylic and allylic regions of the catalytic site. From inspection of steady-state equations which account for the complex binding properties of the enzyme, it was concluded that one could not distinguish between ordered and random mechanisms for addition of substrates.

3.2.3 Non-Steady State Kinetics

The question of binding order was addressed by "hot-trap" experiments (Laskovics and Poulter 1981) in which avian liver farnesyl disphosphate synthetase was pulsed with radioactive substrate, followed by a chase solution containing the second substrate along with a large excess of cold first substrate. Since bound substrate trapped by the chase and converted to product before dissociation will generate 4 with high specific activity, one can determine how much initially bound isopentenyl or geranyl diphosphate reacts prior to dissociation. One half of bound 3 was trapped by 1.4 μM 1; whereas, no trapping of 1 was detected with concentrations of 3 as high as 2.1 mM. These results clearly establish an ordered mechanism in which the allylic diphosphate binds first.

Kinetic studies also revealed another important feature of the 1′−4-condensation. When the reaction was monitored at short reaction times using a multimixing machine, a distinct burst phase was seen. The rate of the burst was about 50 times that of the steady-state rate. Since the acid lability assay only measures formation of 4 from 1 and 2, the burst represents a rapid build-up of an enzyme-4 complex, followed by a rate limiting release of 4. From known values of equilibrium and steady-state kinetic constants, it was possible to calculate individual rate constants for binding, catalysis, and release. It is interesting to note that release of products is rate limiting at

$$
E \; \underset{1.4 \text{ s}^{-1}}{\overset{(2\times10^{6} \text{ M}^{-1} \text{ s}^{-1})\,[\text{GPP}]}{\rightleftarrows}} \; [\text{E}-\text{GPP}] \; \underset{5 \text{ s}^{-1}}{\overset{(5\times10^{-6} \text{ M}^{-1} \text{ s}^{-1})\,[\text{IPP}]}{\rightleftarrows}} \; [\text{E}-\text{GPP}-\text{IPP}]
$$

$$
\downarrow 4.7 \text{ s}^{-1}
$$

$$
E+\text{PP}_i+\text{FPP} \; \xleftarrow{\;0.1 \text{ s}^{-1}\;} \; [\text{E}-\text{PP}_i-\text{FPP}]
$$

steady state and that both substrate additions are considerably slower than anticipated for a diffusion controlled process. The most straightforward explanation is that conformational changes accompany binding of substrates and release of products.

3.3 Mechanistic Studies

3.3.1 Farnesyl Diphosphate Synthetase Mediated Solvolysis

During binding experiments, Poulter and Rilling (1978) discovered an autocatalytic hydrolysis of the allylic substrates. It was subsequently found that PP_i was a stimulatory factor. At optimal concentrations of PP_i, V_{max} was 2% that of the normal reaction. The major product was geraniol along with a small amount of hydrocarbon. The hydrolysis reaction could be suppressed by addition of 2-fluoroisopentenyl diphosphate, a potent reversible inhibitor against 1. These experiments suggest that in the absence of the normal homoallylic substrate, PP_i and a molecule of water serve as an alternate in an enzyme-catalyzed solvolysis reaction where the water becomes the prenyl acceptor. This hypothesis was supported by two experiments. When the reaction was conducted in ^{18}O-labelled water, the oxygen in geraniol was found to come exclusively from solvent. In addition, stereochemical studies showed that C1 of 3 was inverted upon hydrolysis in the same manner found for the normal 1'−4-condensation.

3.3.2 Linear Free Energy Studies

A mechanism for 1'−4-condensation, also consistent with enzyme-mediated hydrolysis of allylic substrates, is an electrophilic alkylation of the double bond in isopentenyl diphosphate by the carbocation generated from the allylic substrate. Poulter et al. (1981) obtained evidence for a stepwise electrophilic alkylation in linear-free energy studies which correlated a solvolytic reaction known to produce allylic carbocations

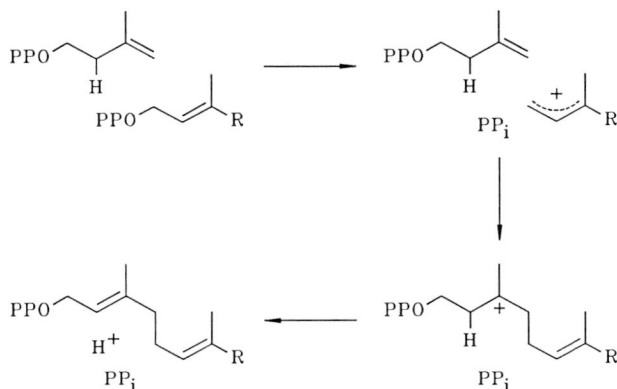

with the enzyme catalyzed 1'−4-condensation with 1 for a series of geranyl derivatives. First order rates for the solvolysis of allylic methanesulfonates derived from 3'-desmethylgeraniol (12), 2-fluorogeraniol (13), 3'-(fluoromethyl)geraniol (14), 3'-(difluoromethyl)geraniol (15), and 3'-(trifluoromethyl)geraniol (16) were measured in aqueous acetone to ascertain the effect of substitutions, which destabilize the allylic intermediates, on the rate of their formation in a simple model system (Table 4). The maximal rates of 1'−4-condensation between 1 and the dipshophate derivatives of alcohols 12−16 were then measured for avian liver enzyme. A Hammett plot of the data shown in Fig. 2 is particularly informative. The rate ratios for 1'−4-condensation and solvolysis correlate linearly for the entire range of reactivities. Also, the slope of the plot (0.8) demonstrates that 1'−4-condensation was slightly more sensitive to substitution than the model solvolysis reaction in water. These results strongly support an electrophilic alkylation mechanism for 1'−4-condensation in which heteroallylic cleavage of the CO linkage in the allylic substrate precedes formation of the 1'−4-bond.

Table 4. Kinetic parameters for model and farnesyl diphosphate synthetase catalyzed reactions

Allylic substrate	Model		Enzyme catalyzed	
	k_I^{rel}	$k_{c\text{-}at}^{rel}$	K_M^{IPP} (μM)	K_I (μM)
3	1	1	0.1	–
12	1.9×10^{-3}	2.3×10^{-4}	0.3	0.68
13	4.4×10^{-3}	1.7×10^{-5}	0.8	0.44
14	7.7×10^{-4}	3.7×10^{-4}	0.5	0.63
15	2.2×10^{-6}	3.7×10^{-8}	0.5	1.8
16	4.0×10^{-7}	7.2×10^{-9}	0.3	3.6

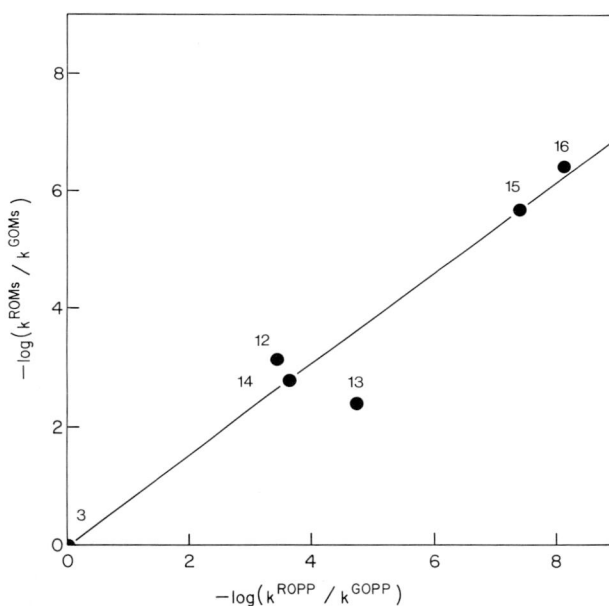

Fig. 2. Hammett plot of relative rates of solvolysis of methanesulfonates (ROMs) in aqueous acetone versus relative rates for catalysis (k_{cat}) for condensation of allylic diphosphates 3 and 12–16 (ROPP) with isopentenyl diphosphate

4 Squalene Synthetase

4.1 General Considerations

Squalene synthetase (EC 2.5.1.21) catalyzes the first two pathway specific reactions in sterol biosynthesis. The first is the condensation of two molecules of farnesyl diphosphate (4) to form presqualene diphosphate (5). The reaction involves loss of PP$_i$ and the pro-S proton from C1 of 4 to give $(1\,R, 2\,R, 3\,R)$–5. In a second reaction, which requires NADH or NADPH, the second diphosphate is expelled, both newly formed cyclopropane bonds are ruptured, and the farnesyl residues are rejoined to form a 1'−1-linkage. The stereochemistry of the 1'−1-condensation was elucidated by Cornforth and Popjak (Cornforth 1968) in their pioneering work on prochiral centers. Inversion is seen at the farnesyl carbon which retains both hydrogens, and retention occurs at the carbon which requires a hydrogen from NADPH, as the result of single inversions during the cyclopropane-forming step and during rearrangement. When NADPH is withheld from a cell-free extract from yeast, 5 accumulates and is eventually hydrolyzed by phosphatases to the corresponding alcohol.

The mechanism for formation of 1'−1-bonds in carotenoids is similar to squalene. During carotenoid biosynthesis, a C_{40} cyclopropylcarbinyl diphosphate, prephytoene diphosphate, is formed, and this compound undergoes a rearrangement identical to that of squalene (Gregonis and Rilling 1974). However, at the final step a proton is eliminated to produce the 1'−1-double bond in phytoene rather than the reduction by NADPH which leads to the 1'−1-methylenes in squalene.

4.2 Enzymology

4.2.1 Purification

Squalene synthetase is an intrinsic microsomal protein which is difficult to handle, and the only significant purifications are from baker's yeast. Early attempts to solubilize the enzyme gave aggregates with mol wts between 100 000 and 300 000. The first major breakthrough was reported by Popjak and Agnew (1979). These investigators used deoxycholate to solubilize the enzyme, and then removed the detergent with cholestyramine. They reported squalene synthetase was a monomer (mol wt 55 000). Unfortunately, the detergent significantly inhibited the enzyme, and removal of the detergent resulted in precipitation of the protein. Recently Rilling and coworkers (Kuswik-Rabiega and Rilling 1987; Sasiak and Rilling 1988) made major progress in solubilizing the yeast enzyme in a catalytically active form and determined a mol wt of 47 000 by SDS gel electrophoresis. The Rilling procedure consists of solubilization of squalene sythetase from yeast microsomes using a combination of octylglucose and Lubrol PX. The solubilized material was chromatographed on DEAE and hydroxylapatite, followed by isoelectric focusing. The final steps were performed at −6 °C with buffer containing 10% methanol and 10% glucose to preserve activity.

Using 4 labeled with tritium at C-1, it is possible to measure the rates of both steps in the conversion of 4 to 6. During cyclopropanation a proton is lost to water from one of the farnesyl residues. The second step produces 6 which is readily separated from 4 or 5 by a simple extraction followed by rapid chromatography. Thus, the rate of synthesis of 5 can be determined from the accumulation of radioactivity in the buffer, and the rate of squalene synthesis, from the accumulation of radioactivity in hexane-soluble material. The specific activities of enzyme were $0.8 \, \mu\text{mol min}^{-1} \, \text{mg}^{-1}$ and $0.15 \, \mu\text{mol min}^{-1} \, \text{mg}^{-1}$ in the presqualene and squalene assays, respectively.

4.2.2 Kinetic Studies

Kinetic studies with squalene synthetase are not as advanced as for the isomerase or farnesyl diphosphate synthetase. $S_{0.5}$ is $10-12 \, \mu M$ for 4 and K_M for NADPH is $4 \, \mu M$ (Sasiak and Rilling 1988). In contrast to previous reports with microsomal preparations, it was also observed by Sasiak and Rilling (1988) that NADH is utilized equally well by solubilized enzyme. None of these studies, however, adequately accounts for the local concentration of the increasingly hydrophobic substrates in detergent systems. As for the isomerase and farnesyl diphosphate synthetase, a divalent metal (Mg^{2+}) is necessary for catalysis. At high concentrations, 4 inhibits synthesis of 6 but not production of 5.

Sasiak and Rilling (1988) demonstrated that a single protein catalyzes both presqualene diphosphate and squalene formation. Through five different purification steps, including examination of individual chromatographic fractions with differing specific activities within a single step, the relative rates of presqualene to squalene synthesis remained constant at 5.4 ± 0.3. The ratio of presqualene to squalene activity indicates that the intermediate is free to leave the catalytic site and that release is preferable to reduction to squalene. In contrast, trace amounts of 5 are found in vivo.

4.3 Mechanistic Studies

4.3.1 General

Among several plausible mechanisms proposed for the cyclopropanation step are electrophilic alkylation, carbenoid insertion, and direct displacement (Poulter and Rilling 1981 c). Unfortunately, there are no experiments that permit one to conclusively distinguish among these suggestions. There is, however, only one mechanism that ade-

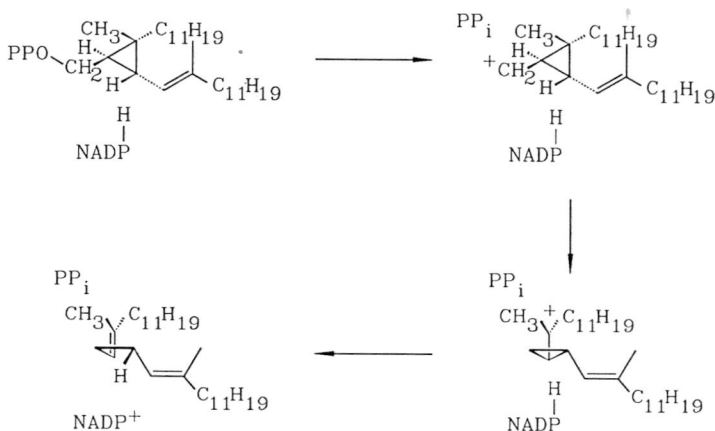

quately rationalizes the rearrangement of 5 to 6, and as observed for the two previous enzymes in the pathway, carbocations are implicated. In this instance the mechanism involves cyclopropylcarbinyl cations, a stable class of intermediates well-known for their ability to rearrange.

4.3.2 Model Studies

Although cyclopropylcarbinyl rearrangements can be used to rationalize the conversion of 5 to 6, the regiochemistry of the cyclopropylcarbinyl system is extremely sensitive to substitution. Because of the unusual pattern of alkyl and vinyl substituents in 5, Poulter et al. (1977) decided to study the behavior of chrysanthemyl derivatives in a simple C_{10} model of 5. They found that the primary cation 17 (R = CH_3) initially produced upon ionization can rearrange to isomeric cations 18–20 and that only a

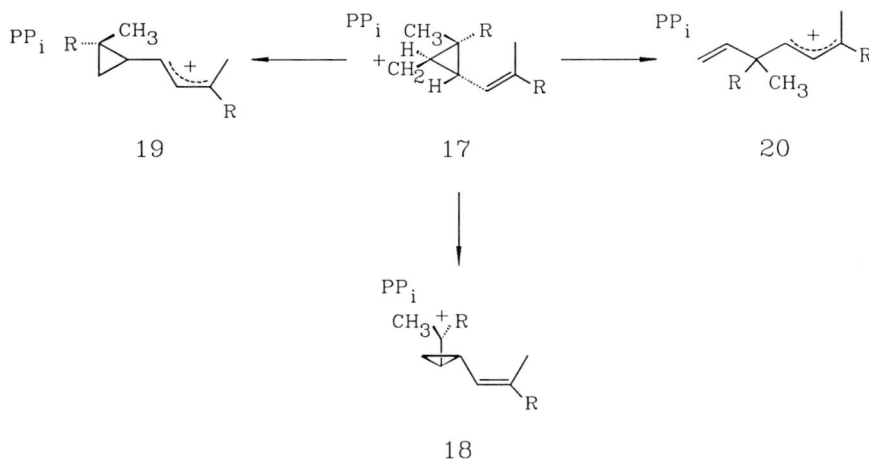

very small percentage (0.04%) of 18, the 1′−1 precursor, was detected. Poulter et al. (1974) suggested that the regiochemistry of the enzyme-catalyzed rearrangement was controlled by the geometry of the intimate ion pair generated upon cleavage of the cyclopropane ring. The relative orientation of the diphosphate moiety and the cyclopropane ring of 5 in the enzyme substrate complex shown can be deduced from observations that the carbinyl position in 5 is inverted in 6 and the bond migrations in cyclopropylcarbinyl rearrangements are suprafacial. Thus, the initial intimate ion pair has the PP$_i$ fragment adjacent to the quaternary cyclopropane carbon, and rearrangement to the tertiary cation does not require separation of charge. In contrast, both of the two competing rearrangements require large charge separation. It is estimated that selective stabilization of the primary to tertiary rearrangement of up to 20 kcal/mol may be possible in the intimate ion pair.

4.3.3 Reactive Intermediate Analogs

Sandifer et al. (1982) reported the synthesis of the ammonium analog 21 designed to mimic the electrostatic and topological properties of the tertiary cyclopropylcarbinyl cation. The inhibitor was observed to have no effect on the production of 6 from 4 using a microsomal preparation of yeast squalene synthetase at concentrations of up to 170 μM. In a related experiment, the effect of PP_i was studied at concentrations between 0 and 2 mM. As PP_i levels were increased to 1 mM, the rate of squalene syn-

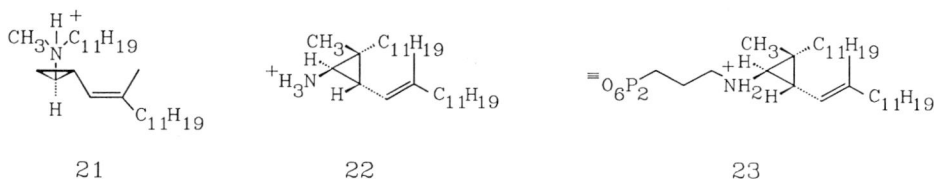

21   22   23

thesis almost doubled (Fig. 3). The origin of this stimulatory effect is unknown, and the activities of other microsomal enzymes were not affected in a parallel manner. Concentrations of PP_i above 1 mM were inhibitory, as expected for product inhibition. The combination of 21 and PP_i, however, resulted in a potent synergistic inhibition of the enzyme. At 1 mM PP_i, a concentration which causes a two-fold stimulation of squalene synthesis, 21 inhibited the reaction by 50% at 3 μM and by >90% at 20 μM.

Recently we synthesized 22, an ammonium analog of the primary cation, and inhibition studies gave results almost identical to those found for 21. We also investigated the effect of inhibitors on the rate of synthesis of 5 from 4 and 6 from 5. In each instance the results were similar to those found for the conversion of 4 to 6. We studied the effect of the inhibitors on synthesis of 5 and 6 from 4 in the same reaction by combining the proton release (synthesis of 5) and direct squalene synthesis assays. It was found that the inhibitor had parallel effects on both steps. These results strongly suggest that both reactions are catalyzed at the same or substantially overlapping catalytic sites and conflicts with the conclusions of Popjak and Agnew (1979) based on kinetic studies.

The ion pair hypothesis received additional support from inhibition studies with 23, an ammonium analog of the primary cyclopropylcarbinyl cation in which the inorganic

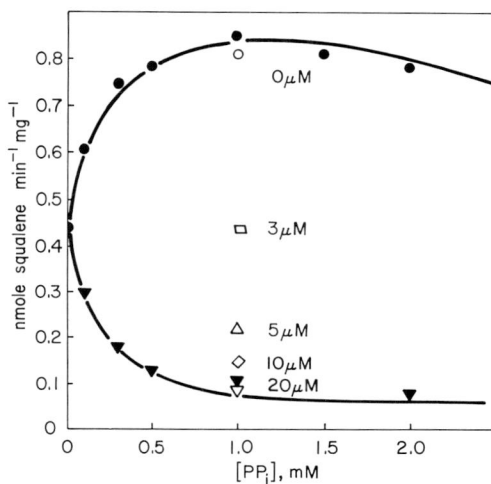

Fig. 3. Rate of squalene synthesis from farnesyl diphosphate in the presence of tertiary ammonium analog 21 (9−20 μM) and or inorganic pyrophosphate (0−2 mM)

diphosphate fragment is covalently tethered. Phosphonatephosphate 23 was a potent inhibitor of squalene synthesis ($S_{0.5}$ = 3 µM) which did not depend on PP_i. As found for 21 and 22, 23 exerted similar effects on the rates of synthesis of both 5 and 6.

It is evident that duplication of the electrostatic and topological features of either fragment of the cyclopropylcarbinyl diphosphate-ion pair is insufficient for inhibition. The synergism seen for inorganic pyrophosphate with 21 and 22 and the ability of 23 to inhibit without added PP_i demonstrates that the enzyme binds the ion pair much more tightly than either fragment.

5 Conclusions

During the past 25 years the collective results from several laboratories have provided a clear picture of the chemistry of all but one of the steps between isopentenyl diphosphate and squalene. It is now evident that isomerase, farnesyl diphosphate synthetase, and squalene synthetase employ carbocations during catalysis. These potent electrophiles are highly reactive, typically indiscriminate in their reactions with nucleophiles, and are rarely encountered during biological reactions. Yet, isoprenoid enzymes have harnessed normally recalcitrant intermediates in efficient electrophilic alkylation reactions.

Unfortunately, complementary progress at the biological front has been painfully slow. Except for farnesyl diphosphate synthetase, the enzymes in the central part of the pathway have only recently been purified to homogeneity. Again with the exception of farnesyl diphosphate synthetase, none have been well characterized kinetically. Only the most rudimentary structural characteristics such as molecular weight and subunit composition are available for any of the enzymes. One possible exception is the recent report by Clarke et al. (1987) of a clone from a rat liver cDNA library which may contain the gene for farnesyl diphosphate synthetase, which may provide the first primary sequence of any of these enzymes. These deficiencies are obvious challenges for the future.

References

Banthorpe DV, Doonan S, Gutowski JA (1977) Isopentenyl pyrophosphate isomerase from pig liver. Arch Biochem Biophys 184:381–390

Barnard GF (1985) Prenyltransferase from human liver. In: Law JH, Rilling HC (eds) Methods in enzymology, vol 110. Academic Press, London, pp 155–167

Barnard GF, Langto NB, Popjak G (1978) Pseudo-isozyme forms of liver prenyltransferase. Biochem Biophys Res Commun 85:1097–1103

Bartlett DL, King C-HR, Poulter CD (1985) Purification of farnesylpyrophosphate synthetase by affinity chromatography. In: Law JH, Rilling HC (eds) Methods in enzymology, vol 110. Academic Press, London New York, pp 171–184

Bruenger E, Chayet L, Rilling HC (1986) Isopentenyl pyrophosphate: dimethylallyl pyrophosphate isomerase: isolation from *Claviceps sp.* SD 58 and comparison to the mammalian enzyme. Arch Biochem Biophys 248:620–625

Clarke CF, Tanaka RD, Svenson K, Wamsley M, Fogelman AM, Edwards PA (1987) Molecular cloning and sequence of a cholesterol-repressible enzyme related to prenyltransferase in the isoprene biosynthetic pathway. Mol Cell Biol 7:3138–3146

Cornforth JW (1968) Olefin alkylation in biosynthesis. Angew Chem Int Ed 7:903–911

Cornforth JW, Clifford K, Mallaby R, Phillips GT (1972) Stereochemistry of isopentenyl pyrophosphate isomerase. Proc R Soc London Ser B 182:277–295

Dihanich ME, Najarian D, Clark R, Gillman EC, Martin NC, Hopper AK (1987) Isolation and characterization of *MOD5*, a gene required for isopentenylation of cytoplasmic and mitochondrial tRNAs of *Saccharomyces cerevisiae*. Mol Cell Biol 1:177–182

Eberhardt NL, Rilling HC (1975) Prenyltransferase from *Saccharomyces cerevisiae*. J Biol Chem 250:863–866

Gregonis DE, Rilling HC (1974) The Stereochemistry of *trans*-phytoene synthesis. Some observations on lycopersene as a carotene precursor and a mechanism for the synthesis of *cis*- and *trans*-phytoene. Biochemistry 13:1538–1542

Huang Z, Poulter CD, Wolf FR, Somers TC, White JD (1988) A novel cyclic C_{32} isoprenoid from *Botryococcus braunii*. J Am Chem Soc 110:3959–3964

Kuswik-Rabiega G, Rilling HC (1987) Squalene synthetase. Solubilization and partial purification of squalene synthetase. Copurification of presqualene pyrophosphate and squalene synthetase activities. J Biol Chem 262:1505–1509

Laskovics FM, Poulter CD (1981) Prenyltransferase. Determination of the binding mechanism and individual kinetic constants for farnesylpyrophosphate synthetase by rapid quench and isotope partitioning experiments. Biochemistry 20:1893–1901

Laskovics FM, Krafcik JM, Poulter CD (1979) Prenyltransferase. Kinetic studies of the 1′–4-coupling reaction with avian liver enzyme. J Biol Chem 254:9458–9463

Megregian S (1978) Fluorine. In: Boltz DF, Howell JA (eds) Chemical analysis, Wiley, New York, p 109

Muehlbacher M, Poulter CD (1988) Isopentenyl diphosphate isomerase. Inactivation of the enzyme with active-site-directed inhibitors. Biochemistry 27:7315–7328

Muehlbacher M, Poulter CD (1985) J Am Chem Soc 107:8307–8308

Popjak G, Agnew WS (1979) Squalene synthetase. Mol Cell Biochem 27:97–116

Poulter CD, Epstein WW (1973) A survey of some irregular monoterpenes and their biogenetic analogies to presqualene alcohol. Phytochemistry 12:737–747

Poulter CD, Rilling HC (1978) The prenyl transfer reaction. Enzymatic and mechanistic studies of the 1′–4-coupling reaction in the terpene biosynthetic pathway. Acc Chem Res 11:307–313

Poulter CD, Rilling HC (1981a) Prenyl transferases and isomerase. In: Porter JW (ed) Biosynthesis of isoprenoid compounds. Wiley, New York, pp 209–216

Poulter CD, Rilling HC (1981b) Prenyl transferases and isomerase. In: Porter JW (ed) Biosynthesis of isoprenoid compounds. Wiley, New York, pp 162–209

Poulter CD, Rilling HC (1981c) Conversion of farnesyl pyrophosphate to squalene. In: Porter JW (ed) Biosynthesis of isoprenoid compounds. Wiley, New York, pp 415–441

Poulter CD, Muscio OJ, Goodfellow RJ (1974) Biosynthesis of head-to-head terpenes. Carbonium ion rearrangements which lead to head-to-head terpenes. Biochemistry 13:1530–1538

Poulter CD, Marsh LL, Hughes JM, Argyle JC, Satterwhite DM, Goodfellow RJ, Moesinger SG (1977) Model studies of the biosynthesis of non-head-to-tail terpenes. Rearrangements of the chrysanthemyl system. J Am Chem Soc 99:3816–3823

Poulter CD, Wiggins PL, Le AT (1981) Farnesyl pyrophosphate synthetase. A stepwise mechanism for the 1′–4-condensation reaction. J Am Chem Soc 103:3926–3927

Reardon JE, Abeles RH (1986) Mechanism of action of isopentenyl pyrophosphate isomerase: evidence for a carbonium ion intermediate. Biochemistry 25:5609–5616

Reed BC, Rilling HC (1975) Crystallization and partial characterization of prenyltransferase from avian liver. Biochemistry 14:50–54

Reed BC, Rilling HC (1976) Substrate binding of avian liver prenyltransferease. Biochemistry 15:3739–3745

Sandifer RM, Thompson MD, Gaughan RG, Poulter CD (1982) Squalene synthetase. Inhibition by an ammonium analogue of a carbocationic intermediate in the conversion of presqualene pyrophosphate to squalene. J Am Chem Soc 104:7376–7378

Sasiak K, Rilling HC (1988) Purification to homogeneity and some properties of squalene synthetase. Arch Biochem Biophys 260:622–627

Satterwhite DM (1985) Isopentenyldiphosphate Δ-isomerase. In: Law JH, Rilling HC (eds) Methods in enzymology, vol 110. Academic Press, London New York, pp 92−99
Segal IH (1975) Enzyme kinetics. Wiley, New York, pp 506−515
Yeh LS, Rilling HC (1977) Purification and properties of pig liver prenyl transferase: interconvertible forms of the enzyme. Arch Biochem Biophys 183:718−724

Chapter 13

Squalene Epoxidase – Enzymology and Inhibition

N. S. RYDER [1]

1 Introduction

1.1 Historical Background

Squalene occupies a key position in the biosynthetic sequence from acetate to ergosterol (Fig. 1), being the first lipophilic intermediate in the pathway. The conversion of this 30-carbon hydrocarbon to the tetracyclic lanosterol structure was originally thought to occur in one step without intermediates. Tchen and Bloch (1957) showed that rat liver extracts require oxygen and reduced pyridine nucleotide to perform cyclisation of squalene. Yeast cells were also shown to require oxygen for ergosterol biosynthesis (Andreasen and Stier 1953) and to accumulate squalene under anaerobic conditions (Klein 1955). Subsequently, the intermediate 2,3-oxidosqualene was discovered (Corey et al. 1966; Van Tamelen et al. 1966) and squalene is now known to be converted by a two-step sequence of epoxidation and subsequent cyclisation involving two separate enzymes. The first of these enzymes, squalene epoxidase (EC 1.14.99.7), was initially described by Yamamoto and Bloch (1970) using rat liver extracts.

1.2 Occurrence

Being the first enzyme in the sterol pathway to require molecular oxygen, squalene epoxidase is of considerable evolutionary significance, an aspect which has been reviewed by Bloch (1979) and Poralla (1982). Under anaerobic conditions, the sterol pathway can proceed only as far as squalene, which thus represents the limit to which the pathway could have developed in primitive organisms. This is indeed the case in present-day prokaryotes, which lack sterols but have anaerobic mechanisms of squalene cyclisation to produce pentacyclic triterpenoids which appear to fulfil some of the functions of sterols (Rohmer et al. 1984). An exception to this is the bacterium *Methylococcus capsulatus* which contains sterols (Bird et al. 1971) and can epoxidate

---

[1] Sandoz Forschungsinstitut, 1235 Vienna, Austria

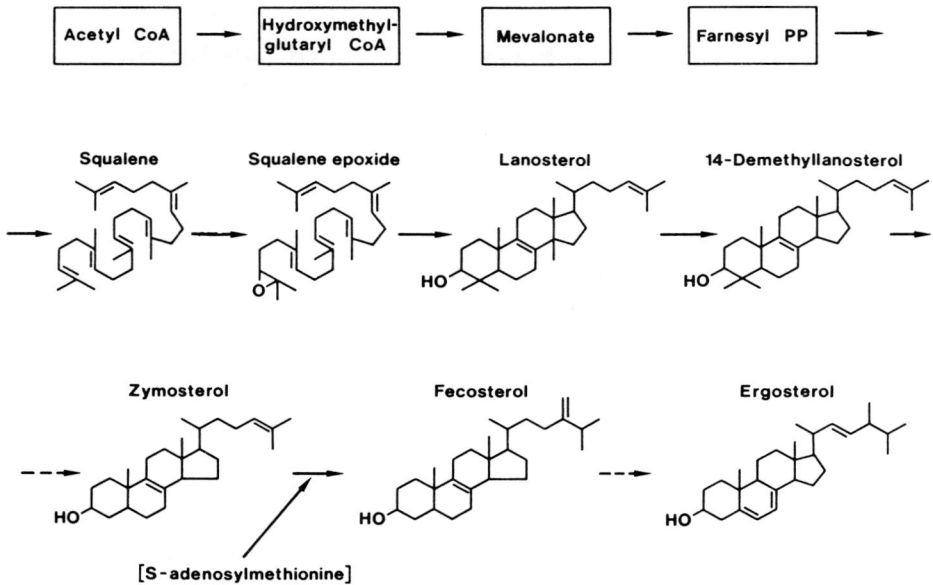

Fig. 1. Simplified pathway of ergosterol biosynthesis as occurring in *Candida*

squalene (Rohmer et al. 1980). In contrast, the great majority of eukaryotes (with the exception of insects) can synthesise sterols and possess squalene epoxidase. The enzyme has been found in various mammalian tissues and fungi, as described below, and occurs in plants as evidenced by the formation of squalene epoxide (Goodwin 1979), and in protozoa (Goad et al. 1985). Fungi of the family Pythiaceae are unable to synthesise sterols, apparently due to the absence of squalene epoxidase (Wood and Gottlieb 1978).

1.3 Regulatory Function

In fungi it is clear that squalene epoxidase is the site at which ergosterol biosynthesis is modulated by oxygen availability. Very little is otherwise known about regulation of sterol biosynthesis in fungi. There is some evidence that squalene epoxidase may be a site of regulation of mammalian cholesterol biosynthesis (Astruc et al. 1977; Eilenberg and Shechter 1984). In addition, squalene 2,3,22,23-dioxide formed in liver acts as a source of 24,25-epoxysterols which may play an important role in regulation of hepatic cholesterol metabolism (Taylor et al. 1986). Squalene dioxide also occurs in fungi (Buttke and Pyle 1982) but is of unknown function.

2 Enzymology

2.1 Rat Liver Squalene Epoxidase

This is the only squalene epoxidase enzyme system which has been extensively investigated and purified. The enzyme is membrane-bound and recovered in the microsomal fraction. For epoxidase activity the microsomes require molecular oxygen, NADPH and the soluble cytoplasmic fraction (Yamamoto and Bloch 1970). Two components of the latter are necessary for full activity, soluble protein(s) and a low molecular weight factor which was identified as an anionic phospholipid such as phosphatidylglycerol (Tai and Bloch 1972). The enzyme could be solubilised by treatment with the detergent Triton X-100 which also eliminated the need for the soluble fraction (Ono and Bloch 1975). The soluble enzyme was separated into two fractions, both of which were required for epoxidase activity. One fraction possessed NADPH-cytochrome c reductase activity and was later purified by Ono et al. (1977). This flavoprotein (EC 1.6.2.3) appears to be identical with that involved in a variety of microsomal oxidations such as those mediated by cytochrome P-450 enzymes. The second component of the epoxidase system, the terminal oxidase was purified to homogeneity by Ono et al. (1982). Fig. 2 shows a diagrammatic representation of the microsomal squalene epoxidase system.

The purified squalene epoxidase (Ono et al. 1982) was a single polypeptide chain with molecular weight 51 000. Enzyme activity could be reconstituted by addition of NADPH-cytochrome c reductase, NADPH, FAD and oxygen in the presence of Triton X-100. There was no evidence for the presence of heme in this enzyme and activity was unaffected by cyanide or carbon monoxide and other inhibitors of cytochrome P-450, indicating that squalene epoxidase is not an enzyme of the cytochrome P-450 type. The purified epoxidase had an absolute requirement for FAD but did not appear to contain tightly bound FAD. There was no evidence for involvement of superoxide, peroxide or hydroxyl radicals in the epoxidase reaction. The Km of the purified enzyme for squalene was 13 µM, similar to that (11 µM) measured with the microsomal epoxidase (Ryder and Dupont 1985).

SQUALENE EPOXIDASE COMPLEX

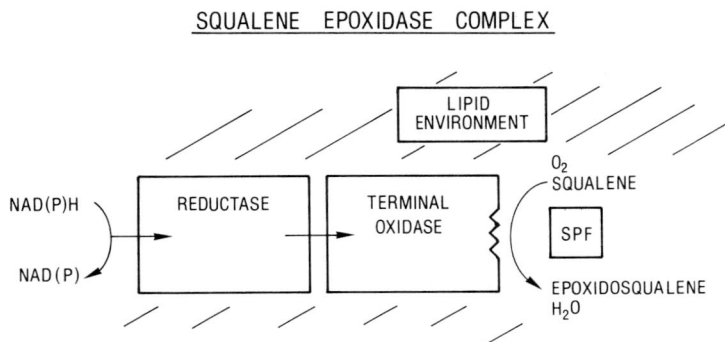

Fig. 2. Diagrammatic representation of squalene epoxidase enzyme system. (From Ryder 1987a)

Rat liver microsomal squalene epoxidase shows an almost complete requirement for the soluble cytoplasmic fraction (Yamamoto and Bloch 1970; Ryder and Dupont 1985). This requirement could be satisfied by a combination of anionic phospholipid and a non-catalytic protein termed "Supernatant Protein Factor" (SPF) which was purified by Ferguson and Bloch (1977). The SPF, molecular weight about 45000, did not bind squalene or its epoxide (Caras et al. 1980) but appeared rather to facilitate transfer of squalene between membrane fractions and its uptake into microsomes, the anionic phospholipid being essential for this activity (Friedlander et al. 1980; Chin and Bloch 1984). Srikantaiah et al. (1976) have described the isolation from liver cytosol of another protein activator of the microsomal epoxidase, termed "Sterol Carrier Protein" (SCP), with molecular weight about 16000. The role played by these soluble proteins in cholesterol biosynthesis in vivo is still not clear, since squalene is formed within the membrane system in vivo rather than being added exogenously as in the in vitro assay. However, as pointed out by Chin and Bloch (1984), exogenous squalene may be delivered within lipoproteins to cells and play a regulatory role there. In addition, labelling studies (Etemadi et al. 1969) suggest that newly synthesised squalene dissociates from the area of synthesis before being epoxidized, so that access to the epoxidase could be of importance. Senjo et al. (1985) have recently suggested a possible regulatory mechanism involving the inactivation of SPF by adenine nucleotides.

2.2 Fungal Squalene Epoxidases

The fungal enzyme has been much less studied than its mammalian counterpart, but has attracted interest in recent years as a target of antifungal compounds. Klein et al. (1967) demonstrated the conversion of squalene to lanosterol by a yeast membrane fraction. Microsomal squalene epoxidase activity has been described and partially characterised in *Saccharomyces cerevisiae* (Jahnke and Klein 1983; M'Baya and Karst 1987), *Candida albicans* (Ryder and Dupont 1984) and in *C. parapsilosis* (Ryder and Dupont 1985). A squalene epoxidase-deficient mutant of *S. cerevisiae* has been identified by Karst and Lacroute (1974). There are no reports of the enzyme from filamentous fungi and up to now we have failed to detect epoxidase activity in cell-free extracts of such fungi (N. S. Ryder, unpublished results) although squalene epoxidation undoubtedly occurs in whole cells. The enzyme may be particularly unstable or is perhaps in some way dependent on maintenance of cellular structure in filamentous fungi, which might be connected with the great sensitivity of the latter to squalene epoxidase inhibition (see below).

In general, squalene epoxidase enzymes from fungal and mammalian sources have similar properties, all being membrane bound and requiring oxygen, FAD and a reduced pyridine nucleotide. In common with the mammalian enzyme, fungal squalene epoxidase is not inhibited by carbon monoxide or other cytochrome P-450 inhibitors and indeed some stimulation of activity was observed (Ryder and Dupont 1984), probably by inhibiting other microsomal enzymes competing for reducing equivalents. The absence of heme from the epoxidase is indicated by the fact that porphyrin-deficient yeast mutants lacking heme could still synthesise lanosterol (Bard et

al. 1974). The *Candida* epoxidase differs from the others in preferring NADH as co-
factor, whereas the yeast and especially the rat liver enzyme have a strong preference
for NADPH. The most obvious difference between the fungal and mammalian epoxidases is in the requirement for soluble protein factors. The rat liver microsomal epoxidase shows almost total dependence on the soluble fraction (Yamamoto and Bloch
1970; Ryder and Dupont 1985), while the yeast enzyme is completely unaffected
(Jahnke and Klein 1983; M'Baya and Karst 1987). The *C. albicans* microsomal epoxidase prepared by cell breakage was stimulated up to 100% by addition of the soluble
fraction (Ryder and Dupont 1984), while enzyme prepared by the gentler process of
sphaeroplast lysis had a higher specific activity and no requirement for soluble factors
(Ryder 1987 a). Cytosol probably facilitated substrate access to the cruder membrane
preparation and also appeared to permit higher squalene concentrations in the assay
(Ryder and Dupont 1984). Interestingly, a protein with SCP activity for rat liver
squalene epoxidase has been isolated from yeast (Dempsey and Meyer 1977) and *C.
albicans* cytosol could also fully satisfy the requirement of the rat liver enzyme (Ryder
and Dupont 1985). These factors may however have a completely different function
in fungal cells, or may play a role in later stages of the ergosterol biosynthesis
pathway.

Microsomes prepared by lysis of *C. albicans* sphaeroplasts were used to investigate
the effects of detergents and fatty acids on squalene epoxidase activity. The majority
of detergents drastically reduced enzyme activity (Table 1). The irreversible inactivation of yeast squalene epoxidase by Triton X-100 has been described previously by
Hata et al. (1982). The non-ionic detergents octyl glucoside, Mega-8 and Mega-9 at
a concentration of 1 mg/ml were strongly stimulatory, but inhibited the enzyme at
higher concentrations (5 mg/ml). This inhibition was reversible upon dilution, thus
permitting use of these detergents for solubilisation of the epoxidase. Best results were
obtained with Mega-8 at 5 to 10 mg/ml which, after centrifugation and dilution pro-

Table 1. Effect of detergents on squalene epoxidase activity of *C. albicans* microsomes derived from
sphaeroplasts, prepared and assayed as described by Ryder (1987 a)

Detergent (Conc.)	Epoxidase activity (% control)	
	(0.1%)	(0.5%)
Tween 80	29.9	14.1
Triton X-100	5.4	0.4
Lubrol WX	11.3	16.7
Polyoxyethylene ether W-1	13.5	17.7
Tergitol NP-40	6.1	5.5
Deoxycholate	13.6	10.0
Taurodeoxycholate	4.0	4.3
Na dodecylsulphate	10.0	2.6
Digitonin	9.6	7.1
Octyl glocoside	189.9	8.6
Mega-9	155.5	22.9
Mega-8	156.7	15.2
CHAPS	3.6	3.4

Table 2. Effect of fatty acids on *C. albicans* squalene epoxidase from sphaeroplast microsomes

Fatty acid (Conc. mM)	Epoxidase activity (% control)	
	0.01 mM	0.1 mM
14:0	103.1	156.9
14:1	109.5	151.7
16:0	106.7	147.6
16:1	96.0	131.8
18:0	103.1	130.6
18:1	101.3	135.1
18:1, 12-OH	105.5	140.6
18:2	104.7	156.9
18:2 (trans)	107.4	140.5
18:3 (alpha)	115.1	173.7
18:3 (gamma)	119.9	204.6
20:0	102.4	120.0
20:4	124.9	231.8

vided a soluble squalene epoxidase (Ryder 1987a). A range of fatty acids were also found to stimulate the epoxidase (Table 2), the polyunsaturates being most effective. This is in contrast to the rat liver epoxidase which was inhibited by unsaturated fatty acids (Yamamoto and Bloch 1970) and may be indicative of differences in the membrane lipid environment of the two enzymes. There is some evidence (Buttke and Pyle 1982) that a source of unsaturated fatty acid is essential for squalene epoxidase activity in yeast cells.

3 Squalene Epoxidase Inhibitors

3.1 Squalene Analogues

This is an obvious approach to development of inhibitors. A range of azasqualene and related nitrogen-containing structures have been synthesized by Cattel and coworkers as analogues of a carbocationic high energy intermediate of the 2,3-oxidosqualene cyclase reaction. Many of these compounds are effective inhibitors of the cyclase from both animals and plants (Duriatti et al. 1985), while others have been shown to inhibit both growth and ergosterol biosynthesis in yeast (Balliano et al. 1988). Two such compounds, 2-aza-2,3-dihydrosqualene (Fig. 3) and a quaternary ammonium derivative thereof, were found to inhibit *C. albicans* ergosterol biosynthesis at several points including squalene epoxidase (Ryder et al. 1986b). Azasqualene was however much more effective against the rat liver squalene epoxidase than against the fungal enzyme (Table 3). Although this inhibitor would appear to be an obvious competitor with the substrate squalene, no clear kinetic pattern could be obtained with the rat liver epoxidase. Duriatti et al. (1985) observed apparent non-competitive kinetics for inhibition

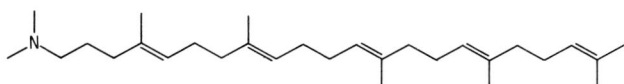

2-AZA-2,3-DIHYDROSQUALENE

Fig. 3. Structure of 2-aza-2,3-dihydrosqualene

Table 3. Inhibitors of *C. albicans* and rat liver microsomal squalene epoxidase

Compound	Conc. (µM) for 50% inhibition	
	C. albicans	Rat liver
2-Aza-2,3-dihydrosqualene	32.9	2.4
Naftifine (*1*)	1.1	144
Terbinafine (*2*)	0.03	77
SDZ 87-469 (*3*)	0.011	43
Benzylamine 880-349 (*4*)	0.045	–
Tolnaftate	1.04	215
Tolciclate	0.12	145

of the cyclase, most likely due to the problems introduced when both substrate and inhibitor are highly lipophilic compounds. A variety of isoprenoid compounds was found to inhibit conversion of squalene to sterols (Morin and Srikantaiah 1982), probably by competing with squalene.

3.2 Allylamine Derivatives

Naftifine (*1*) was discovered during a screening programme for antifungal compounds (Georgopoulos et al. 1981) and is used clinically as a topical antimycotic. Preliminary biochemical studies identified naftifine as an ergosterol biosynthesis inhibitor acting on squalene epoxidation (Paltauf et al. 1982; Ryder and Troke 1982). An extensive programme of chemical derivatisation (Stuetz and Petranyi 1984; Stuetz et al. 1986; Stuetz 1987) has produced over one thousand related compounds including the orally active antimycotic terbinafine (*2*) (Petranyi et al. 1984, 1987a, b). Naftifine and terbinafine (Fig. 4) have been subjected to intensive biochemical studies which have shown that squalene epoxidase is the primary target of these compounds in a number of yeast-like and filamentous fungi (Ryder 1984, 1986, 1987a, b; Ryder and Dupont 1985).

Naftifine and terbinafine are potent inhibitors of the *Candida* squalene epoxidase (Tables 3, 4) at concentrations reflecting their relative antifungal activities. The inhibition is specific to squalene epoxidase and no significant effect on other enzymes of the sterol pathway has been found (Ryder et al. 1984; Ryder 1985a). Both compounds show reversible inhibition and apparent non-competitive kinetics with respect to the

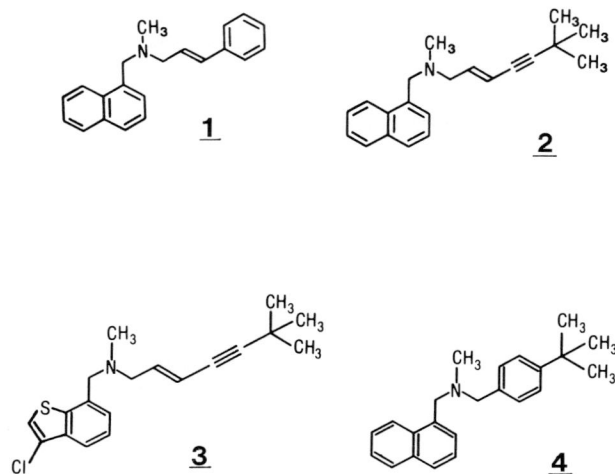

Fig. 4. Structures of epoxidase inhibitors of the allylamine class: *1* naftifine; *2* terbinafine; *3* SDZ 87-469; *4* 880-349

Table 4. Comparative properties of inhibition by allylamines of microsomal squalene epoxidases from *C. albicans* and rat liver

Property	*C. albicans*	Rat liver
Ki naftifine (μM)	1.1	—
Ki terbinafine (μM)	0.03	77
Type	Reversible	Reversible
Kinetics/squalene	Non-competitive	Competitive
Effect cytosol	Non-competitive	Competitive
Effect NAD(P)H	None	None
Effect fatty acids	None	None

substrate squalene (Fig. 5). Inhibition is unaffected by varying the concentrations of the cofactors FAD and NAD(P)H, or by addition of the soluble cytoplasmic fraction (Ryder and Dupont 1985), or by solubilising the epoxidase with detergents (Ryder 1987a). Many of the allylamine type compounds synthesised up to now also have antifungal activity (Stuetz 1987), and on the basis of our biochemical investigations of representative compounds it may be assumed that all of these active structures are squalene epoxidase inhibitors. Figure 4 shows the structures of two further compounds, SDZ 87-469 (*3*) (Stuetz et al. 1988; Petranyi et al. 1988) and the benzylamine 880-349 (*4*) (Kaken 1985; Sandoz 1987). Both compounds have epoxidase inhibitory activity comparable to that of terbinafine (Table 3).

The molecular mechanism of epoxidase inhibition is still not clear. The allylamines show no obvious resemblance to squalene and the non-competitive kinetics are also incompatible with the action of substrate analogues. Studies on the substrate specificity of rat liver squalene epoxidase (Van Tamelen and Heys 1975) indicated that the enzyme bound the entire squalene molecule fairly specifically; major alterations of the molecule; including partial cyclisation, were not accepted as substrates. The

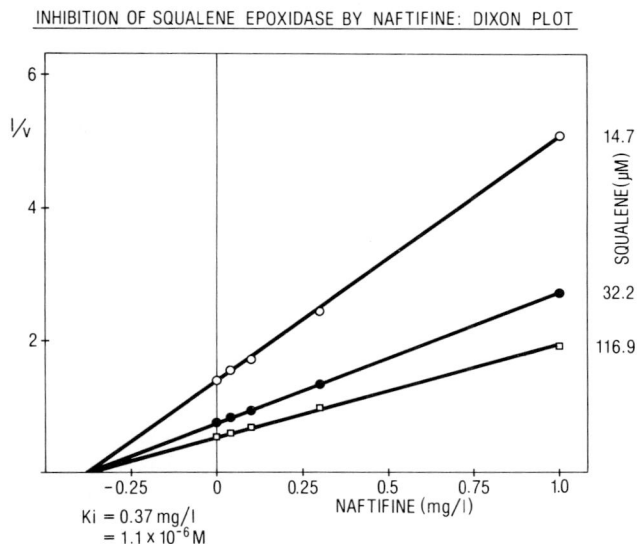

Fig. 5. Inhibition by naftifine of *C. albicans* squalene epoxidase (Dixon analysis). Data from Ryder and Dupont (1985)

kinetic properties and particularly reversibility of inhibition rule out the possibility of mechanisms-based "suicide" inhibition involving the acetylene function of terbinafine, such as described for inhibition of monoamine oxidase by propargyline derivatives (Maycock et al. 1976). The lack of interaction with co-factors and the specificity of action tend to rule out an influence on the electron transport supply to the epoxidase. The available evidence thus indicates that the allylamines act directly on the epoxidase enzyme, perhaps binding to a regulatory site leading to allosteric inhibition. Attachment to a lipid-binding site would be plausible in view of the lipophilic nature of the allylamines. We have recently found that various allylamines including the types shown in Figure 4 interact specifically with anionic phospholipids, which must contain an unsaturated acyl chain for binding to occur (N. S. Ryder, unpublished results). Rat liver squalene epoxidase is known to have specific phospholipid requirements (Tai and Bloch 1972), while unsaturated fatty acids, perhaps in the form of phospholipids, are essential for squalene epoxidation in yeast (Buttke and Pyle 1982). Disruption of a specific lipid domain of the epoxidase system would thus be a possible mechanism for these inhibitors, compatible with all the experimental evidence.

The effects of allylamines, in particular terbinafine, on the rat liver microsomal squalene epoxidase have also been examined in detail (Ryder and Dupont 1985). Terbinafine is a weak reversible inhibitor of this enzyme (Ki = 0.1 mM) showing, in contrast to the *Candida* epoxidase, apparently competitive kinetics with respect to squalene. The very high degree of selectivity of action of terbinafine cannot be explained on the basis of our present knowledge of the enzymes involved. The hypothetical mechanism of inhibition described above would account for selectivity on the basis of variation in the lipid requirements of the respective epoxidases. The most obvious known difference between the fungal and rat liver epoxidases is the requirement of the latter for a soluble protein factor. The rat liver cytosolic fraction was found to interact competitively with terbinafine in the epoxidase assay, suggesting that it might protect the enzyme against inhibition (Ryder and Dupont 1985). The detergent Triton X-100 was reported to substitute for the soluble fraction (Ono and Bloch 1975) but we were unable to reproduce this effect, and used octyl glucoside, which could partially substitute. In the absence of cytosol, the rat liver epoxidase became about 5-fold more sensitive to terbinafine (Ryder 1987a) but this in no way approaches the sensitivity of the fungal epoxidase. Similarly, although the *Candida* cytosol could substitute for rat liver cytosol it did not affect the sensitivity of the rat liver epoxidase to terbinafine, and rat liver cytosol did not protect the *Candida* epoxidase from inhibition. We recently found the guinea pig liver squalene epoxidase to be particularly sensitive to terbinafine (50% inhibition at about 5 µM) indicating a considerable degree of variation even among mammalian epoxidase enzymes, all of which, however, appear to be orders of magnitude less sensitive than fungal squalene epoxidase. The difference in sensitivity between rat and guinea pig liver epoxidases was not connected with properties of the respective cytosolic fractions (Ryder 1987a). In summary, liver cytosol partially protects the epoxidase against inhibition by allylamines but selectivity is primarily due to intrinsic differences between the respective epoxidase enzymes.

3.3 Thiocarbamates

Naftifine was the first specific inhibitor of squalene epoxidase to be identified and discovery of this novel mode of action led to testing of a wide range of other compounds as potential ergosterol biosynthesis inhibitors. In this way, the thiocarbamate antifungal agents tolnaftate and tolciclate were also found to inhibit fungal squalene epoxidase (Ryder et al. 1986a), this property of tolnaftate being discovered independently by Morita and Nozawa (1985) and Barrett-Bee et al. (1986). Figure 6 shows the structures of these two compounds which although chemically quite distinct from the allylamines, have some similarity to naftifine in terms of overall molecular shape. Like the allylamines, tolnaftate and tolciclate inhibit the *C. albicans* microsomal squalene epoxidase, being somewhat less active than terbinafine (Ryder et al. 1986a), and have weak inhibitory activity against the rat liver epoxidase (Table 3). The kinetic properties of inhibition by thiocarbamates have not been reported and it is not certain whether the mechanism is identical to that of the allylamines.

THIOCARBAMATES

TOLNAFTATE   TOLCICLATE

Fig. 6. Structures of thiocarbamate antifungals

3.4 Other Inhibitors

A wide range of known metabolic and respiratory inhibitors, antifungal agents and natural products has been tested for inhibition of the *Candida* squalene epoxidase. Up to now, significant inhibition has been observed only in the case of two known electron transport inhibitors, rotenone and antimycin A, and at the relatively high concentration of 0.1 mM (Ryder and Dupont 1984). Thus at present, the allylamines and thiocarbamates are the only known classes of specific squalene epoxidase inhibitors. No specific inhibitors of the mammalian enzyme have been reported.

4 Squalene Epoxidase Inhibitors as Antifungal Agents

Since ergosterol biosynthesis is essential for growth of virtually all fungi, squalene epoxidase is clearly a potential target for antifungal action, its importance being underlined by the existence of two classes of antifungals acting at this site. Several features contribute to making squalene epoxidase an attractive target in fungi:

1) The enzyme is clearly not cytochrome P-450-dependent, so clinical use involves no danger of interfering with human P-450 enzymes which have many important physiological functions. This has been confirmed in the case of the allylamines (Schuster 1985).
2) High selectivity of action against the fungal enzyme is provided, although this could not have been predicted from our knowledge of the respective enzymes. The weak inhibition of mammalian epoxidases by terbinafine in vitro does not affect squalene or cholesterol levels in vivo, as confirmed by studies in experimental animals and with patients.
3) Squalene epoxidase inhibition leads to strictly fungicidal action in many fungi (Petranyi et al. 1987b). The mechanism of cell death appears to be related to accumulation of very high levels of intracellular squalene (Ryder et al. 1985).

Growth inhibition of various pathogenic fungi correlates to some extent with the degree of ergosterol biosynthesis inhibition (resulting from squalene epoxidase inhibition) by allylamines, as shown in Table 5. However, it is clear from these and other data that physiological factors specific to each fungus play an important role in determining susceptibility. For example, the epoxidases of *C. albicans* and *C. parapsilosis* show almost identical sensitivity to inhibition by allylamines (Ryder and Dupont 1985) although the latter species is much more susceptible to growth inhibition by the compounds. In general, filamentous fungi seem particularly susceptible and require only partial ergosterol biosynthesis inhibition for full inhibition of growth (Table 5). This principle is also illustrated by the dimorphic fungus *C. albicans*, the filamentous form of which is much more susceptible than the yeast form to allylamines (Schaude et al. 1987; Ryder 1987a), while ergosterol biosynthesis is equally affected in the two forms (N. S. Ryder, unpublished results). Further important variables include the rapidity of onset of inhibition, which correlates with fungal susceptibility (Ryder

Table 5. Concentrations of naftifine and terbinafine causing inhibition of fungal growth (MICs) and of sterol biosynthesis in cells of various fungi

Fungus	Compound	MIC (mg/litre)	Inhibitory concentrations[a] (mg/litre)	
			50%	95%
Trichophyton rubrum	Naftifine	0.05	0.005	0.11
T. mentagrophytes	Naftifine	0.05	0.006	0.12
Candida parapsilosis	Naftifine	1.6	0.23	4.0
Candida albicans	Naftifine	50	0.35	7.5
Candida glabrata	Naftifine	100	0.34	11.3
Trichophyton rubrum	Terbinafine	0.003	0.0005	0.02
T. mentagrophytes	Terbinafine	0.003	0.002	0.04
Aspergillus fumigatus	Terbinafine	0.8	0.07	1.2
Candida parapsilosis	Terbinafine	0.4	0.006	0.3
Candida albicans	Terbinafine	3.1	0.008	0.2
Candida glabrata	Terbinafine	100	0.040	0.9

[a] Concentration of drug causing respectively 50% and 95% inhibition of sterol biosynthesis compared with untreated controls (mean of 3 separate experiments). Sterol biosynthesis was measured by [^{14}C]acetate incorporation. Data from Ryder (1987b).

1985b), and ease of uptake of inhibitor into the fungal cell. Poor penetration of the fungal cell envelope accounts for the lack of activity of the thiocarbamates against yeasts (Ryder et al. 1986a).

The data reviewed here demonstrate that squalene epoxidase is a very appropriate biochemical target for attack by antifungal agents. Development of novel and more potent epoxidase inhibitors may therefore lead to the discovery of improved antifungals for both clinical and agricultural applications.

References

Andreasen AA, Stier TJB (1953) Anaerobic nutrition of *Saccharomyces cerevisiae*. I. Ergosterol requirement for growth in a defined medium. J Cell Physiol 41:23−36

Astruc M, Tabacik C, Descomps B, Crastes de Paulet A (1977) Squalene epoxidase and oxidosqualene lanosterol-cyclase activities in cholesterogenic and non-cholesterogenic tissues. Biochim Biophys Acta 487:204−211

Balliano G, Viola F, Ceruti M, Cattel L (1988) Inhibition of sterol biosynthesis in *Saccharomyces cerevisiae* by *N,N*-diethylazasqualene and derivatives. Biochim Biophys Acta 959:9−19

Bard M, Woods RA, Haslam JM (1974) Porphyrin mutants of *Saccharomyces cerevisiae*: correlated lesions in sterol and fatty acid biosynthesis. Biochem Biophys Res Commun 56:324−330

Barrett-Bee KJ, Lane AC, Turner RW (1986) The mode of action of tolnaftate. J Med Vet Mycol 24:155−160

Bird CW, Lynch JM, Port FJ, Reid WW, Brooks CJW, Middleditch BS (1971) Steroids and squalene in *Methylococcus capsulatus* grown on methane. Nature (London) 230:473

Bloch KE (1979) Speculations on the evolution of sterol structure and function. CRC Crit Rev Biochem 7:1−5

Buttke TM, Pyle AL (1982) Effects of unsaturated fatty acid deprivation on neutral lipid synthesis in *Saccharomyces cerevisiae*. J Bacteriol 152:747–756

Caras IW, Friedlander EJ, Bloch K (1980) Interactions of supernatant protein factor with components of the microsomal squalene epoxidase system. Binding of supernatant protein factor to anionic phospholipids. J Biol Chem 255:3575–3580

Chin J, Bloch K (1984) Role of supernatant protein factor and anionic phospholipid in squalene uptake and conversion by microsomes. J Biol Chem 259:11735–11738

Corey EJ, Russey WE, Ortiz de Montellano PR (1966) 2,3-Oxidosqualene, an intermediate in the biological synthesis of sterols from squalene. J Am Chem Soc 88:4750–4751

Dempsey ME, Meyer GM (1977) Purification of yeast sterol carrier protein. Fed Proc 36:779

Duriatti A, Bouvier-Nave P, Benveniste P, Schuber F, Delprino L, Balliano G, Cattel L (1985) In vitro inhibition of animal and higher plants 2,3-oxidosqualene-sterol cyclases by 2-aza-2,3-dihydrosqualene and derivatives, and by other ammonium-containing molecules. Biochem Pharmacol 34:2765–2777

Eilenberg H, Shechter I (1984) A possible regulatory role of squalene epoxidase in Chinese hamster ovary cells. Lipids 19:539–543

Etemadi AH, Popjak G, Cornforth JW (1969) Assay of the possible organisation of particle-bound enzymes with squalene synthetase and squalene oxidocyclase systems. Biochem J 111:445–451

Ferguson JB, Bloch K (1977) Purification and properties of a soluble protein activator of rat liver squalene epoxidase. J Biol Chem 252:5381–5385

Friedlander EJ, Caras IJ, Lin LFH, Bloch K (1980) Supernatant protein factor facilitates intermembrane transfer of squalene. J Biol Chem 255:8042–8045

Georgopoulos AG, Petranyi G, Mieth H, Drews J (1981) In vitro activity of naftifine, a new antifungal agent. Antimicrob Agents Chemother 19:386–389

Goad LJ, Holz GG, Beach DH (1985) Effect of the allylamine antifungal drug SF 86-327 on the growth and sterol synthesis of *Leishmania mexicana mexicana* promastigotes. Biochem Pharmacol 34:3785–3788

Goodwin TW (1979) Biosynthesis of terpenoids. Ann Rev Plant Physiol 30: 369–404

Hata S, Nishino T, Ariga N, Katsuki H (1982) Effect of detergents on sterol synthesis in a cell-free system of yeast. J Lip Res 23:803–810

Jahnke L, Klein HP (1983) Oxygen requirements for formation and activity of the squalene epoxidase in *Saccharomyces cerevisiae*. J Bacteriol 155:488–492

Kaken Pharmaceutical Co Ltd (1985) European Patent Application No 0164697

Karst F, Lacroute F (1974) Yeast mutant requiring only a sterol as growth supplement. Biochem Biophys Res Commun 59:370–376

Klein HP (1955) Synthesis of lipids in resting cells of *Saccharomyces cerevisiae*. J Bacteriol 69:620–627

Klein HP, Volkmann CM, Leaffer MA (1967) Subcellular sites involved in lipid synthesis in *Saccharomyces cerevisiae*. J Bacteriol 94:61–65

Maycock AL, Abeles RH, Salach JI, Singer TP (1976) The structure of the covalent adduct formed by the interaction of 3-dimethylamino-1-propyne and the flavine of mitochondrial amine oxidase. Biochemistry 15:114–125

M'Baya B, Karst F (1987) In vitro assay of squalene epoxidase of *Saccharomyces cerevisiae*. Biochem Biophys Res Commun 147:556–564

Morin RJ, Srikantaiah MV (1982) Inhibition of rat liver sterol formation by isoprenoid and conjugated ene compounds. Pharmacol Res Commun 14:941–947

Morita T, Nozawa Y (1985) Effects of antifungal agents on ergosterol biosynthesis in *Candida albicans* and *Trichophyton mentagrophytes*: differential inhibitory sites of naphthiomate and miconazole. J Invest Dermatol 85:434–437

Ono T, Bloch K (1975) Solubilization and partial characterization of rat liver squalene epoxidase. J Biol Chem 250:1571–1579

Ono T, Ozasa S, Hasegawa F, Imai Y (1977) Involvement of NADPH-cytochrome c reductase in the rat liver squalene epoxidase system. Biochim Biophys Acta 486: 401–407

Ono T, Nakazono K, Kosaka H (1982) Purification and partial characterization of squalene epoxidase from rat liver microsomes. Biochim Biophys Acta 709:84–90

Paltauf F, Daum G, Zuder G, Hoegenauer G, Schulz G, Seidl G (1982) Squalene and ergosterol biosynthesis in fungi treated with naftifine, a new antimycotic agent. Biochim Biophys Acta 712:268–273

Petranyi G, Ryder NS, Stuetz A (1984) Allylamine derivatives: new class of synthetic antifungal agents inhibiting fungal squalene epoxidase. Science 224:1239–1241

Petranyi G, Meingassner JG, Mieth H (1987a) Antifungal activity of the allylamine derivative terbinafine in vitro. Antimicrob Agents Chemother 31: 1365–1368

Petranyi G, Stuetz A, Ryder NS, Meingassner JG, Mieth H (1987b) Experimental antimycotic activity of naftifine and terbinafine. In: Fromtling RA (ed) Recent trends in the discovery, development and evaluation of antifungal agents. JR Prous Science Publishers, Barcelona, p 441–450

Petranyi G, Meingassner JG, Schaude M (1988) Experimental chemotherapeutic activity of SDZ 87-469, a new allylamine antimycotic. Poster presentation at 10th ISHAM Congress, Barcelona 1988

Poralla K (1982) Considerations on the evolution of steroids as membrane components. FEMS Microbiol Lett 13:131–135

Rohmer M, Bouvier P, Ourisson G (1980) Non-specific lanosterol and hopanoid biosynthesis by a cell-free system from the bacterium *Methylococcus capsulatus*. Eur J Biochem 112:557–560

Rohmer M, Bouvier-Nave P, Ourisson G (1984) Distribution of hopanoid triterpenes in prokaryotes. J Gen Microbiol 130:1137–1150

Ryder NS (1984) Selective inhibition of squalene epoxidation by allylamine antimycotic agents. In: Nombela C (ed) Microbial cell wall synthesis and autolysis. Elsevier, Amsterdam, pp 313–321

Ryder NS (1985a) Specific inhibition of fungal sterol biosynthesis by SF 86-327, a new allylamine antimycotic agent. Antimicrob Agents Chemother 27:252–256

Ryder NS (1985b) Effect of allylamine antimycotic agents on fungal sterol biosynthesis measured by sterol side-chain methylation. J Gen Microbiol 131:1595–1602

Ryder NS (1986) Biochemical mode of action of the allylamine antimycotic agents naftifine and terbinafine. In: Iwata K, Vanden Bossche H (eds) In vitro and in vivo evaluation of antifungal agents. Elsevier, Amsterdam, pp 89–99

Ryder NS (1987a) Squalene epoxidase as the target of antifungal allylamines. Pestic Sci 21:281–288

Ryder NS (1987b) Mechanism of action of the allylamine antimycotics. In: Fromtling RA (ed) Recent trends in the discovery, development and evaluation of antifungal agents. JR Prous Science Publishers, Barcelona, pp 451–459

Ryder NS, Dupont MC (1984) Properties of a particulate squalene epoxidase from *Candida albicans*. Biochim Biophys Acta 794:466–471

Ryder NS, Dupont MC (1985) Inhibition of squalene epoxidase by allylamine antimycotic compounds: a comparative study of the fungal and mammalian enzymes. Biochem J 230:765–770

Ryder NS, Troke PF (1982) The activity of naftifine as a sterol synthesis inhibitor in *Candida albicans*. In: Periti P, Grassi GG (eds) Current chemotherapy and immunotherapy. American Society for Microbiology, Washington, pp 1016–1017

Ryder NS, Seidl G, Troke PF (1984) Effect of the antimycotic drug naftifine on growth of and sterol biosynthesis in *Candida albicans*. Antimicrob Agents Chemother 25:483–487

Ryder NS, Seidl G, Petranyi G, Stuetz A (1985) Mechanism of the fungicidal action of SF 86-327, a new allylamine antimycotic agent. In: Ishigami J (ed) Recent advances in chemotherapy. University of Tokyo Press, Tokyo, pp 2558–2559

Ryder NS, Frank I, Dupont MC (1986a) Ergosterol biosynthesis inhibition by the thiocarbamate antifungal agents tolnaftate and tolciclate. Antimicrob Agents Chemother 29:858–860

Ryder NS, Dupont MC, Frank I (1986b) Inhibition of fungal and mammalian sterol biosynthesis by 2-aza-2,3-dihydrosqualene. FEBS Lett 204:239–242

Sandoz Ltd (1987) German Patent DE 3702039 A1

Schaude M, Ackerbauer H, Mieth H (1987) Inhibitory effect of antifungal agents on germ tube formation in *Candida albicans*. Mykosen 30:281–287

Schuster I (1985) The interaction of representative members from two classes of antimycotics – the azoles and the allylamines – with cytochromes P-450 in steroidogenic tissues and liver. Xenobiotica 15:529–546

Senjo M, Ishibashi T, Imai Y (1985) Regulation of rat liver microsomal squalene epoxidase: inactivation of the supernatant protein factor by nucleotides. Arch Biochem Biophys 238:584–587

Srikantaiah MV, Hansbury E, Loughran ED, Scallen TJ (1976) Purification and properties of sterol carrier protein 1. J Biol Chem 251:5496–5504

Stuetz A (1987) Allylamine derivatives – a new class of active substances in antifungal chemotherapy. Angew Chem Int Ed Engl 26:320–328

Stuetz A, Petranyi G (1984) Synthesis and antifungal activity of (E)-N-(6,6-demethyl-2-hepten-4-ynyl)-N-methyl-1-naphthalene-methanamine (SF 86-327) and related allylamine derivatives with enhanced oral activity. J Med Chem 27:1539–1543

Stuetz A, Georgopoulos A, Granitzer W, Petranyi G, Berney D (1986) Synthesis and structure-activity relationships of naftifine-related allylamine antimycotics. J Med Chem 29:112–125

Stuetz A, Nussbaumer P, Petranyi G (1988) SDZ 87-469: synthesis and structure-activity relationships of a novel allylamine antimycotic. Poster presentation at 10th ISHAM Congress, Barcelona 1988

Tai HH, Bloch K (1972) Squalene epoxidase of rat liver. J Biol Chem 247:3767–3773

Taylor FR, Kandutsch AA, Gayen AK, Nelson JA, Nelson SS, Phirwa S, Spencer TA (1986) 24,25-Epoxysterol metabolism in cultured mammalian cells and repression of 3-hydroxy-3-methylglutaryl-CoA reductase. J Biol Chem 261:15039–15044

Tchen TT, Bloch K (1957) On the conversion of squalene to lanosterol in vitro. J Biol Chem 226:921–939

Van Tamelen EE, Heys JR (1975) Enzymic epoxidation of squalene variants. J Amer Chem Soc 97:1252–1253

Van Tamelen EE, Willett JD, Clayton RB, Lord KE (1966) Enzymic conversion of squalene 2,3-oxide to lanosterol and cholesterol. J Amer Chem Soc 88:4752–4754

Wood SG, Gottlieb D (1978) Evidence from cell-free systems for differences in the sterol biosynthetic pathway of *Rhizoctonia solani* and *Phytophthora cinnamomi*. Biochem J 170:355–363

Yamamoto S, Bloch K (1970) Studies on squalene epoxidase of rat liver. J Biol Chem 245:1670–1674

Chapter 14

Inhibition of Sterol Biosynthesis in Higher Plants by Analogues of High Energy Carbocationic Intermediates

A. Rahier, M. Taton, and P. Benveniste [1]

1 Introduction

Cholesterol (1) is the major sterol in vertebrates and ergosterol (2) in fungi whereas plant cells contain a mixture of C-24-alkylated sterols, mainly campesterol (3), stigmasterol (4) and sitosterol (5). The sterol biosynthesis pathway in plants has been extensively investigated and rewieved in detail recently (Benveniste 1986). After a biosynthetic pathway from acetate to 2,3-oxidosqualene, common to animals, fungi and vascular plants, sterol biosynthesis differs in these three classes of organisms (Fig. 1): First: 2(3)-oxidosqualene (6) cyclases produce lanosterol (7) in non-photosynthetic eukaryotic organisms (animals, fungi), whereas they yield cycloartenol (8), the $9\beta,19$-cyclopropyl isomer of (7) in photosynthetic eukaryotes (algae, higher plants) (Fig. 1); second, this $9\beta,19$-cyclopropane ring is cleaved by an enzyme specific to photosynthetic eukaryotes, namely cycloeucalenol (9)-obtusifoliol (10)-isomerase (COI); third, most higher plants contain two C-methyl-transferases responsible for the introduction of the two extra carbon atoms at C-24 in the plants sterols side chain. These three enzymatic reactions are of interest both for their intricate mechanisms and as phylogenetic markers in higher plants.

The Conversion of Squalene-Oxide into Δ^5-Sterols is Predominantly Performed by Reactions Involving Carbocationic Intermediates. During recent years we have been engaged in studying the membrane-bound enzymes isolated from higher plants systems responsible for converting oxidosqualene into the final Δ^5-sterols (sitosterol, campesterol) and our discussion will only consider those enzymes. Among the 12 steps in the synthesis of sitosterol from squalene-oxide, 8 enzymatic steps involve postulated or demonstrated carbocationic high energy intermediates (HEIs). We shall now discuss some mechanistical and experimental arguments which strongly support this proposal. For this purpose we shall not discuss the 8 enzymes in the order of expected biosynthetic sequence, but rather consider the different types of reaction they catalyze.

[1] Institut de Botanique, Laboratoire de Biochimie Végétale, UA CNRS No. 1182, 28, rue Goethe, 67083-Strasbourg Cédex, France

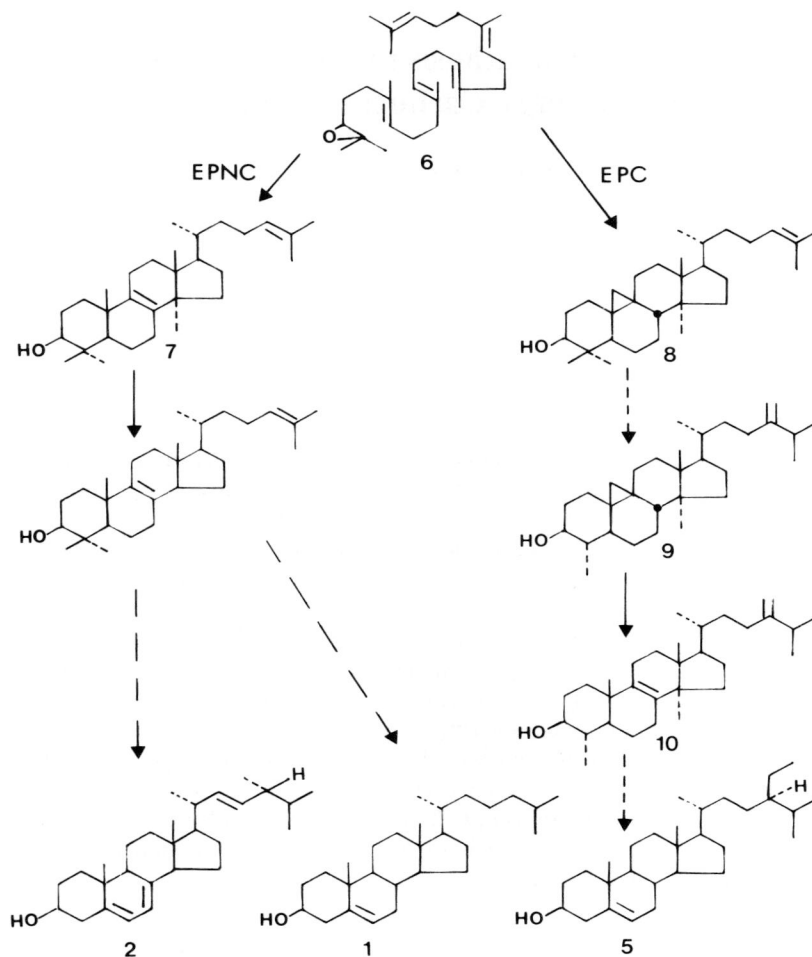

Fig. 1. Comparative biosynthesis of sterols in photosynthetic and non-photosynthetic eukaryotes

1.1 Cyclization and Alkylation Reactions

This first group includes the 2(3)-oxidosqualene cyclases (OSC) and the S-adeno-syl-L-methionine-sterol C-24 and C-28 methyltransferases (C24 and C28 MeTr).

1.1.1 2(3)-Oxido-Squalene-Cycloartenol (7) and -Lanosterol (6) Cyclases (OSC)

Following the pioneering theoretical model proposed by the Zürich School (Eschen-moser et al. 1955), the mechanism of this enzymatic reaction which cyclizes the chair-boat-chair-boat folded 3S-2(3)-O.S (6) into (7) or (8 EC 5.4.99.8 and EC 5.4.99.7),

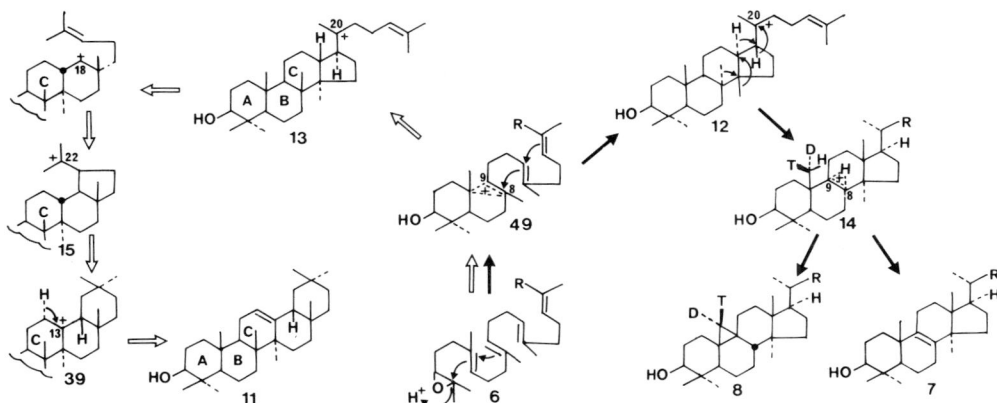

Fig. 2. Mechanism of the enzyme-catalysed cyclization of 2,3-oxidosqualene to lanosterol (7), cycloartenol (8) and β-amyrin (11)

possessing seven asymmetric centers, has been the subject of numerous studies (Van Tamelen 1982; Johnson et al. 1987). Because of subtle changes in the catalytic pathway, 3(S)-2,3-O.S can be cyclized into (7) or (8), or in the case of all chair folded O.S into various tetracyclic and pentacyclic triterpenes widely distributed in higher plants such as β-amyrine (11) (Fig. 2).

Enzymatic cyclization of the all-*trans*-2,3-oxidosqualene is believed to be triggered by a general acid-catalyzed epoxide ring opening assisted by the neighbouring π-bond. The concertedness of the ensuing overall annelation and backbone rearrangements is a matter of debate (Dewar and Reynolds 1984). However, for entropic reasons and from experimental evidence based upon biogenetic synthesis (Van Tamelen and James 1977; Van Tamelen 1981; Johnson et al. 1987) or isolation of intermediate cyclization products (Boar et al. 1984), it has been suggested that the cyclization process is likely to proceed through a series of discrete conformationally rigid carbocationic intermediates leading to the tetracyclic protosteryl (12) or dammarenyl (13) intermediates (Fig. 2). Assuming this last proposal, in the case of cycloartenol (lanosterol) cyclase, the C-20 carbocation (12), ending the annelation process, should probably be strongly stabilized by the active site of the cyclase since a 120° rotation of the alkyl chain around the C17−C20 bond must precede the series of migrations in order to achieve the 20(R) stereochemistry found in cycloartenol (lanosterol). After a series of hydride and methyl transpositions, in the case of cycloartenol formation, the final cyclopropane ring closure, i.e. the 9β,19-carbon-carbon bond formation, *cannot* be concerted with the migration of the 9β-hydrogen. Therefore the rearrangement process of the reaction should involve at least the passage through a transient C9 carbocationic intermediate (14) before the last elimination of a proton occurs. This C9 carbocation is not compulsory in lanosterol formation, for which, a totally concerted rearrangement-elimination process is stereochemically possible. The stereochemistry of the cyclopropane ring closure in the case of cycloartenol (8) formation has been shown to occur with retention of configuration suggesting that the basic subsite of the cyclase involved in the C-19 hydrogen abstraction is situated above the C-ring of (14) (Blättler 1978; Altman et al. 1978), a situation probably comparable to that found for the 8β-proton abstraction in the case of lanosterol (7) formation. In the case of β-amyrin formation, after the annelation process leading to the dammarenyl (13) intermediates, a series of cycle enlargements and of cyclizations leads to the oleanyl carbocation (15) via the dammarenyl, baccharenyl, lupenyl carbocationic intermediates. Then, a series of hydride-1,2-shifts results in a carbocationic species at C-13 which undergoes elimination of the 12-αH, yielding (11). In this case also, a totally concerted rearrangement-elimination process is stereochemically possible (Eschenmoser et al. 1955).

1.1.2 S-Adenosyl-L-Methionine-Sterol C-24 and C-28 Methyltransferases (C24MeTr and C28MeTr)

The mechanism of the SAM cycloartenol-C-24 methyl transferase (EC 2.1.1.41) has been thoroughly studied in *Trebouxia* sp. (Mihailovic 1984) and the major results of this study are shown in Fig. 3a.

After nucleophilic attack of the methyl group of SAM with inversion of configuration of this methyl group by the $\Delta^{24(25)}$-double bond of cycloartenol (8) a stereospecific hydride migration from C24 to C25 occurs. Then proton loss from the C28 position leads to C24-methylene-cycloartanol (16) whereas a direct loss of proton at C26 leads to cyclolaudenol (17). Both elimination steps are stereospecific and display a strong kinetic isotopic effect ($k_H/k_D = 3.5$), underlining that this last elimination step is mediated by a basic subsite of the enzyme. Moreover, the alkylation step and the proton loss occur on opposite faces of the nodal plane of the substrate double bond. In the context of this paper, the major point is that the C24 methyl insertion and the C24−C25 hydride migration occur on opposite faces of the original double bond. This rules out the involvement of covalent stabilization of the intermediates (18) by a nucleophilic group of the active site (Fig. 3b) which would give an opposite stereochemistry at C-25 in the case of a substrate stereospecifically labelled at one of the terminal methyl groups (C26 or C27) (Fig. 3). Therefore, the involvement of a C25-carbocationic intermediate in this reaction is indicated.

1.2 Allylic Isomerizations

Plant sterol biosynthesis involves two demonstrated allylic isomerization steps, catalyzed by cycloeucalenol-obtusifoliol isomerase (COI) (EC 5.5.19) and $\Delta^8 \rightarrow \Delta^7$-sterol isomerase (SI) (EC 5.3.3.5); both activities have been isolated from a cell-free system of higher plant material for the first time in our laboratory (Heintz and Benveniste 1974; Rahier et al. 1985).

1.2.1 Cycloeucalenol-Obtusifoliol Isomerase (COI) (Fig. 4)

Results of mechanistic studies are consistent with a general acid-catalyzed cyclopropane ring opening with incorporation of one proton from the medium at C-19

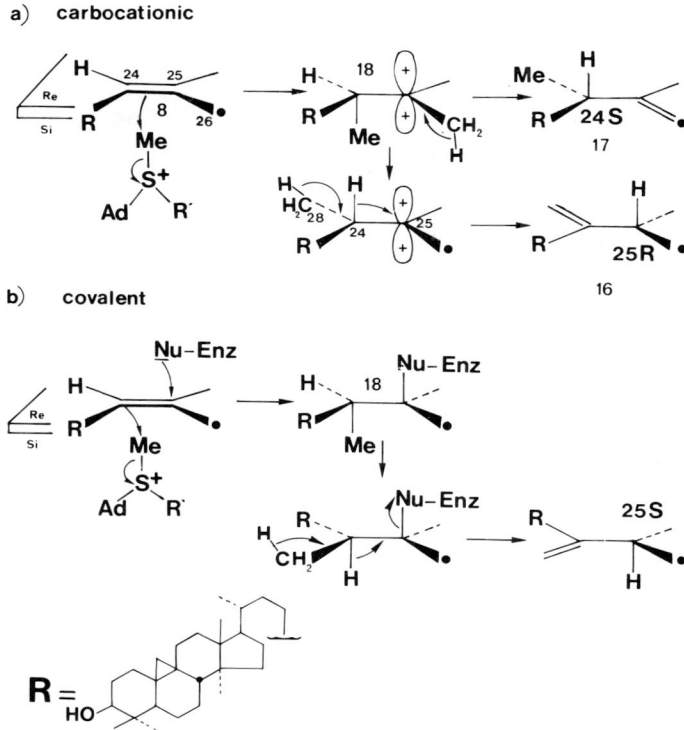

Fig. 3a, b. Possible mechanisms for the C-24 methylation of cycloartenol by the AdoMet-cycloartenol-C24-methyltransferase. Adapted from the work of Arigoni (Mihailovic 1984)

(Rahier 1980). The cleavage of the C19-C9β bond of the cyclopropane occurs with retention of configuration on the C19 carbon atom. This result suggests that the enzyme subsite which is involved in the acid-catalyzed cyclopropane ring opening is located above the C ring of the substrate (Blättler 1978; Rahier 1980), a situation analogous to that occupied by the basic subsite eliminating the C-19 proton in the OS-cycloartenol cyclase. Moreover, since the cleavage of the C9-C19β bond and the elimination of the C8β-H are *cis*, the reaction cannot be concerted and has to go through the C9, most probably carbocationic, intermediate.

1.2.2 $\Delta^8 \rightarrow \Delta^7$-Sterol Isomerase (SI) (Fig. 4)

Although little is known about the mechanism of this isomerization, it has been shown that in rat liver the reaction involves an antarafacial mechanism, i.e. 9α-protonation and 7β-elimination, whereas in yeast the 7α hydrogen atom is eliminated (Akhtar et al. 1970). Antarafacial transfers are mostly encountered for allylic isomerizations of non-activated double bonds (for which allylic deprotonation is not facilitated), and must involve a two-base mechanism (Cornforth 1974) which is most probably non

C.O.I.

Δ^{14}– Red.

Δ^8– Δ^7 S.I.

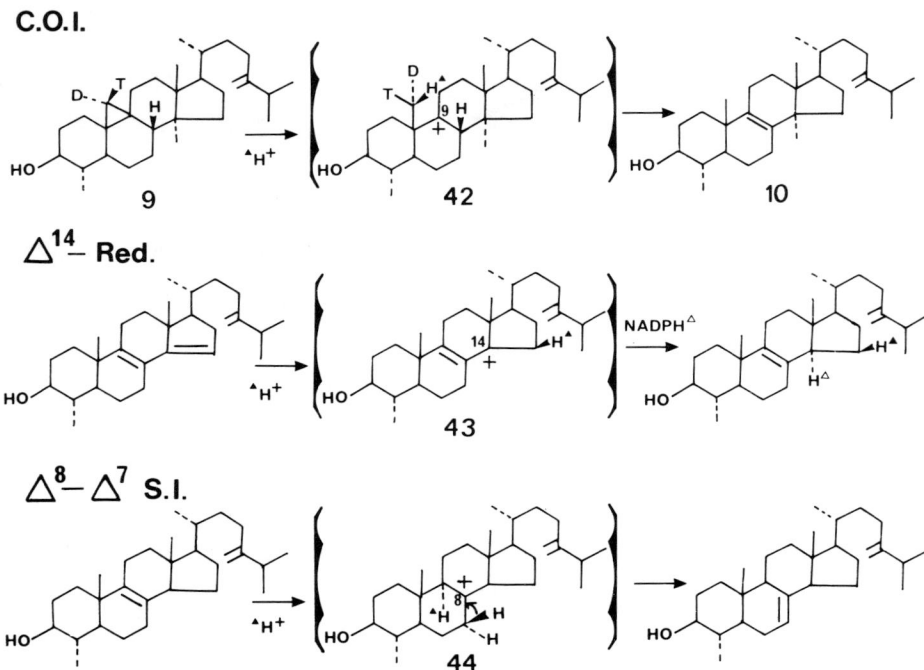

Fig. 4. Comparative mechanisms of action of COI, Δ^{14}-Red. and $\Delta^8 - \Delta^7$-SI involving high energy carbocationic intermediates

concerted, i.e. there must be protonation of the double bond, leading to a carbocationic intermediate from which a proton is eliminated. Such an antarafacial allylic isomerization involving a carbocationic intermediate has been demonstrated in the case of isopentenyl-pyrophosphate isomerase (Cornforth 1974; Reardon and Abeles 1986). In contrast, a number of allylic isomerizations of activated double bonds involve a suprafacial and carbanionic mechanism; $\Delta^5 \rightarrow \Delta^4$-3-cetosteroid-isomerase is a well known example of this class of reaction (Wang et al. 1963).

These considerations led us to consider that the most probable mechanism for the plant $\Delta^8 \rightarrow \Delta^7$ SI would involve a C8 carbocationic intermediate.

1.3 Ene-Reductases (Fig. 4)

$\Delta^{8,14}$-sterol Δ^{14}-reductase has been described in animals (Wilton et al. 1970) and fungi (Bottema and Parks 1978); Δ^7-sterol reductase and Δ^{24}-sterol reductase have also been described in animals. The results of mechanistic studies performed in the rat liver system clearly show that the first step of the reaction is the stereospecific protonation of the double bond generating the most stable carbocation, which is then reduced stereospecifically in a *trans* manner by a hydride ion from NADPH (Wilton et al. 1970).

Besides the involvement of carbocationic HEIs, the three types of reactions we have described share the following important properties:

1) They involve tight enzymatic control since they are regio- and stereo-specific, although this stereospecificity generally does not control the structure of the reaction product.

2) In the first two types of reaction, an electron deficient species triggers the reaction which is terminated by a stereospecific elimination of a proton by a basic subsite. Therefore, one important function of the active site of such enzymes is probably to maintain the substrate and carbocationic intermediates in an hydrophobic, slightly basic but compelling environment which is relaxed by elimination of a proton helping the product to leave the active site. The structure of the final product of the reaction is probably determined by the regiospecificity of the base responsible for the last elimination. Moreover the lack of racemization during the passage through the carbocationic HEIs reflects tight Van der Waals and electrostatic interactions between the charged intermediate species and the enzymatic surface during catalysis.

2 Inhibition of Phytosterol Biosynthesis by Carbocationic HEI and Transition State Analogues

The transition state (TS) analogue concept is very useful for the design of potent and specific enzyme inhibitors since TS analogue inhibitors may have a much higher affinity for the enzymatic active site than traditional ground state analogue inhibitors (Pauling 1946; Wolfenden 1976).

We have synthesized stable analogues with a positive charge at a position identical to that of the sp^2 carbon atom in the carbocationic HEI. In most cases we used a nitrogen atom; the tertiary amine having a pK_A, close to 10 is protonated at physiological pH (7.4) and generates a stable ammonium derivative displaying a positive charge at the expected position. In our early work, we decided to first apply this strategy to SAM C24 Me Tr in higher plant systems since it had been previously observed in yeast that 24-dihydro-25 azazymosterol was a good inhibitor of the in vivo and in vitro C-24 methylation of zymosterol (Avruch et al. 1976). Moreover, aza-steroid analogues containing a nitrogen atom in the side chain had been already used as potential hypocholesterolemic agents although their precise mode of action had not been elucidated (Thompson et al. 1963).

2.1 Inhibition of SAM C24 Me Tr (for structures of compounds see Fig. 6)

We confirmed that 25-aza-cycloartanol (19) was a potent inhibitor of the plant SAM C-24 Me Tr ($K_i = 3.10^{-8}$ M) (Km for cycloartenol = 2×10^{-5} M) (Rahier et al. 1980). It was then important to ascertain whether or not the charge of the nitrogen atom was responsible for the affinity. We showed that the quaternary ammonium derivative (20) (Table 1) was as potent an inhibitor as the corresponding amine (19), whereas the cor-

Table 1. Inhibition of C24 Me Tr by 25 azasteroids: influence of the charge in the C25 position

Compounds	Ki (nM)
19	30
20	35
21	no inhibition
22	200 000

Table 2. Inhibition of C24 Me Tr by various heteroatom substituted sterols displaying onium ion in the C25 position

Compounds	Ki (nM)
23	35
24	50
25	25

Table 3. Inhibition of C24 Me Tr by azasteroids. Influence of the location of the positive charge

Compounds	Ki (nM)
26	45
27	180
28	90
29	11 000

Table 4. Inhibition of C24 Me Tr by an *N*-oxide derivative

Compounds	Ki (nM)
31	15
32	100 000
23	35

responding electrostatically isosteric analogues (21) and (22) were totally devoided of affinity (Table 1). Moreover, replacement of the nitrogen atom by other positively charged heteroatoms leading to sulfonium (24) or arsonium (25) ions produced highly potent inhibitors (Table 2). These results provide strong evidence that the charged species mimics the cationic intermediates, and argues against the possibility that the C24 Me Tr binds the neutral form of (19) as a substrate ground state analogue which then undergoes N-methylation leading to the active species as it has been proposed elsewhere (Oehlschlager et al. 1984). Moreover these results emphasize that tetrahedral species such as ammonium, arsonium or sulfonium ions can take the place of a planar carbocation. The fact that these compounds are inhibitors shows that the interaction between such species and the enzyme active site is probably mostly electrostatic.

In the case of nitrogen-substituted analogues, we showed by assaying azasterols substituted at various positions in the side chain (Table 3) that the 25 position was indeed optimal, though a shift of the charged atom from one to two carbon-carbon distance along the side chain was possible without loss of inhibitory activity. This could be due either to the flexibility of the sterol side chain which however must have precise interactions with the active site since the reaction is totally regio- and stereospecific, or to the nature of the charge of such ammonium species which is thought to be delocalized on the neighbouring carbon and hydrogen atoms (Pullman and Armbruster 1974; Greenberg et al. 1982).

Finally a series of analogues possessing different tetracyclic moieties or without a steroidal nucleus shows that a steriod-like structure for the inhibitor was also important, though some features which are compulsory for the substrate to be transformed (such as a free 3β-hydroxy group or the bent conformation of cycloartenol) were no longer required for inhibition by the 25-azacycloartanol series (Rahier et al. 1984).

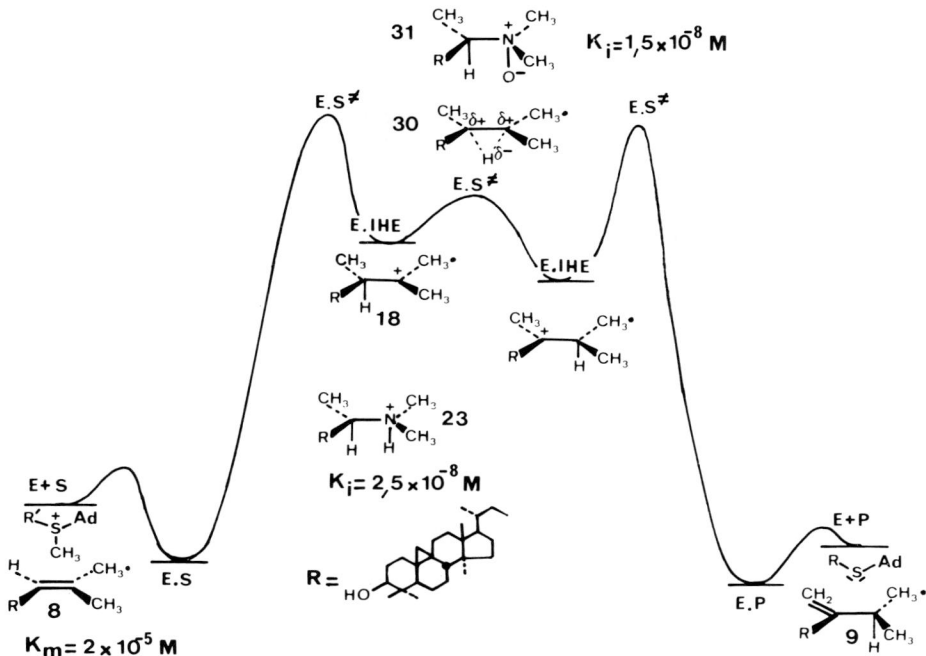

Fig. 5. Free energy diagram for AdoMet-cycloartenol-C24-methyltransferase

In the case of SAM C24 Me Tr beside the involvement of a carbocationic HEI, a dipolar transition state (30) can be postulated for the C24−C25-hydride migration (Fig. 5). A good mimic of this species is the stable analogue (31) (Figs. 5 and 6) possessing a *tertiary* amine N-oxide function, which is globally neutral but displays a very high dipolar moment. Indeed (31) was shown to strongly inhibits the C24 Me Tr, whereas the corresponding neutral isosteric analogue (32) displayed a very low affinity (Table 4); (31) was an even more potent inhibitor than the parent amine (23).

Similar results have been obtained in the case of inhibition of OS-cyclase by 2-aza-2,3-dihydrosqualene-*N*-oxide designed to mimic the transition state involved in the initial epoxide ring opening step of the reaction (Cerutti et al. 1985).

2.2 Inhibition of 2(3)-Oxidosqualene Cyclases by HEI Analogues Inhibitors

2,3-oxidosqualene cyclase is a key target for manipulating sterol contents in animals and plants cells. The design of potent and novel inhibitors of this enzyme is, therefore, of great interest, as well as the challenge of elucidating the reaction mechanism.

We synthesized *N*-substituted-8-azadecalins (33−37) in order to mimic the postulated C8 carbocationic HEI (49) occurring in the annelation process brought about by the three cyclases (cycloartenol-, lanosterol- and β-amyrin cyclases) (Fig. 2) or the C8 or C9 HEI (14) occurring in the last step of cycloartenol (lanosterol) cyclase.

Table 5. Comparative inhibition of 2(3)-oxidosqualene-cycloartenol and $\beta(\alpha)$-amyrin cyclases by N-substituted 8-aza-decalins and analogues

	33	34	35	36	37	38
	ID$_{50}$ (µM)					
OS-cycloartenol cyclase	1	4	50	NI	NI	50
OS-$\beta(\alpha)$-amyrin cyclase	NI	ND	ND	ND	NI	NI

ID$_{50}$: concentration of inhibitor required to reduce the reaction velocity by 50% at a given substrate concentration. Under the assay conditions used in this study where the concentration of the substrate was close to its Km value (125 µM), ID$_{50}$ values are of the order of the inhibition constants.
NI: not inhibitory at 10^{-4} M (the highest concentration tested).
ND: not determined.
The β- and α-amyrins have not been separated in our enzymatic assay (Taton et al. 1986).

Table 6. Comparative inhibition of 2(3)-oxidosqualene cyclases from different origins by the aza-decalin (33)

		OS-cycloartenol (lanosterol) cyclase	OS-$\beta(\alpha)$-amyrin cyclase
		ID$_{50}$ (µM)	
Maize seedlings	A	1	NI
	B	1	NI
Suspension cultures of	A	100	NI
bramble cells	B	10	NI
Pea cotyledons	A	ξ	NI
Rat liver	A	2	ξ

ID$_{50}$: concentration of inhibitor required to reduce the reaction velocity by 50%. The concentration of the substrate (100 µM) is close to the Km values (125 µM) of both OS-cyclases.
NI: not inhibitory at 10^{-4} M (the highest concentration tested).
ξ: enzyme not present in this material.
A: microsomal enzyme; B: solubilized enzyme.

Amongst the molecules (33–38) tested in vitro in cell-free systems prepared from plants or animals, only (33) and (34) strongly inhibited the OS-cycloartenol cyclase, underlining the importance of both a branched isoprenoid side chain and a bicyclic decalin skeleton for activity (Taton et al. 1986) (Table 5). We then showed that the N-[(1,5,9)-trimethyl-decyl]-4α,10-dimethyl-8-aza-trans-decal-3β-ol (33) strongly inhibited the OS-cycloartenol cyclase obtained in microsomal suspensions isolated from various species of plant whereas at the highest concentration tested (10^{-4} M) it failed to inhibit OS-β-amyrin cyclase present in the same systems. The same results were obtained with solubilized enzymes, thus excluding the possibilities that this difference results from partitioning phenomena or topological differences between the two cyclases in the microsomal preparation (Table 6). Independently (33) was shown to potently inhibit OS-lanosterol cyclase isolated from rat liver (Table 6).

In the present case, the use of an analogue of a postulated carbocationic HEI led to a new potent inhibitor of OS-cycloartenol (lanosterol) cyclase. This inhibitor may be a mechanistic probe for the intricate and debated reaction pathway of this reaction. Compound (33) failed to inhibit OS-β-amyrin cyclase despite the postulated passage through a C-8 carbocationic HEI occurring in the annelation process brought about by all three cyclases (Fig. 2). This lack of inhibition could reflect a concerted annelation and backbone rearrangement process for the three cyclizing reactions (Eschenmoser et al. 1955) without the intervention of stabilizing basic subsites of the active site, until the reaction coordinates reach the last step, i.e. the unconcerted elimination of a proton from a stabilized carbocation by a basic residue in the active site (see 1.1.1). Such a proposal would leave open the question as to how the enzyme manages to maintain the stereochemistry at the C20 carbocationic HEI during the 120° rotation around the C17−C20 bond preceeding the rearrangement process in the case of the cycloartenol (lanosterol) cyclase. Some recent work would suggest that no direct stabilization occurs in the case of C20 intermediate involved in β-amyrine cyclase since 20-aza-dammarenol (40), an aza-analogue of this intermediate, did not inhibit β-amyrin cyclase (Delprino et al. 1984). However, one can argue that in this special case no rotation around the C17−C20 bond has to occur before the next cycle enlargement occurs and that there is no reason to stabilize the progression of the electron deficient wave at this point. Synthesis and assay of a 20-azaprotosteryl analogue might provide an answer to this question. It has been proposed recently (Johnson et al. 1987) that a strong active site point charge stabilization of the C8 cationic intermediate by the α-face could be involved in the synchronous enzymatic annelation process of the chair-boat-chair-boat, unfolded OS into cycloartenol (lanosterol). This was postulated by analogy with the chemical biomimetic cyclization of OS. Thus, another possibility for the lack of inhibition of OS-β-amyrin cyclase by (33) would be this C8-α point charge stabilization to be attenuated by the β-face point charge stabilizations at C-10 and C-13 in the case of cyclization of the all chair unfolded OS into β-amyrin (Johnson personal communication). However, in the case of a concerted annelation process, we suggest that (33) could interact strongly either with a basic subsite involved in the elimination of the 19H or the 8βH from (14) to give (8) or (7) (Fig. 2); in these conditions the N-8 ammonium ion of (33) could not interact with the basic residue responsible for the abstraction of the C-12α proton of (39) during β-amyrin formation.

Examination of the structure-activity relationship for the 8-azadecalin analogues shows the importance for affinity of substitution of the nitrogen atom by a flexible isoprenoid-like chain able to mimic the C, D rings and side chain of the sterol molecule. The effectiveness in the 8-azadecalin series of only (33) and (34) could reflect their relative conformational flexibility confering on these HEI analogues the ability to bind the OS-cyclase in its ground state and to trigger a rapid conformational change of the active site close to the target HEI, where the ammonium ion would be at the vicinity of its enzymatic anionic counterpart. The importance of both the complete carbon skeleton and of the length of the hydrocarbon chain attached to C17 has recently been observed in the enzymic cyclization of Z $\Delta^{18,19}$-oxidosqualene analogues with or without a truncated substituent of this double bond. In this case the results underline the importance of the complete framework in controlling the nature of the cyclization products probably through conformational folding (Krief et al. 1987).

The fact that 20-aza-dammarenol (40) is completely inactive on all OS cyclases (Delprino et al. 1984), whereas 2-aza-2-dihydrosqualene (41) (Delprino et al. 1983) and N-substituted-8-aza decalin (33) are strongly active, could reflect the inability of the cyclase active site, to rapidly attain in the absence of catalysis, conformations complementary to HEI such as (12) or (13) which have structures far from these of substrates or products (Fig. 2) (Frieden et al. 1980). Thus analogues of TS or HEI close to the structure of substrate, such as (41), or close to the structure of product, such as (33), would be in rapid equilibrium whereas those with a structure far from that of product or substrate, such as (40) would not.

According to these views, if there is a pronounced difference in structure between the substrate in the ground state and in the transition state, there is no reason why the active site should rapidly bind such ammonium derivatives since perfect complementarity does not exist between enzyme and these transition-state analogue inhibitors when they first encounter each other.

2.3 Inhibition of Cycloeucalenol-Obtusifoliol Isomerase (COI) Δ^{14}-Reductase (Δ^{14}-Red) and $\Delta^8 - \Delta^7$-Sterol Isomerase (SI)

2.3.1 N-Substituted Azadecalins

The N-substituted 8-azadecalins (33 – 37) (Fig. 6) have electronic and structural similarities with the carbocationic HEIs (42), (43) and (44) involved in the thre enzymatic steps shown in Fig. 4 and could, therefore, be considered as potential inhibitors of COI, Δ^{14}-Red and SI. Indeed these compounds were shown to strongly inhibit COI, SI (Table 7) (Rahier et al. 1985) and more recently Δ^{14}-Red (Taton et al. 1989) in a cell-free system from maize seedlings.

We were able to show that a positive charge at C8 is the major basis of the affinity, since an electrostatic neutral isosteric analogue (46) of (45) does not inhibit these reactions. The affinities measured for the different analogues in this series also show the importance of the presence of a N-substituent able to mimic the hydrophobic part of the C, D rings and lateral chain of the steroid nucleus. The strongest inhibitors in this series show a ID$_{50}$/K$_m$ ratio of 2×10^{-4} (Table 7). However, in this series of inhibitors the structure-activity relationship for the N-alkyl substituent is not so tight as for the inhibition of OS-cyclase. This could reflect the smaller conformational changes occurring for substrate and active site during the catalysis brought about by COI and SI in comparison to OS-cyclase.

In the case of COI, comparison of the structure-affinity relationship for substrates and for compounds of the azadecalin series, shows that structural features which are compulsory for binding of substrates or ground state analogues, affect the binding of the related analogue in the azadecalin series to a much lesser extent. Thus, substrates and ground state analogues inhibitors on one hand and azadecalins on the other probably interact with two different conformations of the active site of COI; this would be in accordance with the nature of HEI analogue postulated for the azadecalins (Rahier et al. 1989). Moreover, these results stress the major contribution of the electrostatic interaction during the binding of these stable HEI analogues or the stabilization of the unstable carbocationic HEIs.

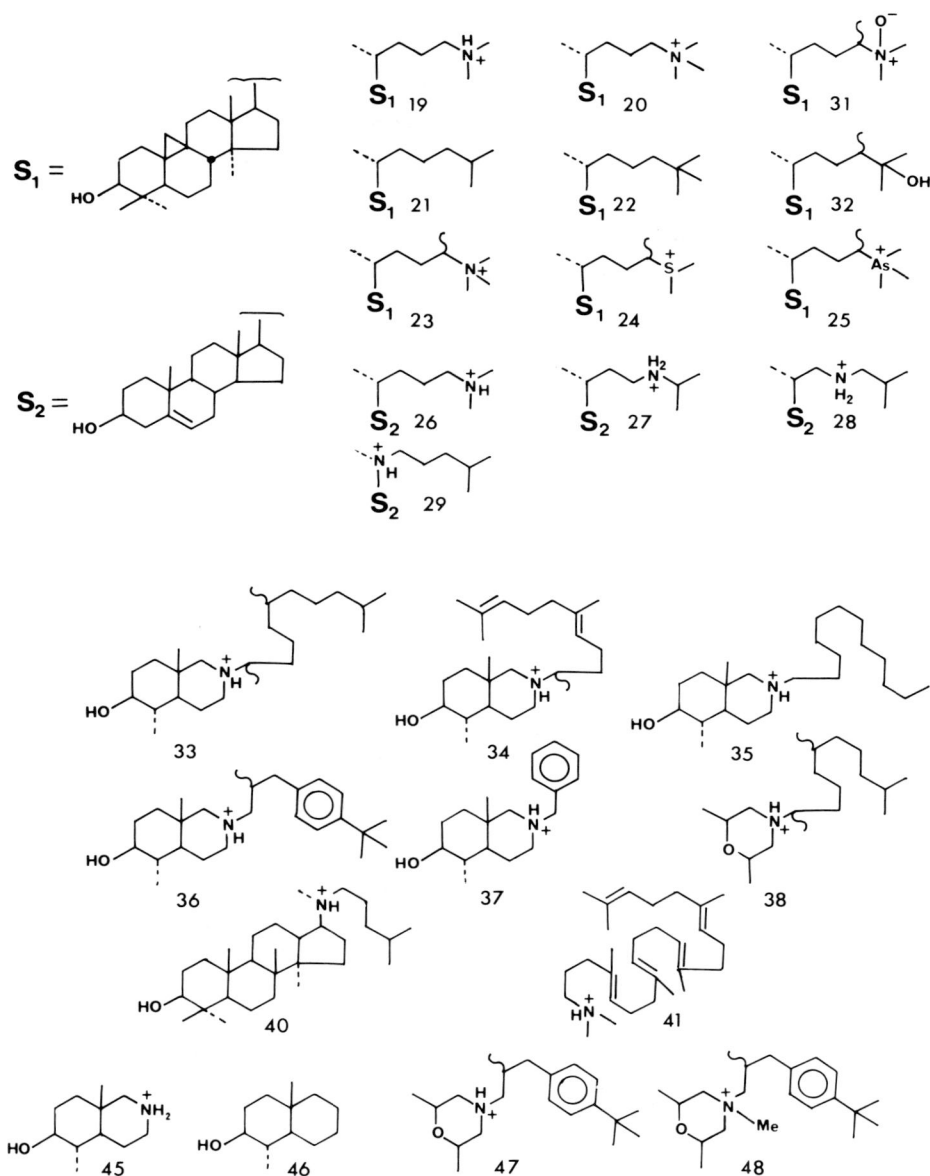

Fig. 6. Chemical structures of the compounds considered in the present work

It is interesting to consider how the same ammonium derivative is able to mimic three different HEI in the C9 (42), C14 (43) and C8 (44) positions. As already discussed, this might de due to the delocalized charge of such ammonium ions. The relative freedom within the active site of model analogues of HEI with a flexible N-substituent could also account for the inhibition of these three enzymes by the same

Table 7. Inhibition of the cycloeucalenol-obtusifoliol isomerase (COI) and $\Delta^8 \rightarrow \Delta^7$-sterol isomerase (SI) by N-substituted-8-aza-decalins and analogues

	33	34	35	36	37	38	45	46
	ID_{50} (μM)							
COI	0.025	0.035	0.1	0.1	0.1	0.4	17	NI
SI	0.2	ND	ND	ND	0.13	0.4	10	NI

ID_{50}: concentration of inhibitor required to reduce the reaction velocity by 50%. The concentration of the substrates: cycloeucalenol and 4α-methyl-5α-ergosta-8,24-dien-3β-ol of the COI and the SI respectively are close to the Km values (100 μM) of both enzymes, therefore ID_{50} values are of the order of the Ki.
ND: not determined.
NI: not inhibitory.

N8-ammonium derivatives, stressing again the importance of the electrostatic interaction of such compounds with the active site.

2.3.2 N-Substituted Morpholines

One important group of fungicides that act as inhibitors of sterol biosynthesis are the morpholine derivatives such as tridemorph (38) and fenpropimorph (47). In fungi, $\Delta^8 - \Delta^7$-SI is the most sensitive target for tridemorph whereas fenpropimorph or fenpropidin generally strongly inhibit Δ^{14}-Red (Baloch and Mercer 1987). However, under some experimental conditions the major target of fenpropimorph has been found to be $\Delta^8 \rightarrow \Delta^7$-SI (Kerkenaar 1987).

Using an enzyme assay in a microsomal preparation isolated from corn embryos, we have recently shown that tridemorph and fenpropimorph strongly inhibit COI (Rahier et al. 1986) and SI (Taton et al. 1987b).(Ki = 4.10^{-7} M). Based on our previous results obtained with the azadecalin series we investigated the importance of the charge on the nitrogen atom of the morpholine ring in theses molecules.

We first showed that the quaternary ammonium derivative (48) of fenpropimorph (47) displays a strong activity relative to the parent amine probably binding in its protonated form.

We then demonstrated that the potency of inhibition of COI and SI is strongly pH dependent. Assuming that the observed strong affinity is due mainly to the interaction of the morpholine cation with a complementary negatively charged surface of the enzyme, the variation in its affinity with pH is not only dependent on the fraction of inhibitor in the charged form but also on the protonation state of the residue(s) of the enzyme interacting with it. Results of the study of the variation of affinity with pH for fenpropimorph (47) and its ammonium derivative (48) for both COI and SI show that changing pH alters the affinity of the enzyme for fenpropimorph in a manner consistent with exclusive binding of the morpholinium cation to both COI and SI. The results suggest the existence of an electrostatic interaction between this cation and a negatively charged residue of the enzyme with a more basic pK_A (>8) in the case of COI than in the case of SI ($pK_A < 6$). An attractive possibility would be that these residues are those stabilizing the HEIs involved in both reactions. N-substituted alkyl-morpholines could then be considered as in the case of azadecalins, as transition-state analogues inhibitors (Taton et al. 1987b). The physiological consequences of these findings may be important if one considers the in vivo effects of the morpholines (38) and (47) on sterol biosynthesis. Since the pH optimum for SI inhibition is around 6 whereas COI inhibition does not vary in the pH range 6 to 8.5, a change in intracellular pH may strongly modify the inhibition of one enzyme with respect to the other. This could explain why, for instance fenpropimorph at low concentrations, mostly inhibits SI in bramble suspension cultures whereas COI is the major target in maize seedlings. Such considerations could also account for the only slight inhibition in vivo of $\Delta^{8,14}$-sterol Δ^{14}-reductase in bramble suspension culture by (47) whereas we have recently demonstrated that (47) is a potent inhibitor of this enzyme in vitro (Taton et al. 1989) in accordance with recent data obtained in several fungi (Baloch and Mercer 1987). Finally this study emphasizes the fact that the effectiveness of such systemic xenobiotic compounds, including fenpropidin, in plants and possibly in fungi, could depend on the pH of the cellular compartments containing the target enzymes.

3 Conclusions

The studies described in this article stress the overwhelming importance of electrostatic interactions during the binding of inhibitory analogues of HEIs to the enzymatic site. The nature of the negatively charged group on the enzyme interacting with these inhibitors or stabilizing the carbocationic intermediate in the catalytic process is not known, but it could be a delocalized carboxylate anion as shown in glycosidases or a sulphhydryl group as recently shown in the case of isopentenyl pyrophosphate isomerase (Reardon and Abeles 1986). Moreover, the major importance of such electrostatic interactions between charged substrates and complementary charged amino acids, besides steric and hydrophobic effects, has recently been demonstrated in the case of subtilisin by protein engineering of such ion-pair interactions. It was clearly demonstrated that substrate preference was not dominated by steric repulsion but correlated best with the charge and not the size of the interacting amino-acid side chains. In that case, tailoring of electrostatic interactions could increase k_{cat}/K_m toward complementary charged substrates by up to 1900 times (Wells et al. 1987).

The inhibitors discussed here permit the control in vitro of several key enzymes involved in sterol biosynthesis. Moreover the results we obtained in the use of these inhibitors in in vivo systems, such as bramble cells suspension cultures or maize seedlings, are in full agreement with the in vitro effects. Profound qualitative and quantitative changes in sterol profiles were obtained using these inhibitors (Taton et al. 1987c). Recent data confirm that the resulting abnormal sterols are distributed in pericellular and intracellular membranes (mitochondria, endoplasmic reticulum and plasma membrane) (Benveniste et al. 1984).

The effects of these inhibitors on cell growth were quite different depending on the enzymatic target. The link between phytotoxicity and the nature of the target is still unresolved except in the case of a complete inhibition of OS-cyclase, leading to accumulation of precursors laching the tetracyclic nucleus and unable to fulfill the role of sterols in the membrane. The divergent hological effects may be due to the existence of other, as yet unknown, cellular targets for these molecules.

Finally, in addition to their value as probes to study enzymatic reaction mechanisms and the role of sterols in plant membranes, these inhibitors could lead to applications as potential antifungal or hypocholesterolemic agents (Gerst et al. 1988).

Acknowlegments. We are indebted to Drs. C. Anding and P. Place (Rhône-Poulenc Agrochimie) for their participation in the synthesis of the 8-aza-decalins series and Dr. A. S. Narula for his participation in the synthesis of the 25-aza-cycloartanol series. We acknowledge BASF Agrochemical Station (Limburgerhof, Federal Republic of Germany) for having provided the *N*-alkyl-morpholines used in the present work. Many thanks to B. Bastian for kindly typing the manuscript.

References

Akhtar M, Rahimtula AD, Wilton DC (1970) The stereochemistry of hydrogen elimination from C7 in cholesterol and ergosterol biosynthesis. Biochem J 117:539–542

Altman LJ, Hau CY, Bertolino A, Haudy G, Laugani D (1978) Stereochemistry of the 1,3-proton loss from a chiral methyl group in the biosynthesis of cycloartenol as determined by tritium NMR spectroscopy. J Am Chem Soc 100:3235–3237

Avruch L, Fisher S, Pierce HD Jr, Oehlschlager AC (1976) The induced biosynthesis of 7-dehydrocholesterols in yeast: potential sources of new provitamin D_3 analogs. Can J Biochem 54:657–665

Baloch RI, Mercer EI (1987) Inhibition of $\Delta^8 - \Delta^7$-isomerase and Δ^{14}-reductase by fenpropimorph, tridemorph and fenpropidin in cell-free enzyme systems from *Saccharomyces cerevisiae*. Phytochemistry 26:3663–3668

Benveniste P (1986) Sterol Biosynthesis. Annu Rev Plant Physiol 37:275–308

Benveniste P, Bladocha M, Costet MF, Ehrhardt A (1984) Membranes and compartmentalisation in the regulation of plant functions. (Boudet AM, Alibert G, Margio G, Lea PJ (eds). Annu Proc Phytochem Soc Eur 24:283–300

Blättler WA (1978) Sterischer Verlauf der Bildung und Öffnung des Cyclopropanringes in der Biosynthese von Phytosterinen. PhD Thes, Eidgenöss Tech Hochsch, Zürich

Boar RB, Couchman LA, Jaques AJ, Perkins MJ (1984) Isolation from *Pistacia* resins of a bicyclic triterpenoid representing an apparent trapped intermediate of squalene 2,3 epoxide cyclization. J Am Chem Soc 106:2476–2477

Bottema CK, Parks LW (1978) Δ^{14}-reductase in *Saccharomyces cerevisiae*. Biochem Biophys Acta 531:301–307

Cerutti M, Delprino L, Cattel L, Bouvier-Navé P (1985) N-oxide as potential function in the design of enzyme inhibitors. Application to 2,3-oxidosqualene sterol cyclases. J Chem Soc Chem Commun 1985:1054–1055

Corey EJ, Lin K, Yamamoto H (1969) Separation of the cyclisation and rearrangement processes of sterol biosynthesis. Enzymic formation of a protosterol derivation. J Am Chem Soc 91:2132–2134

Cornforth JW (1974) Enzyme and stereochemistry. Tetrahedron 30:1515–1524

Delprino L, Balliano G, Cattel L, Benveniste P, Bouvier P (1983) Inhibition of Higher Plant 2,3-Oxidosqualene Cyclase by 2-Aza-2,3-dihydrosqualene and its Derivatives. J Chem Soc Chem Commun 1983:381–382

Delprino L, Caputo O, Balliano G, Berta S, Bouvier P, Cattel L (1984) Biosynthesis of β-amyrin. Part 3. Synthesis and biological evaluation of 17(βH)- and 17(αH)-azadammaran-3β-ol. J Chem Res 5:254–255

Dewar MJS, Reynolds CH (1984) Ground-states of molecules. π-complexes as intermediates in reactions. Biomimetic cyclization. J Am Chem Soc 106:1744–1750

Eschenmoser A, Ruzicka L, Jeger O, Arigoni D (1955) Zur Kenntnis der Triterpene. Eine stereochemische Interpretation der biogenetischen Isoprenregel bei den Triterpenen. Helv Chim Acta 38:1890–1904

Frieden C, Kurz LC, Gilbert HR (1980) Adenosine Deaminase and Adenylate Deaminase: Comparative kinetic studies with Transition State and Ground State Inhibitors. Biochemistry 19:5303–5309

Gerst N, Duriatti A, Schuber F, Taton M, Benveniste P, Rahier A (1988) Potent inhibition of cholesterol biosynthesis in 3T3 fibroblasts by N-[(1,5,9)-trimethyldecyl]-4α,10-dimethyl-8-aza-trans-decal-3β-ol, a new 2,3-oxidosqualene cyclase inhibitor. Biochem Pharmacol 37 (10):1955–1964

Goodwin TW (1979) Biosynthesis of terpenoids. Annu Rev Plant Physiol 30:369–404

Greenberg A, Winkler R, Smith BL, Liebman JF (1982) The negatively charged nitrogen in ammonium ion and derived concepts of acidity, basicity, proton affinity and ion energetics. J Chem Educ 59(5):367–370

Heintz R, Benveniste P (1974) Plant sterol metabolism. Enzymatic cleavage of the 9β,19-cyclopropane ring of cyclopropyl sterols in bramble tissue cultures. J Biol Chem 249:4267–4274

Johnson SW, Lindell SD, Steele J (1987) Rate enhancement of biomimetic polyene cyclizations by a cation-stabilizing auxililary. J Am Chem Soc 109:2517–2518

Kerenaar A (1987) Modern selective fungicides. In: Lyr H (ed). Fischer, Jena

Krief A, Schander JR, Guitte E, Herve du Penhoat C, Lalemand JY (1987) About the mechanism of sterol biosynthesis. J Am Chem Soc 109:7910–7911

Mihailovic MM (1984) Biosynthesis of phytosterols in *Trebouxia* spp.: Steric course of the C-alkylation step. PhD Thes, Eidgenöss Techn Hochsch, Zürich

Oehlschlager AC, Augus RH, Pierce AM, Pierce HD Jr, Srinivasan R (1984) Azasterol inhibition of Δ^{24}-sterol methyl transferase in *Saccharomyces cerevisiae*. Biochemistry 23:3582–3589

Pauling L (1946) Molecular architecture and biological reactions. Chem Eng News 24:1375–1377

Pullman A, Armbruster AM (1974) An ab initio study of the hydration and ammoniation of ammonium ions. Int J Quant Chem Symp 8:169–176

Rahier A (1980) Biosynthèse des stérols chez les plantes. PhD Thes, Univ Strasbourg, France

Rahier A, Narula AS, Benveniste P, Schmitt P (1980) 25-aza-cycloartanol, a potent inhibitor of S-adenosyl-L-methionine-sterol-C-24 and C-28-methyl-transferase in higher plant cells. Biochem Biophys Res Commun 92:20–25

Rahier A, Genot JC, Schuber F, Benveniste P, Narula AS (1984) Inhibition of S-adenosyl-L-methionine sterol C-24-methyltransferase by analogues of a carbocationic ion high energy intermediate. J Biol Chem 259:15215–15223

Rahier A, Taton M, Schmitt P, Benveniste P, Place P, Anding C (1985) Inhibition of Δ^8–Δ^7-sterol isomerase and of cycloeucalenol-obtusifoliol isomerase by N-benzyl-8-aza-4α,10-dimethyl-trans-decal-3β-ol, an analogue of a carbocation high energy intermediate. Phytochemistry 24:1223–1232

Rahier A, Schmitt P, Huss B, Benveniste P, Pommer EH (1986) Chemical structure activity relationships of the inhibition of sterol biosynthesis by N-substituted morpholines in higher plants. Pestic Biochem Physiol 25:112–124

Rahier A, Taton M, Benveniste P (1989) Cycloencalenol-obtusifoliol isomerase. Structural requirements for transformation or binding of substrates and inhibitors. Eur J Biochem

Reardon JE, Abeles RH (1986) Mechanism of Action of Isopentenyl Pyrophosphate Isomerase: Evidence for a Carbonium Ion Intermediate. Biochemistry 25:5609–5616

Taton M, Benveniste P, Rahier A (1986) N-[(1,5,9)-trimethyl-decyl]-4α,10-dimethyl-8-aza-trans-decal-3β-ol, a novel potent inhibitor of 2,3-oxidosqualene-cycloartenol and -lanosterol cyclases. Biochem Biophys Res Commun 138(2):764–770

Taton M, Benveniste P, Rahier A (1987a) Use of rationally designed inhibitors to study sterol and triterpenoid biosynthesis. Pure Appl Chem 59(3):287–294

Taton M, Benveniste P, Rahier A (1987b) Mechanism of inhibition of sterol biosynthesis enzymes by N-substituted morpholines. Pesti Sci 21:269–280

Taton M, Benveniste P, Rahier A (1987c) Comparative study of the inhibition of sterol biosynthesis in *Rubus fructicosus* suspension cultures and *Zea mays* seedlings by N-[(1,5,9)-trimethyl-decyl]-4α-10-dimethyl-8-aza-trans-decal-3β-ol and derivative. Phytochemistry 26(2):385–392

Taton M, Benveniste P, Rahier A (1989) Microsomal sterol Δ^{14}-reductase in higher plants. Characterization and inhibition by analogues of a phesumptive carbocationic intermediate of the reduction reaction. Eur J Biochem (in press)

Thompson MJ, Dupont J, Robbins WE (1963) The sterols of liver and carcass of 20,25-diaza-choles-terol-fed rats. Steriods 2 (1):99–104

Van Tamelen EE (1981) The role of organic synthesis in bioorganic chemistry. Pure Appl Chem 53:1259–1270

Van Tamelen EE (1982) Bioorganic characterization and mechanism of the 2,3-oxidosqualene-lanosterol conversion. J Am Chem Soc 104:6480–6481

Van Tamelen EE, James DR (1977) Overall Mechanism of terpenoid terminal epoxide polycyclization. J Am Chem Soc 99:950–952

Wang SF, Kawahara FS, Talalay P (1963) The mechanism of the Δ^5-3-cetosteroid isomerase reaction: absorption and fluorescence spectra of enzyme-steroid complexes. J Biol Chem 238:576–585

Wells JA, Powers DB, Bott RR, Graycar TP, Estell DA (1987) Designing substrate specificity by protein engineering of electrostatic interactions. Proc Natl Acad Sci USA 84:1219–1223

Wilton DC, Watkinson IA, Akhtar M (1970) The stereochemistry of hydrogen transfer from reduced nicotinamide adenine dinucleotide phosphate in the reduction of ethylenic linkages during choles-terol biosynthesis. Biochem J 119:673–675

Wolfenden R (1976) Transition state analog inhibitors and enzyme catalysis. Annu Rev Biophys Bioeng 5:271–306

Chapter 15

Lanosterol to Ergosterol – Enzymology, Inhibition and Genetics

S. L. KELLY, S. KENNA, H. F. J. BLIGH, P. F. WATSON, I. STANSFIELD, S. W. ELLIS, and D. E. KELLY [1]

1 Introduction

The biosynthetic steps from lanosterol to ergosterol are of both pure and applied interest in terms of the biotransformations involved and in their inhibition by antifungal compounds. These antifungal compounds have application in medical (Saag and Dismukes 1988) and agricultural (Kuck and Scheinpflug 1986) contexts.

In recent years ergosterol biosynthesis and its inhibition have been subject to a variety of experimental studies and these have been reviewed extensively. The reader is referred to reviews by Vanden Bossche (1985) and Kato (1986) and to the text "Modern Selective Fungicides" edited by Lyr (1987). Studies on ergosterol biosynthesis inhibitors (EBI's) have also shown such compounds to be useful tools in the investigation of the roles of sterols in membranes and in the initiation of molecular cloning studies (Kalb et al. 1986; Kelly et al. 1986). These developments are set to continue, and offer the prospect for producing elegant experimental systems for evaluating the function of and requirements for sterols, as well as for comparative gene sequence determination of target proteins. The latter development will also permit overexpression of such proteins which with subsequent evaluation will be of value in the rational design of novel antifungal compounds.

1.1 Biosynthetic Steps of Ergosterol Biosynthesis

The biosynthesis of ergosterol from lanosterol in yeasts and filamentous fungi is illustrated in Fig. 1 (Barton et al. 1973; Fryberg et al. 1973, 1975; H. D. Pierce et al. 1978). The reactions in the biosynthetic pathway are:

(1) demethylation of the two methyl groups at C-4 and the single methyl group at C-14;
(2) methylation at C-24 with reduction of $\Delta^{24(25)}$ and formation of $\Delta^{24(28)}$;
(3) isomerization of the Δ^8-double bond to Δ^7;
(4) introduction of a Δ^5-double bond;
(5) introduction of Δ^{22}-double bond;
(6) reduction of $\Delta^{24(28)}$.

[1] Department of Pharmacology and Therapeutics, and Department of Molecular Biology and Biotechnology, Sheffield University, Sheffield S10 2TN, UK

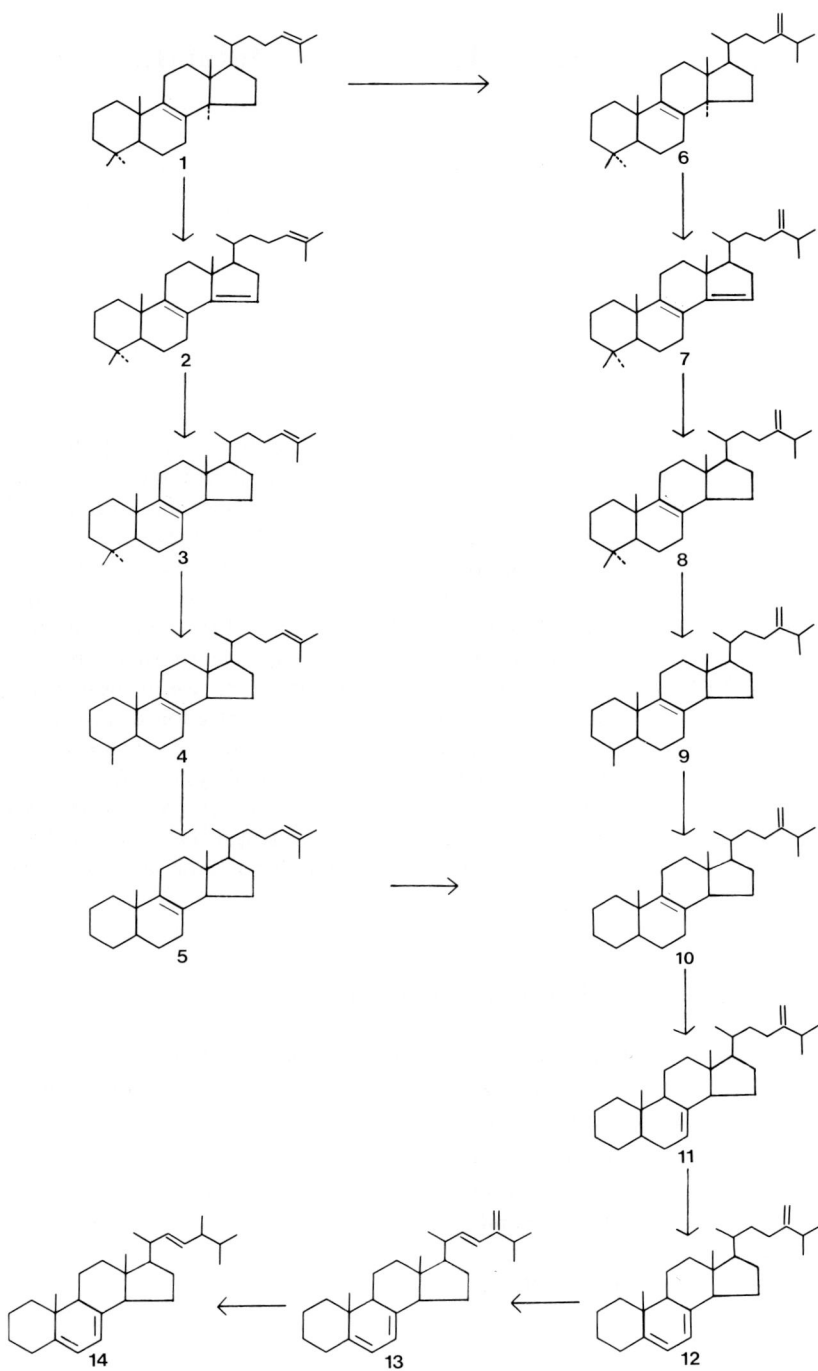

Fig. 1

It appears that in filamentous fungi methylation at C-24 predominates as the first reaction from lanosterol (Goulston et al. 1975). The first demethylation is at C-14 with conversion of the methyl to hydroxymethyl and formyl groups and elimination of the carbon atom as formic acid. This reaction is mediated by a single cyt P-450 (Aoyama et al. 1987) and is followed by the reduction of the Δ^{14}-double bond produced to give a Δ^8 sterol. The two methyl groups at C-4 are then removed by stepwise oxidation (Gibbons et al. 1982).

After demethylation (C-14 and C-4) and transmethylation (C-24) reactions, isomerization of the Δ^8-double bond takes place to give the Δ^7 sterol. This is followed by desaturations at the C-5 and C-22 position, and finally by reduction of the $\Delta^{24(28)}$-double bond. Alterative reaction sequences are possible depending on fungal species and strains (Kato 1986). Some fungi also have sterols other than ergosterol as their abundant sterol (Weete 1973; Huang-Kuang et al. 1972; McCorkindale et al. 1969).

2 Antifungal Compounds and Their Activities

Inhibitors of ergosterol biosynthesis after lanosterol synthesis have been found for the following steps:

(1) C-14 demethylation;
(2) $\Delta^8 \rightarrow \Delta^7$ isomerization;
(3) Δ^{14} reduction;
(4) C-24 transmethylation;
(5) Δ^{22} desaturation;
(6) $\Delta^{24(28)}$ reduction.

2.1 Inhibitors of C-14 Demethylation

Compounds that have been found to be active as inhibitors of the C-14 demethylation of lanosterol include pyrimidines, pyridines, piperazines, imidazoles and triazoles (Fig. 2). Collectively they are known as demethylation inhibitors (DMI's) or sterol demethylase inhibitors (SDI's). This group of compounds represents a large number of economically important antifungal agents. Analysis of the sterols produced following treatment with these compounds shows that the sterols that accumulate contain a methyl group at C-14.

Fig. 1. Biosynthetic sequences from lanosterol to ergosterol (after Kato 1986). *1*, Lanosterol; *2*, 4,4-Dimethyl-cholesta-8,14,24-trienol; *3*, 4,4-Dimethyl-cholesta-8,24-dienol; *4*, 4-Methylcholesta-8,24-dienol; *5*, Zymosterol; *6*, 24-Methylenedihydrolanosterol; *7*, 4,4-Dimethylergosta-8,14,24(28)-trienol; *8*, 4,4-Dimethylergosta-8,24(28)-dienol; *9*, 4-Methylergosta-8,24(28)-dienol; *10*, Fecosterol; *11*, Episterol; *12*, Ergosta-5,7,24(28)-trienol; *13*, Ergosta-5,7,22,24(28)-tetraenol; *14*, Ergosterol

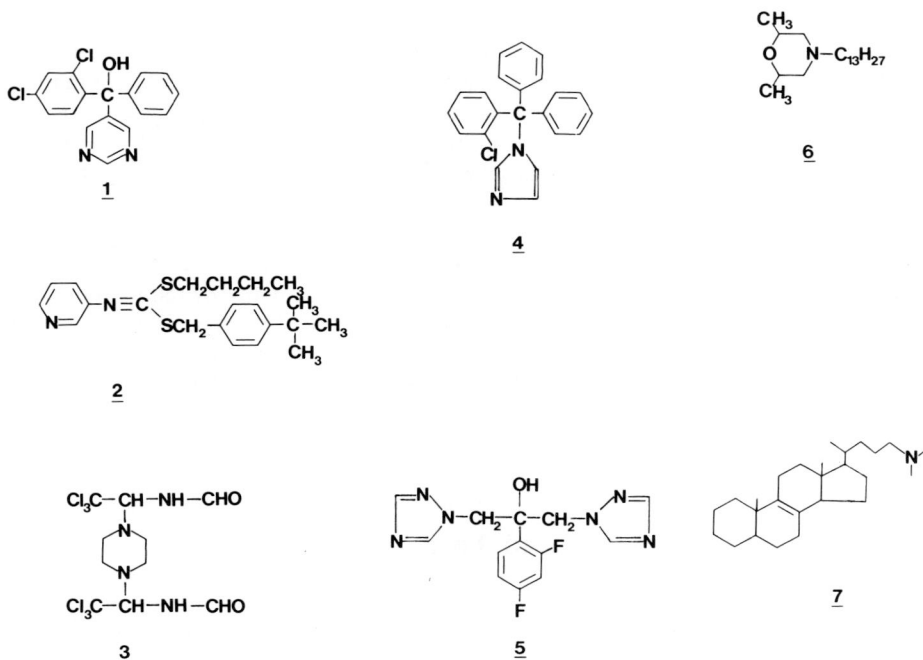

Fig. 2. Examples of EBI's that inhibit enzymic steps between lanosterol and ergosterol: *1*, Triarimol; *2*, Buthiobate; *3*, Triforine; *4*, Clotrimazole; *5*, Fluconazole; *6*, Tridemorph; *7*, 25-Aza-24, 25-dihydro-zymosterol

Of the DMI's it is the imidazoles, and increasingly the triazoles, that have been particularly important commercially, and which have a broad spectrum of activity (Scheinpflug and Kuck 1987). In medical applications the triazoles, fluconazole and itraconazole offer the prospect of effective antifungal treatment with good toxicological properties. Such drugs will become increasingly important with the development of immune-suppressive therapies and with the spread of AIDS. In the latter case the persistence or recurrence of oropharyngeal *Candidosis* is frequent and DMI's will be of value in control of the disease (Dupont and Prouhet 1988).

The structure of a representative from each of the groups of EBI's is shown in Fig. 2. The presence of a heterocyclic ring containing an unhindered nitrogen enables binding of the inhibitor to the haem moiety of the cytochrome P-450 component of the C-14 demethylase enzyme system (Wilkinson et al. 1972; Vanden Bossche 1985). The relative affinity of this binding, which includes interaction with the apoprotein, determines the efficiency of the compound at inhibiting the cyt P-450-mediated transformation.

2.1.1 Pyrimidines, Pyridines and Piperazines

The pyrimidine fungicide triarimol was the first DMI for which inhibition of ergosterol biosynthesis was demonstrated (Ragsdale and Sisler 1972). The vast majority of sterols in treated cultures of *Ustilago maydis* were found to be C-14 methylated. The pyrimidines, fenarimol and nuarimol, have also achieved commercial status and are used mainly on fruits and vegetables.

Buthiobate is the only commercial pyridine DMI and is used mainly against powdery mildews. This compound has been shown to inhibit ergosterol biosynthesis in *Monilinia fructigena* (Kato et al. 1974). Similarly triforine is the only piperazine DMI to be commercialized as an agricultural fungicide. It has a broad spectrum of activity (Schicke and Veen 1969) and has been shown to inhibit ergosterol biosynthesis with the accumulation of C-14-methylated sterols in *Neurospora crassa* (Sherald et al. 1973) and *Aspergillus fumigatus* (Sherald and Sisler 1975).

2.1.2 Imidazoles

Imazalil was the first agricultural compound of this class and is used in seed treatment of cereals, foliar treatment of bananas, vegetables and ornamental plants and for the post-harvest treatment of citrus fruits (Scheinpflug and Kuck 1987). Imazalil inhibits ergosterol biosynthesis in a range of species including *U. avenae, U. maydis, B. cinerea, Penicillium expansium* and *A. nidulans* causing accumulation of C-14 methylated sterols (Kato 1986). Other imidazole DMI's used as agricultural fungicides include prochloraz, while the medical antifungal agents include clotrimazole, miconazole, econazole, tioconazole and ketoconazole. The latter compound has proved the compound of choice for many fungal infections and may be administered orally or intravenously (Saag and Dismukes 1988).

2.1.3 Triazoles

Various triazole compounds have been introduced as agricultural fungicides including fluotrimazole, triadimefon, triadimenol, bitertanol, propiconazole, etaconazole, penconazol, flutriafol, and flusilazol, and others, e.g. BAS 454 06 F are under investigation (Scheinpflug and Kuck 1987). The triazole antimycotics, fluconazole and itraconazole, are important potential improvements for use in medicine and are close to general release. The development of these compounds was in response to the need for DMI's with broad spectrum anti-fungal activity but also with selectivity, so as not to inhibit human cyt P-450. In addition it was desirable to develop a drug which could be administered orally. Fluconazole has been reported to show promising activities against cryptococcal meningitis and urinary tract infection caused by *Candida*. Itraconazole also has good general activity and may be of particular value in treating *A. fumigatus* (Saag and Dismukes 1988).

In the case of the triazoles, both agricultural fungicides and antimycotics, there is again evidence for the accumulation of C-14-methylated sterols following treatment of a range of target organisms (M. J. Henry and Sisler 1981; Marriott and Richardson 1987; Vanden Bossche 1987).

2.2. Inhibitors of $\Delta^8 \rightarrow \Delta^7$ Isomerization and Δ^{14} Reduction

The morpholine antifungal agents, which include tridemorph, fenpropimorph and fenpropidin, inhibit the $\Delta^8 \rightarrow \Delta^7$ isomerization step in ergosterol biosynthesis and are particularly effective against powdery mildews (Kerkenaar 1987). The mode of action of these EBI's has been elucidated in a series of studies since the late 1970s, and it has shown that in addition to inhibition of the $\Delta^8 \rightarrow \Delta^7$ isomerization, the Δ^{14} reductase may also be inhibited (Kerkenaar et al. 1981; Baloch et al. 1984). For example in *S. cerevisiae*, tridemorph, fenpropimorph and fenpropidin were found to inhibit both enzymes, though tridemorph was more active against the isomerase, while fenpropidin and to a lesser extent fenpropimorph were more inhibitory towards the reductase (Baloch and Mercer 1987). In *B. cinerea* tridemorph was found to inhibit the $\Delta^8 \rightarrow \Delta^7$ isomerization (Kato et al. 1980). The azasterol A25822B is another inhibitor of Δ^{14} reduction and is active at concentrations of 1.0 to 10.0 µg/ml. This activity has been detected in *S. cerevisiae* and *U. maydis* (Hays et al. 1977; Bottema and Parks 1978; Kato 1986).

2.3 Inhibitors of Side Chain Modification

Side chain modifications during the biosynthesis of ergosterol from lanosterol include transmethylation at C-24, introduction of a $\Delta^{22(23)}$-double bond and $\Delta^{24(28)}$ reduction. Azasterols with nitrogen substitution in the side chain have been found to inhibit various activities. For instance 25-aza-24,25-dihydrozymosterol inhibits side chain methylation (Avruch et al. 1976), while 23-azacholesterol is active against $\Delta^{24(28)}$ reduction leading to the accumulation of ergosta-5,7,22,24(28)-tetraenol in *S. cerevisiae* (H. D. Pierce et al. 1978). This effect of 23-azacholesterol was observed at concentrations $> 10^{-6}$ M while a higher concentration of 10^{-5} M resulted in inhibition of C-24 transmethylation. Inhibition of C-24 methyltransferase has also been found after treatment with 25-azacholesterol and 25-azacholestanol (A. M. Pierce et al. 1979). The azasterol A25822B was found to inhibit C-24 transmethylation and $\Delta^{24(28)}$ reduction at higher concentrations than that required to inhibit Δ^{14} reduction (Bailey et al. 1976; Hays et al. 1977).

Of the other side chain modification reactions, the inhibition of $\Delta^{24(28)}$ reduction has been reported for the compound DMEA-DHA in *S. cerevisiae* (Field et al. 1979) and triarimol has been found to inhibit Δ^{22} desaturation in addition to C-14 demethylation (Sherald and Sisler 1975). The latter activity probably reflects an inhibitory action on the Δ^{22} desaturase enzyme which is a cyt P-450 though different from the one responsible for C-14 demethylation (Hata et al. 1987).

2.4 Biological Effects of Ergosterol Biosynthesis Inhibitors (EBI's)

The inhibition of ergosterol biosynthesis after treatment with EBI compounds can give rise to diverse effects which may be due to either ergosterol depletion and/or the accumulation of sterol intermediates (see also Chapter 10). Similar phenotypes in mutant strains defective in the same biosynthetic step will support the association of ergosterol-biosynthesis inhibition and cellular perturbations although it must be assumed that for an effective antifungal action that the biosynthetic step inhibited must be vital for growth.

Other secondary effects of EBI's may be unrelated to their effect on sterol biosynthesis. In studies of the antimycotics miconazole and clotrimazole, Sud and Feingold (1981) observed that direct membrane damage resulted from treatment of S. cerevisiae with concentrations in excess of those required to inhibit growth. At the higher concentrations required to produce direct membrane damage, these compounds were fungicidal rather than fungistatic. Effects upon the membranes of C. albicans have been observed following treatment with a fungicidal concentration of clotrimazole that caused membranes to leak cellular constituents (Iwata et al. 1973). Treatment of B. cinerea with high concentrations of tridemorph has also been observed to result in membrane leakage (Kato et al. 1980).

Ketoconazole apparently lacks this secondary effect (Sud and Feingold 1981) although of interest is the difference between strains of S. cerevisiae some of which show a fungicidal response and others a fungistatic effect, after ketoconazole treatment at the MIC (Fig. 3). This may reflect different genetic backgrounds of the strains involved. In similar studies of fungicidal vs fungistatic effects, but between different species rather than strains, Sisler et al. (1984) suggested that a fungicidal response may be associated with an inability to remove the two C-4 methyl groups from sterols retaining the C-14 methyl group. This interpretation is based on sterol feeding experiments with S. cerevisiae sterol auxotrophs where it was found that growth occurred upon supplementing the medium with cholesterol or 14α-methylcholest-7-en-3β-ol, but not with C-4 methyl sterols (Buttke and Bloch 1981). In another case a

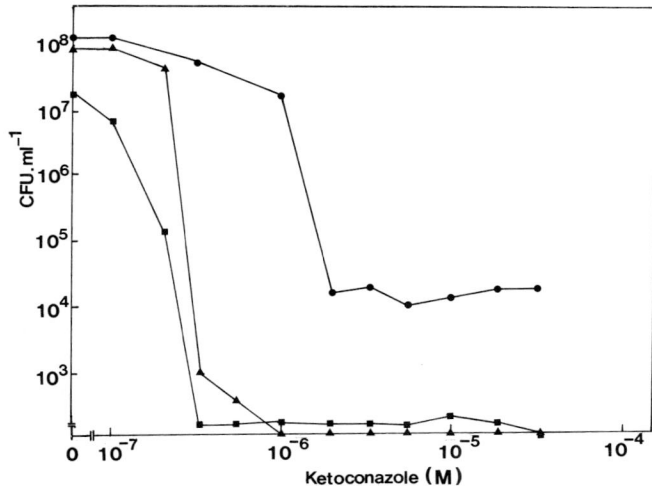

Fig. 3. Response of S. cerevisiae strains BB (▲), XL165B (■) and XY729 5a (●) to treatment with ketoconazole. Colony forming units (CFU) were determined after treatment of 500 cells/ml with ketoconazole for 3 days in liquid YEPD medium

fungicidal rather than fungistatic activity was reported for bifonazole at doses higher than that required for growth inhibiton. This activity may, however, be associated with an inhibition of HMG-CoA reductase (Berg et al. 1984).

Although some azoles interact with the membrane lipid bilayer and this may be the cause of the secondary fungicidal activity of miconazole described above (Vanden Bossche 1985), many of the effects observed on treatment with EBI's are due to the altered sterols produced and/or to the depletion of ergosterol. Bloch (1982) considered the effect of sterol structure on membranes and indicated that methylated sterols were not suitable for membrane stability and function. This was further supported by sterol auxotroph studies with *S. cerevisiae* which showed that lanosterol could not support growth (Nes et al. 1978).

EBI's have also been found to have effects on the activity of membrane-bound enzymes, on respiration, on fatty acid synthesis and on cell morphology. The changes in cell morphology include thickening of the cell wall, formation of small vesicles between the cell wall and plasma membrane and the deposition of vesicular material in the thickened cell wall (Kato 1986). Among the membrane-bound enzymes indicated to be affected is chitin synthase. This enzyme was inhibited by high levels of ergosterol (Chiew et al. 1982) and Barug et al. (1983) observed altered chitin deposition after bifonazole treatment of *C. albicans*. The chitin was found distributed all over the cell instead of being localized at the junction between mother and daughter cell. In addition, the septae of many treated cells did not exhibit chitin staining with diethanol, and cell separation was disturbed. In filamentous fungi the production of branched germ tubes after SDI treatment may have a similar origin (Kato et al. 1974; Sherald et al. 1973).

It has also been observed that the hyphal mode of growth of *C. albicans* was suppressed by ketoconazole, and miconazole (Vanden Bossche et al. 1983). It seems unlikely that depletion of ergosterol and stimulation of chitin synthase were associated with this effect, as the mycelial form contains about three times the chitin level of the yeast form (Chattaway et al. 1968) and the chitin synthase activity found in the hyphae was twice that of the yeast form (Braun and Calderone 1978). It is of interest to note that polyene-resistant mutants of *C. albicans* accumulating C-14-methylated sterols were observed to be defective in hyphal growth. Changes of both phenotypes in revertants suggest that this may be a defect associated with the metabolic block in sterol metabolism (Shimokawa et al. 1986). It is possible that chitin synthase or other enzymes are critically affected in particular cell locations by ergosterol depletion in both mutants and treated cells. Alternatively, these effects may be due to the accumulation of C-14-methylated sterols. The C-14 methyl sterols may be incompatible with the structural integrity of mycelial membranes or the sterol changes may cause regulatory alterations directing the cell from a mycelial to a yeast growth form.

The effect of EBI compounds on membrane-bound enzymes and on respiration in fungi have also been studied as possible sites of action. Ketoconazole has been found to block respiratory chain electron transport in isolated mitochondria and in whole cells of *C. albicans* through an interaction with cytochrome C oxidase (Shigematsu et al. 1982; Uno et al. 1982). These effects were observed at concentrations several orders of magnitude in excess of those required for inhibitory effects on ergosterol biosynthesis (Vanden Bossche 1985). Other studies using a variety of fungi have shown respiration to be relative insensitive to fenarimol (Sisler et al. 1984).

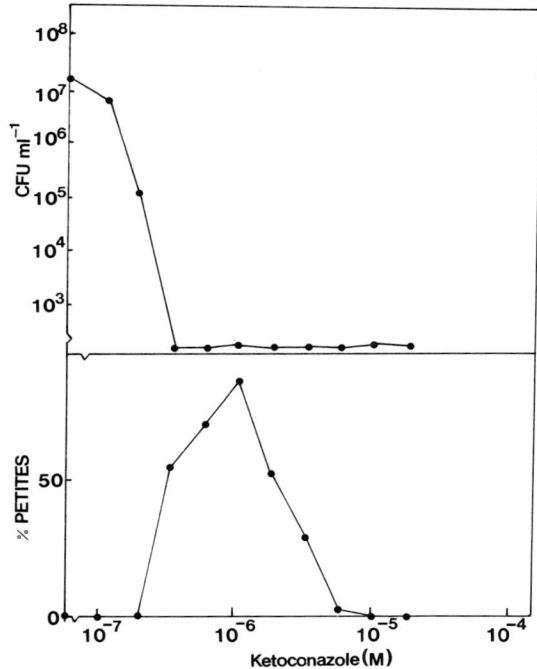

Fig. 4. Petite induction and colony forming units (CFU) observed after treatment of 5000 cells/ml of *S. cerevisiae* strain XL16-5B with ketoconazole for 3 days in YEPD medium

Effects of azole DMI's on other enzyme activities e.g. $Mg^{2+}Na^{+}K^{+}$ ATPase (Vanden Bossche 1985) and cytochrome C peroxidase (Borgers 1980; De Nollin et al. 1977) have also been observed though these effects may be indirect and due to ergosterol depletion in the membranes to which these activities are bound. Studies in our laboratory have shown that ketoconazole and other azoles at sub-MIC concentrations are potent inducers of petite mutations in *S. cerevisiae* (Fig. 4). The level of petite induction has been found to approach 100% of colonies in some strains and result from mitochondrial rather than nuclear mutation (Bligh 1988). This effect does not appear to be associated with DNA damage since DNA repair mutants are not hypersensitive to ketoconazole. The effect is likely to be due to ergosterol depletion, but is of interest since the respiratory deficiency of ergosterol auxotrophs is reversible (Astin et al. 1977; Haslam and Astin 1979). The sensitivity observed in mitochondria from *S. cerevisiae*, if generally applicable, may affect the MIC values for petite negative yeast where petite mutations are not tolerated. This would effectively lower the MIC for these yeasts.

Azole antifungal compounds have also been observed to affect fatty acid metabolism causing accumulation of free fatty acids and increased fatty acid desaturation (Vanden Bossche 1985). Again this effect may be a response by the cell to the inhibition of ergosterol biosynthesis rather than to any direct effect, and these changes in lipids may in turn affect membrane-bound enzymes.

2.4.1 Effects of EBI's on Plants and Animals

In using EBI's as agrochemicals or antimycotics these compounds may potentially also inhibit enzymes in the plants or animals involved. The effects of EBI's on plant metabolism have been investigated, and sterol biosynthesis may be inhibited by SDIs acting against sterol C-14 demethylation (Burden et al. 1987) or by the action of morpholines on cycloeucalenol-obtusifoliol isomerase and $\Delta^8 \to \Delta^7$ isomerase (Rahier et al. 1987; see also Chapter 14). Azole compounds may also affect plant growth due to interference with the cyt P-450 mediated conversion of kaurene to kaurenoic acid in the biosynthesis of gibberellins, the site of action of azole-based plant growth regulators (Shire and Sisler 1976).

In animals, as with plants, the desired effect of the antifungal compounds depends on a differential selectivity towards the fungal systems. The cyt P-450's are a superfamily of enzymes (Nebert et al. 1987) and despite differential levels of activity for target cyt P-450's some compounds have been found to have side effects associated with inhibition of mammalian cyt P-450. Examples are the inhibition of testosterone synthesis and possibly adrenal corticosteroid synthesis by ketoconazole (DeFelice et al. 1981; Pont et al. 1984).

3 Fungal Cytochrome P-450

Fungal cyt P-450 studies are best developed in the yeast *S. cerevisiae* where under conditions of catabolite repression a high level of cyt P-450 is commonly observed (Lindenmayer and Smith 1964; Fig. 5). A repressor gene common in laboratory strains of *S. cerevisae* can, however, cause levels to drop below that detectable by reduced carbon monoxide (CO) difference spectrophotometry used to assay the enzyme (King et al. 1983).

The *S. cerevisiae* cyt P-450 system was analyzed by Yoshida et al. (1974) and found to resemble the hepatic system, with the haemoprotein coupled to NADPH-cyt P-450 reductase in the microsomal fraction together with cyt b_5 and NADH-cyt b_5 reductase. By employing techniques similar to those used in mammalian liver microsomal cyt P-450 purification, the abundant cyt P-450 form present in *S. cerevisiae* grown semi-anaerobically was purified to near homogeneity by Yoshida and Aoyama (1984a, b) and found to catalyse the C-14 demethylation of lanosterol in a reconstituted system with NADPH cyt P-450 reductase. The cyt P-450, named cyt P-450$_{14DM}$ had a molecular weight of 58000 and a reduced CO difference spectrum with a maximum absorbance at 448 nm.

In similar studies using a different strain and 20% w/v glucose in place of 6% (w/v) glucose in the medium, King et al. (1984) have also analyzed the properties of the principal cyt P-450 of *S. cerevisiae*. They found a cyt P-450-mediated activity for benzo[a]pyrene metabolism and determined a molecular weight of 55500 for the purified protein. This showed a Type I substrate-binding spectrum on addition of lanosterol. The enzyme purified by Yoshida and Aoyama (1984a, b) did not show a Type I-binding spectrum with benzo[a]pyrene and the enzymes purified may, there-

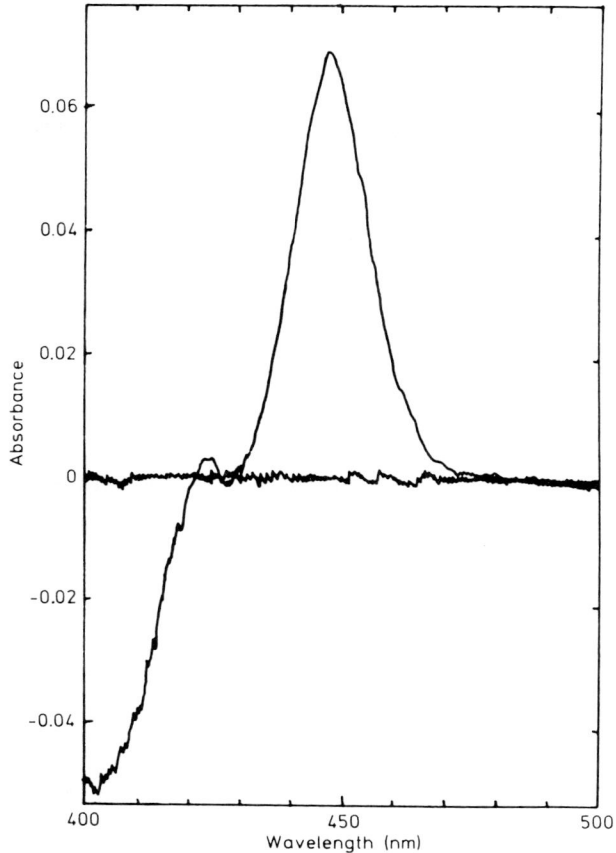

Fig. 5. Reduced carbon monoxide difference spectrum of a microsomal fraction of *S. cerevisiae* strain NCYC 754

fore, represent different forms of cyt P-450. However, in each case only a single predominant cyt P-450 form was found in the microsomal fraction. Furthermore, it is unlikely that these purified enzymes represent different forms resulting from physiological variations in growth. The presence of a mutant cyt P-450 in the strain NYS P-100 when grown in 20% w/v glucose medium (King et al. 1985) and the knowledge that this cyt P-450 is defective in C-14 demethylation of sterols (H. D. Pierce et al. 1978), show that the elevated cyt P-450 present during growth under these conditions is the C-14 demethylase. The differences observed in the preparations are likely to be due to polymorphism or to artefacts associated with purification procedure.

The extent of the superfamily of cyt P-450 genes in fungi remains undetermined. The Δ^{22}-desaturation step in ergosterol biosynthesis in *S. cerevisiae* has recently been indicated to be a cyt P-450-mediated reaction (Hata et al. 1987) and this may represent the minor form observed during purification (King et al. 1984; Kelly et al. 1985). A different form of cyt P-450 is involved in the initial metabolism of alkane in alkane-utilising yeasts (Sanglard et al. 1984), and a cyt P-450 has also been shown to be involved in steroid hydroxylation in filamentous fungi such as in the 11 α-hydroxylation of progesterone by *Rhizopus nigricans*. The latter system is a novel microsomal electron-transport system involving the haemoprotein, rhizaporedoxin and a FAD containing rhizaporedoxin reductase (Brevskar et al. 1987). The sterol C-14 demethylase system of filamentous fungi has yet to be purified, but presumably resembles the yeast and mammalian system with cyt P-450 coupled to NADPH-cyt P-450 reductase.

The presence of cyt P-450 in a variety of yeast species including *C. albicans* has been demonstrated (Karenlampi et al. 1980; Vanden Bossche et al. 1987). Obligate aerobic yeasts do not grow well, if at all, in high glucose concentrations and cytochrome oxidase usually obscures the cyt P-450 peak due to a deep trough at 441–445 mm in reduced CO difference spectra. This interference can be overcome with careful separation of the mitochondrial fraction after lysing of protoplasts (Sanglard et al. 1984). We have observed similar interference with germlings of *A. fumigatus* and *N. crassa* and it may be possible to utilise a similar technique to examine cyt P-450 spectra in such filamentous fungi.

3.1 Interaction of Fungal Cyt P-450 with Sterol Demethylation Inhibitors (DMI's)

The interaction of DMI's with the C-14 demethylase occurs by the binding of the unhindered nitrogen to the haem group as a sixth ligand. The N-3 in the imidazole ring and the N-4 in the triazole ring are the positions which bind to the haem and cause a shift to a low spin state in the iron atom. In addition the interaction of N-1 substitutions in the azole molecule with the apoprotein is important in determining the affinity of binding to the haemoprotein. The binding is characterized by a Type II spectrum with a maximum absorption at 425 to 430 nm and a minimum absorption at 390–410 nm (Jefcoate 1979) and such spectra have been observed between yeast cyt P-450 and DMI compounds including cyt P-450 from *S. cerevisiae* (Aoyama et al. 1983 b; Wiggins and Baldwin 1984; Fig. 6). Similar results may be obtained using cyt P-450 obtained from *C. albicans* which exhibits a cyt P-450 with an absorption maximum at 448 nm (Vanden Bossche et al. 1987). In the latter study comparisons were made between the concentrations of azole required to reduce the peak height for cyt P-450 by 50% in reduced CO difference spectra recorded 45s after addition of carbon monoxide. The DMI's azaconazole, penconazole, propiconazole and imazalil were examined. The results obtained indicated selectivity of the compounds towards inhibition of the fungal enzyme, IC_{50} values of $1-10 \times 10^{-8}$ M were obtained for the azoles with an enzyme preparation from yeast compared with IC_{50} values greater than 1×10^{-5} M in assays using mammalian or plant preparations. A measure of the affinity of SDI's for a cyt P-450 may also be evaluated by determining the extent of displacement by CO with time (Yoshida and Aoyama 1987). In general, however, K_s values calculated from values of absorbance change and substrate concentration are not employed.

Yoshida and Aoyama (1987) have examined further the interaction of a range of DMI's using cyt P-450$_{14DM}$ purified from *S. cerevisiae*. They found a one-to-one stoichiometry of the azole compounds in binding to purified cyt P-450$_{14DM}$. Reduction of the cyt P-450$_{14DM}$ with sodium dithionite prevented interaction of the azole with the haemoprotein when using the compounds 1-methylimidazole, triadimenol, triadimefon, ketoconazole and itraconazole. The absence of binding to the reduced

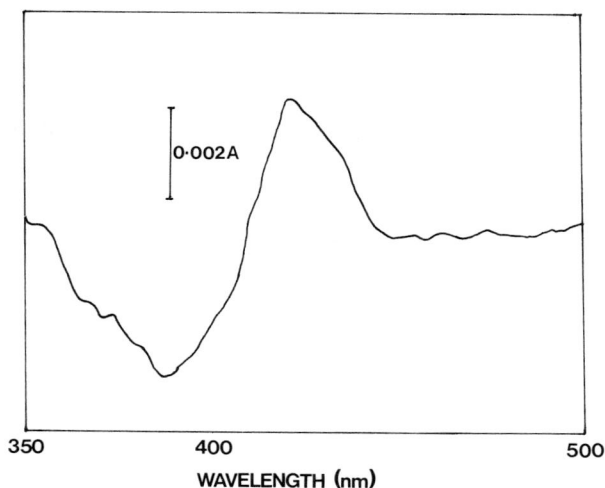

Fig. 6. Type II-binding spectrum obtained on addition of an azole antifungal agent to a *C. albicans* microsomal fraction

cyt P-450$_{14DM}$ may reflect the closing of the haem crevice (Yoshida and Aoyama 1987). Of the azoles tested, ketoconazole and itraconazole were not immediately displaced by carbon monoxide indicating their relatively higher affinities for cyt P-450$_{14DM}$. Yoshida and Aoyama (1987) proposed that the resistance to CO displacement shown by these large azole molecules may be due to low mobility in the haem pocket, together with binding interactions between bulky lipophilic substituents at N-1 and the cyt P-450 apoprotein. In other studies Aoyama et al. (1987) have shown that ketoconazole not only inhibits the first of the monooxygenation reactions performed by cyt P-450$_{14DM}$ in lanosterol metabolism, but also the subsequent steps of the C-14 demethylation reaction.

Interaction of DMI's with the haem iron of cyt P-450 inhibits enzyme activity by preventing the binding of dioxygen at the sixth coordination position. This binding is required for activation of one of the oxygen atoms prior to insertion into a carbon-hydrogen bond in the substrate (Ortiz de Montellano 1987). The evaluation of inhibition as described above has and will continue to be of value in determining structure-activity relationships for inhibitory activity against fungal cyt P-450. An approach complementary to such studies will be the evaluation of binding to cyt P-450 from different sources and to mutant cyt P-450 to investigate the importance to binding of structural features of the proteins.

3.2 Analysis of Fungal Cyt P-450 and Resistance to DMI's

Mutants of the ergosterol-biosynthesis pathway (designated *erg* mutants) were originally isolated on the basis of resistance to polyene-antibiotics. These antifungal agents bind to ergosterol and alter membrane function. In various mutants of *S. cerevisiae* steps from lanosterol to ergosterol were found to be blocked resulting in depletion of ergosterol and thereby nystatin resistance (S. A. Henry 1982). Mutants were found with blocks at C-14 demethylation of lanosterol (*erg* 11), $\Delta^8 \rightarrow \Delta^7$ isomerase (*erg* 2), Δ^5 dehydrogenase (*erg* 3), Δ^{22} desaturase (*erg* 5) and C-24 methyltransferase (*erg* 7). None of these blocks have been found to result in ergosterol auxotrophs. Three isolates have been produced independently with defects in lanosterol C-14 demethylation; the strains are NYS P-100 (H. D. Pierce et al. 1978), SG 1 (Trocha et al. 1977) and JR4 (Taylor et al. 1983). The cyt P-450 from SG 1 has been purified (Aoyama et al. 1987) and shown to have altered spectral properties compared to native wild-type protein. Genetic studies and reduced CO difference spectra have shown that NYS P-100 is a distinct mutant allele from SG 1, the former exhibiting a severly distorted CO difference spectrum with a peak at 445 nm instead of 448 nm in wild types and SG 1 (King et al. 1985). Although Aoyama et al. (1983a) observed no binding of buthiobate to microsomal cyt P-450 from SG 1, in preliminary studies we have observed Type II binding of ketoconazole to NYS P-100 cyt P-450.

Of interest in this regard is the azole resistance found in nystatin-resistant strains of *S. cerevisiae* defective in C-14 demethylase activity (Kenna et al. in preparation). Azole resistance may be due to an altered target cyt P-450 which may nevertheless be "leaky" and synthesize small quantities of ergosterol. Furthermore we have found that these strains may exhibit reduced uptake of azoles (Kenna et al. in preparation) as has been shown previously for a strain of *C. albicans* defective in C-14 demethylase activity (Hitchcock et al. 1986). Genetic analyses of lanosterol C-14 demethylation defective strains and of azole resistant strains of *S. cerevisiae* have indicated another mechanism of resistance.

In the isolation of *S. cerevisiae* strains with defective cyt P-450's, parent strains already had sterol biosynthesis defects in $\Delta^{5,6}$-desaturase activity. This led to the hypothesis that a cyt P-450 defect is lethal except where there is a second and compensatory modification caused by a block in $\Delta^{5,6}$-desaturation (Taylor et al. 1983). The finding that the common mechanism of resistance to ketoconazole in *S. cerevisiae* is a block at $\Delta^{5,6}$-desaturation, further support this as a mechanism. In the case of azole resistance compensation for the accumulation of C-14 methyl sterols in response to SDI treatment may occur with a $\Delta^{5,6}$ desaturation block (Watson et al. 1988). This block does not significantly alter the uptake of azoles. The presence of sterol with C-14 methylation and saturation at C-5 in DMI-treated cells may reflect other mechanisms resulting in the association of C-14-methylation and $\Delta^{5,6}$ desaturation defects for viability in *S. cerevisiae*. It is possible that low levels of $\Delta^{5,6}$ desaturation of C-14 methylated sterol is detrimental to growth. An interesting finding in these studies is the cross-resistance of some polyene-resistant strains to azoles and vice versa.

In the case of resistance to miconazole in *S. cerevisiae*, one resistant mutant has been investigated after screening using glycerol as a carbon source for growth. The mechanism of resistance was based on a mitochondrial gene mutation in subunit 9 of mitochondrial ATPase (Portillo and Gancedo 1985). This mutation may have been selected for as the effect of azoles on petite induction in *S. cerevisiae* has been noted above. Mitochondrial function is required for growth on non-fermentable carbon sources and alterations in mitochondrial enzymes may permit growth on glycerol at higher treatment concentrations of DMI when interference with mitochondrial function occurs.

During the use of azoles as antimycotics only a small number of resistant strains of *C. albicans* have been isolated (Ryley et al. 1984; Smith et al. 1986; Vanden Bossche et al. 1987). Of these, two appear to have a defect in azole permeability or efflux (Hitchcock et al. 1986) and another (Vanden Bossche et al. 1987) has an altered cyt P-450 with reduced affinity for azoles. The latter example represents the first demonstration of azole resistance associated with a structural change in cyt P-450. The differences found between azole resistance mechanisms in *C. albicans* and *S. cerevisiae* may reflect the ease of isolation of recessive mutations in *S. cerevisiae*. A similar explanation, along with physiological variations during mutant isolation, may also apply to resistant mutants isolated in other fungal systems. Of particular note is the isolation of imazalil resistant mutants of *A. nidulans* where eight loci were found to contribute to resistance (Van Tuyl 1977). Amongst filamentous fungi a number of resistant isolates have been produced in the laboratory and in recent years there have also been reports of resistance in the field (Buchenauer 1987). Resistant strains typically show cross-resistance to other DMI's. Resistance may also be associated with reduced pathogenicity, as has been found in *C. albicans* (Smith et al. 1986). In the case of fenarimol resistance in *A. nidulans*, the resistance has been attributed to increased efflux (De Waard and van Nistelrooy 1980).

3.3 Molecular Biology of Yeast Cyt P-450

The molecular cloning of the cyt P-450 genes of yeast has been developed in recent years leading to the publication of the gene sequence for a *S. cerevisiae* cyt P-450 (Kalb et al. 1987). Selection of cyt P-450 genes involved in lanosterol C-14 demethylation have been based on the gene dosage technique (Rine et al. 1983). This technique screens random transformants following gene library transformation for resistance to an inhibitor. In this case transformants exhibiting ketoconazole resistance can be identified. Resistance in the yeast cells results from the overexpression of cyt P-450 from cyt P-450 genes on plasmids present in multiple copies (Kalb et al. 1986; Kelly et al. 1986).

Assignment of a function in lanosterol C-14 demethylation for the gene cloned by Kalb et al. (1987) was made on the basis of 1) ketoconazole resistance for selection of the cloned gene, 2) immunoprecipitation experiments, 3) identical N-terminal sequence data for the gene product and for the cyt P-450 purified by Yoshida and Aoyama (1984a, b) and 4) from the ergosterol auxotrophy produced after gene disruption of the chromosomal locus. Since there is another cyt P-450 involved in ergosterol biosynthesis (Hata et al. 1987), it will be of interest to confirm that the sterols accumulated in the gene-disrupted strains are C-14 methylated. The cyt P-450-amino acid sequence indicates that this protein is the first member of a new cyt P-450 family which has been designated P-450 L1A1 (Kalb et al. 1987). Besides the transmembrane domain, a conserved tridecapeptide was observed and an HR2 or haem-binding domain containing an invariant cysteine which binds to the haem.

In other studies (Kelly et al. 1986) gene libraries have been screened in a *Schizosaccharomyces pombe* host for ketoconazole resistance. Using this approach genes have been isolated from *S. cerevisiae* and *Schizosaccharomyces pombe* gene libraries. This may prove to be a valuable screening procedure for fungal cyt P-450's relying on heterologous function in a host with a capacity to process introns (Kaufer et al. 1985).

The gene isolated by Kalb et al. (1986) has also been used to isolate an orthologous gene from *C. tropicalis* by hydridisation (Chen et al. 1987) and this approach may also prove valuable in isolating cyt P-450 genes from other fungi.

4 Conclusions and Implications for the Discovery of Novel Antifungal Agents

Sites of action for EBI compounds have been outlined with particular emphasis on DMI's. The availability of cyt P-450 genes from yeasts, together with the techniques of molecular genetics may facilitate the identification of orthologous genes from filamentous fungi including those of important pathogens of plants, animals and humans. Using either genomic DNA or expression of cDNA from expression vectors (Oeda et al. 1985) it will be possible to increase the levels of target enzymes for bioassays in vitro (for example Wiggins and Baldwin 1984) and ultimately for macromolecular modelling to improve existing methods of computer-assisted design (for example Marchington and Lambros 1987). For these purposes yeast strains will be available lacking the corresponding endogenous activity (Kalb et al. 1987; King et al. 1985).

The availability of various fungal cyt P-450's, will allow the testing of hypotheses regarding the importance of components of the apoprotein in the binding of inhibitors. This may use in vitro and classical mutagenesis of cyt P-450's and the general approach may be applied to other enzymes in the ergosterol biosynthesis pathway. Study of the X-ray crystal structure of cyt $P-450_{CAM}$ with bound imidazoles has indicated that the inhibitors force the central region of the distal helix of the haem pocket to move away (Poulos 1987; Poulos and Howard 1987). This structure may provide some insight into the interactions important in inhibition, albeit in a soluble cyt P-450 system rather than a membrane-bound eucaryotic cyt P-450. However, deletion of the N terminus of fungal cyt P-450's expressed in yeast may also enable the production of soluble cyt P-450 amenable to crystallisation studies and thereby permit analysis of the relevant protein targets.

Acknowledgements. We are grateful to SERC (PW and IS) and Sheffield University (HFJB and SK) for postgraduate research studentship support and for support from Pfizer (WE), the Wellcome Trust (DK) and to the Sheffield University Medical Research Fund.

References

Aoyama Y, Yoshida Y, Hata S, Nishino T, Katsuki H, Maitra S, Mohan SP, Sprinson DB (1983a) Altered cyt P-450 in a yeast mutant blocked in demethylating C-32 of lanosterol: J Biol Chem 258:9040–9042
Aoyama Y, Yoshida Y, Hata S, Nishino J, Katsuki H (1983b) Buthiobate: a potent inhibitor for yeast cyt P-450 catalysing 14-demethylation of lanosterol. Biochim Biophys Res Commun 115:642–647

Aoyama Y, Yoshida Y, Sonoda Y, Sato Y (1987) Metabolism of 32-hydroxy-24, 25-dihydrolanosterol by purified cytochrome P-450 from yeast. J Biol Chem 262:1239–1243

Astin A, Haslam J, Woods RA (1977) The manipulation of cellular cytochrome and lipid composition in a haem mutant of *Saccharomyces cerevisiae*. Biochemistry 166:275–285

Avruch I, Fischer S, Pierce A Jr, Oehlschlager AC (1976) The reduced biosynthesis of 7-dehydrocholesterols in yeast; potential sources of provitamin D_3 analysis. Can J Biochem 54:567–577

Bailey RB, Hays PR, Parks LW (1976) Homoazasterol-mediated inhibition of yeast sterol biosynthesis. J Bacteriol 128:730–734

Baloch RI, Mercer EI (1987) Inhibition of $\Delta^{8,7}$isomerase and Δ^{14}reductase by fenpropimorph, tridemorph and fenpropidin in cell-free systems from *Saccharomyces cerevisiae*. Phytochemistry 26:663–668

Baloch RI, Mercer EI, Wiggins TE, Baldwin BC (1984) Inhibition of ergosterol biosynthesis by tridemorph, fenpropimorph and fenpropidin. Phytochemistry 23:2219–2276

Barton DHR, Corrie JEJ, Marshall PJ, Widdowson DA (1973) Biosynthesis of terpenes and steroids. VII. Unified scheme for the biosynthesis of ergosterol in *Saccharomyces cerevisiae*. Bioorg Chem 2:363–373

Barug D, Samson RA, Kerkenaar A (1983) Microscopic studies of *Candida albicans* and *Torulopsis glabrata* after in vitro treatment with bifonazole. Light and scanning electron microscopy. Arzneim Forsch 33:528–537

Berg D, Draemer W, Regel E, Buechel K-H, Holmrood G, Plembel M, Scheinpflug H (1984) Mode of action of fungicides: studies on ergosterol biosynthesis inhibitors. Br Crop Protect Conf Pests Dis, vol 3, pp 887–892

Bligh HFJ (1988) Analysis of cytochrome P-450 in yeast. PhD thesis University of Sheffield

Bloch K (1982) Sterols and membranes. In: Mertonosi AN (ed) Membranes and transport. Plenum, New York, London, pp 25–35

Borgers M (1980) Mechanism of action of antifungal drugs, with special reference to the imidazole derivatives. Rev Infect Dis 2:520–534

Bottema CK, Parks LW (1978) Δ^{14}-sterol reductase in *Saccharomyces cerevisiae*. Biochim Biophys Acta 531:301–307

Braun PC, Calderone RA (1978) Chitin synthesis in *Candida albicans*: comparison of yeast and hyphal forms. J Bacteriol 135:1472–1477

Brevskar KK, Cresnar B, Hudnik-Plevnik T (1987) Resolution and reconstitution of cyt. P-450 containing steroid hydroxylation system of *Rhizopus nigricans*. J Steroid Biochem 26:499–501

Buchenauer H (1987) Mechanism of triazoyl fungicides and related compounds. In: Lyr H (ed) Modern selective fungicides — properties, application, mechanism of action. Longman, London, and Fischer, Jena, pp 205–231

Burden RS, Clark T, Holloway PJ (1987) Effects of sterol biosynthesis inhibiting fungicide and plant growth regulators on the sterol composition of barley plants. Pestic Biochem Physiol 27:289–300

Buttke TM, Bloch K (1981) Response of yeast mutant strain GL7 to lanosterol, cycloartenol and cyclolaudenol. Biochem Res Commun 92:229–276

Chattaway FW, Holmes MR, Barlow AJF (1968) Cell wall composition of the mycelial and blastospore forms of *Candida albicans*. J Gen Microbiol 51:367–376

Chen C, Turi TG, Sanglard D, Loper JC (1987) Isolation of the *Candida tropicalis* gene for P-450 lanosterol demethylation and its expression in *S. cerevisiae*. Biochem Biophys Res Commun 146:1311–1317

Chiew YY, Sullivan PA, Shepherd MG (1982) The effects of ergosterol and alcohols on germ-tube formation and chitin synthase in *Candida albicans*. Can J Biochem 60:15–20

De Felice R, Johnson DG, Galgiani JN (1981) Gynecomastia with Ketoconazole. Antimicrob Agent Chemother 19:1073–1074

De Nollin S, Van Belle H, Goossens F, Thone F, Borgers M (1977) Cytochemical and biochemical studies of yeasts after in vitro exposure to miconazole. Antimicrob Agent Chemother 11:500–513

De Waard MA, Nistelrooy JGM van (1980) An energy dependant efflux mechanism for fenarimol in a wild-type strain and fenarimol resistant mutants of *Aspergillus nidulans*. Pestic Biochem Physiol 13:255–266

Dupont B, Prouhet E (1988) Fluconazole in the management of oropharyngeal candidosis in a predominantly HIV antibody-positive group of patients. J Med Vet Mycol 26:67–71

Field RD, Holmlund CE, Whittaker NF (1979) The effects of the hypocholeterimic compound 3β-(β-dimethylaminoethoxy)-andros-5-ene-17-one on the sterol and steryl ester composition of S. cerevisiae. Lipids 14:741–747

Fryberg M, Oehlschlager AC, Unrau AM (1973) Biosynthesis of ergosterol in yeast. Evidence for multiple pathways. J Am Chem Soc 95:5747–5757

Fryberg M, Oehlschlager AC, Unrau AM (1975) Sterol biosynthesis in antibiotic sensitive and resistant Candida. Arch Biochem Biophys 173:171–177

Gibbons GF, Mitropoulos KA, Myant NB (1982) Biochemistry of Cholesterol. Elsevier/North-Holland Biomed Press, Amsterdam

Goulston G, Mercer EL, Goad LJ (1975) The identification of 24-methylene-24, 25-dihydrolanosterol and other possible ergosterol biosynthesis precursors in Phycomyces blakesleeanus and Agricus campestris. Phytochemistry 14:457–462

Haslam J, Astin A (1979) The use of haem deficient mutants to investigate mitochondrial function and biogenesis in yeast. Methods Enzymol 56:558–567

Hata S, Nishino T, Katsuki A, Aoyama Y, Yoshida Y (1987) Characterisation of Δ^{22} desaturation in ergosterol biosynthesis of yeast. Agric Biol Chem 51:1349–1354

Hays PR, Parks LW, Pierce HD Jr, Oehlschlager AC (1977) Accumulation of ergosta -1,14-dien 3-β-ol by Saccharomyces cerevisiae cultured with azasterol antimycotic agents. Lipids 12:666

Henry MJ, Sisler HD (1981) Inhibition of ergosterol biosynthesis in Ustilago maydis by the fungicide 1-[2-(2,4-dichlorophenyl)-4-ethyl-1,3-dioxolan-2-methyl-1H-1, 2,4-triazole. Pestic Sci 12:98–102

Henry SA (1982) Membrane lipids of yeast: Biochemical and genetic studies. In: Strathern J, Jones E, Broach J (eds) The molecular biology of the yeast Saccharomyces cerevisiae, vol II. Cold Spring Harbor Press, Cold Spring Harbor, pp 101–158

Hitchcock CA, Barrett-Bee KJ, Russell NJ (1986) The lipid composition and permeability to azole of an azole and polyene resistant mutant of Candida albicans. J Med Vet Mycol 25:29–37

Huang-Kuang L, Lagenbach RJ, Knoche HW (1972) Sterols of Uromyces phaseoli usedospores. Phytochemistry 11:2319–2322

Jefcoate JR (1979) Measurement of substrate and inhibitor binding to microsomal cyt. P-450 by optical-difference spectroscopy. In: Colowick SP, Kaplan NO (eds) Methods in enzymology, vol 52. Academic Press, London New York, pp 258–279

Kalb VF, Loper JC, Dey CR, Woods CV, Sutter TR (1986) Isolation of a cytochrome P-450 structural gene from Saccharomyces cerevisiae. Gene 45:237–245

Kalb VF, Woods CW, Turi TG, Dey CR, Sutter TR, Loper JC (1987) Primary structure of the P-450 lanosterol demethylase gene from Saccharomyces cerevisiae. DNA 6:528–529

Karenlampi SO, Marin E, Hanninen OOP (1980) Occurrence of cyt P-450 in yeast. J Gen Microbiol 120:529–532

Kato T (1986) Sterol biosynthesis in fungi, a target for broad spectrum fungicides. In: Haug G, Hoffman H (eds) Chemistry of plant protection, vol 1. Springer, Berlin Heidelberg New York Tokyo, pp 1–24

Kato T, Tanaka S, Ueda M, Kawase Y (1974) Effects of the fungicide S1358 on general metabolism and lipid biosynthesis in Monilia fructigena. Agric Biol Chem 15:597–682

Kato T, Shaomi M, Kawase Y (1980) Comparison of tridemorph with buthiobate in antifungal mode of action. Pestic Sci 5:69–79

Kaufer NF, Simanis V, Nurse P (1985) Fission yeast Schizosaccharomyces pombe correctly excises a mammalian RNA transcript intervening sequence. Nature (London) 318:78–80

Kelly SL, Kelly DE, King DJ, Wiseman A (1985) Interaction between yeast cyt. P-450 and chemical carcinogens. Carcinogenesis 6:1321–1325

Kelly SL, Bligh HFJ, Kenna S, Watson PF, Kelly DE (1986) Molecular genetic analysis of yeast cytochrome P-450. Mutagenesis 1:392–393

Kerkenaar A (1987) Mechanism of action of morpholine fungicides. In: Lyr H (ed) Modern selective fungicides – properties, application, mechanism of action. Longman, London, and Fischer, Jena, pp 159–171

Kerkenaar A, Uchiyama M, Versluis GG (1981) Specific effect of tridemorph on sterol biosynthesis in Ustilago maydis. Pestic Biochem Physiol 16:97–104

King DJ, Wiseman A, Wilkie D (1983) Studies on genetic regulation of cytochrome P-450 production in *Saccharomyces cerevisiae*. Mol Gen Genet 192:466–470

King DJ, Azari MR, Wiseman A (1984) Studies on the properties of highly purified cytochrome P-448 and its dependant activity in benzo(α) pyrene hydroxylase from *Saccharomyces cerevisiae*. Xenobiotica 14:187–206

King DJ, Wiseman A, Kelly DE, Kelly SL (1985) Differences in the cytochrome P-450 enzymes of sterol C-14 demethylase mutants of *Saccharomyces cerevisiae*. Curr Genet 10:261–267

Kuck KH, Scheinpflug H (1986) Biology of sterol-biosynthesis inhibiting fungicides. In: Haug H, Hoffman H (eds) Chemistry of plant protection, vol 1. Springer, Berlin Heidelberg New York Tokyo, pp 65–96

Lindenmayer A, Smith L (1964) Cytochromes and other pigments of bakers yeast grown aerobically and anaerobically. Biochim Biophys Acta 93:445–461

Lyr H (ed) (1987) Modern selective fungicides — properties, applications and mechanisms of action. Longman, London, and Fischer, Jena

Marchington AF, Lambros SA (1987) Computer design of fungicides. In: Lyr H (ed) Modern selective fungicides — properties, applications, mechanisms of action. Longman, London, and Fischer, Jena, pp 325–326

Marriott MS, Richardson K (1987) The discovery and mode of action of fluconazole. In: Frontling RA (ed) Recent trends in the discovery, development and evaluation of antifungal agents. Prous, Barcelona, pp 81–92

McCorkindale NJ, Hutchinson SA, Pursey BA, Scott WJ, Wheeler R (1969) A comparison of the types of sterol found in species of the *Saprolegniales* and *Leptomitales* with those found in other Phycomycetes. Phytochemistry 8:861–867

Nebert DW, Adesnick M, Coon MJ, Estabrook RW, Gonzales FJ, Guenguerich FP, Gunsalus IC, Johnson EF, Kemper B, Levin W, Philips IR, Sato R, Waterman MR (1987) The P-450 gene superfamily. Recommended nomenclature. DNA 6:1–11

Nes W, Sekula B, Nes D, Adler J (1978) The functional importance of structural features of ergosterol in yeast. J Biol Chem 253:6218–6225

Oeda K, Sakaki T, Ohkawa H (1985) Expression of rat liver cytochrome P-450 MC cDNA in *Saccharomyces cerevisiae*. DNA 4:203–210

Ortiz de Montellano PR (1987) Oxygen activation and transfer. In: Ortiz de Montellano PR (ed) Cytochrome P-450. Structure, mechanism and function. Plenum, New York London, pp 217–271

Pierce AM, Mueller RB, Unrau AM, Oehlschlager AC (1978) Metabolism of Δ^{24} sterols by yeast mutants blocked in the removal of the C-14 methyl group. Can J Biochem 56:794–800

Pierce AM, Unrau AM, Oehlschlager AC, Woods RA (1979) Azasterol inhibition in yeast. Inhibition of Δ^{24} sterol methyltransferase and 24 methylene sterol. Can J Biochem 57:201–208

Pierce HD Jr, Pierce AM, Srinivasan R, Unrau AM, Oehlschlager AC (1978) Azasterol inhibitors of yeast. Inhibition of 24 methylene sterol, $\Delta^{24(28)}$ reductase and Δ^{24} sterol methyltransferase of *S. cerevisiae* by 23-Azacholesterol. Biochim Biophys Acta 529:429–437

Pont A, Graybill R, Craven PL, Calgiani JW, Dismukes WE, Reitz RE, Stevens DA (1984) High dose ketoconazole therapy and adrenal and testicular function in humans. Arch Intern Med 142:2137–2140

Portillo F, Gancedo C (1985) Mitochondrial resistance to miconazole in *Saccharomyces cerevisiae*. Mol Gen Genet 199:493–499

Poulos TL (1987) The crystal structure of P-450$_{CAM}$. In: Ortiz de Montellano PR (ed) Cytochrome P-450. Structure, mechanism, biochemistry. Plenum, New York London, pp 505–524

Poulos TL, Howard AZ (1987) Crystal structure of metapyrone and phenylimidazole-inhibited complexes of cytochrome P-450$_{CAM}$. Biochemistry 26:8165–8174

Ragsdale NH, Sisler HD (1972) Inhibition of ergosterol biosynthesis in *Ustilago maydis* by the fungicide triarimol. Biochem Biophys Res Commun 46:2048–2053

Rahier A, Schmitt P, Huss B, Beneviste P, Pommer EH (1987) Chemical structure-activity relationships of the inhibition of sterol biosynthesis by N-substituted morpholines in higher plants. Pestic Biochem Physiol 25:112–124

Rine J, Hansen W, Hardeman E, Davis RW (1983) Targeted selection of recombinant clones through gene dosage effect. Proc Natl Acad Sci USA 80:6750–6754

Ryley JF, Wilson RG, Barrett-Bee KJ (1984) Azole resistance in *Candida albicans*. Sabouraudia 22:53–63

Saag MS, Dismukes WE (1988) Azole antifungal agents: emphasis on new triazoles. Antimicrob Agents Chemother 32:1 – 8

Sanglard D, Kappeli O, Fiechter A (1984) Metabolic conditions determining the composition and catalytic activity of cyt. P-450 monooxygenases in *Candida tropicalis*. J Bacteriol 157:297 – 302

Scheinpflug H, Kuck KH (1987) Sterol biosynthesis inhibiting piperazine, phridine, pyrimidine, and azole antifungals. In: Lyr H (ed) Modern selective fungicides – properties, applications, mechanism of action. Longman, London, and Fischer, Jena, pp 173 – 204

Schicke P, Veen KH (1969) A new systemic CELA W524 (N,N'-bis[1-formamide-2,2,2-trichloroethyl]-piperazine) with action against powdery mildew, nut and apple scab. Proc 5th Br Insectic Fungic Conf, pp 819 – 833

Sherald JL, Sisler HD (1975) Antifungal mode of section of triforine. Pestic Biochem Physiol 5:477 – 488

Sherald JL, Ragsdale NN, Sisler HD (1973) Similarities between the systemic fungicide triforine and triarinol. Pestic Sci 4:719 – 727

Sherald JL, Ragsdale NN, Sisler HD (1979) Antifungal mode of action of triforin. Pestic Biochem Physiol 5:477 – 488

Shigematsu ML, Uno J, Arai T (1982) Effects of Ketoconazole on isolated mitochondria from *Candida albicans*. Antimicrob Agents Chemother 21:919 – 924

Shimokawa O, Kato Y, Nakoyara H (1986) Accumulation of 14-methyl sterols and defective hyphal growth in *Candida albicans*. J Med Vet Mycol 24:327 – 336

Shire JB, Sisler HD (1976) Effect of ancymidol (a growth retardant) and triarimol (a fungicide) on growth, sterols and giberellins of *Phaseolus vulgaris* (L.). Plant Physiol 57:640 – 644

Sisler HD, Ragsdale NN, Waterfield WF (1984) Biochemical aspects of fungitoxic and growth regulatory action of fenarimol and other pyridine methanols. Pestic Sci 13:167 – 176

Smith KJ, Warnock DW, Kennedy CTC, Johnson EM, Hopwood V, Van Cutsen J, Vanden Bossche H (1986) Azole resistance in *Candida albicans*. J Med Vet Mycol 24:133 – 144

Sud IJ, Feingold DS (1981) Heterogeneity of action mechanisms among antimycotic imidazoles. Antimicrob Agents Chemother 20:71 – 74

Taylor FR, Rodrigues RJ, Parks LW (1983) Requirement for a second sterol biosynthetic mutation for viability of a sterol C-14 demethylation defect in *Saccharomyces cerevisiae*. J Bacteriol 155:64 – 68

Trocha PJ, Jasne SJ, Sprinson DB (1977) Yeast mutants blocked in removing the methyl group of lanosterol at C-14. Separation of sterol by high pressure liquid chromatography. Biochemistry 16:4721 – 4726

Uno J, Shigematsu ML, Arai T (1982) Primary site of action of ketoconazole on *Candida albicans*. Antimicrob Agents Chemother 21:912 – 918

Vanden Bossche H (1985) Biochemical targets for antifungal azole derivatives: Hypothesis on the mode of action. In: McGinnis MR (ed) Current topics in medical mycology, vol 1. Springer, Berlin Heidelberg New York Tokyo, pp 313 – 315

Vanden Bossche H (1987) Itraconazole: a selective inhibitor of the cytochrome P-450 dependant ergosterol biosynthesis. In: Frontling RH (ed) Recent trends in discovery, development and evaluation of antifungal agents. Prous, Barcelona, pp 207 – 221

Vanden Bossche H, Ruysschaert JM, Defrise-Quertain F, Willemsens G, Cornelissen F, Marichal P, Cools W, Van Cutsern J (1983) Biochemical differences between yeast and mycelia. Do they determine the antimycotic activity of ketoconazole? In: Spitzy H, Karrer K (eds) Proc 13th Int Congr Chemother, Vienna, pp 3 – 9

Vanden Bossche H, Marichal P, Gorrens J, Bellens D, Verhooven H, Coene ML, Lauwers W, Janssen PAJ (1987) Interaction of azole derivatives with cyt. P-450 isozymes in yeast, fungi, plants and mammalian cells. Pestic Sci 21:289 – 306

Van Tuyl JM (1977) Genetics of fungal resistance to systemic fungicides. Meded Landbouwhogesch Wageningen 77-2:1 – 136

Watson PF, Rose ME, Kelly SL (1988) Isolation and characterisation of ketoconazole resistant mutants of *Saccharomyces cerevisiae*. J Med Vet Mycol 26:153 – 162

Weete JD (1973) Sterols of fungi: distribution and biosynthesis. Phytochemistry 12:1843 – 1864

Wiggins T, Baldwin BC (1984) Binding of azole fungicides related to diclobutrazol to cytochrome P-450. Pestic Sci 15:206 – 209

Wilkinson CF, Hetnarski K, Yellis TO (1972) Imidazole derivatives – a new class of microsomal enzyme inhibitors. Biochem Pharmacol 21:3187 – 3192

Yoshida Y, Aoyama Y (1984a) Yeast cytochrome P-450 catalysing lanosterol $^{14}\alpha$ demethylation. 1. Purification and spectral properties. J Biol Chem 259:1655–1660

Yoshida Y, Aoyama Y (1984b) Yeast cytochrome P-450 catalysing lanosterol $^{14}\alpha$ demethylation. II. Lanosterol metabolism by purified P-450$_{DM}$ and by intact microsomes. J Biol Chem 254:1661–1666

Yoshida Y, Aoyama Y (1987) Interaction of azole antifungal agents with cyt. P-450$_{14DM}$ purified from *Saccharomyces cerevisiae* microsomes. Biochem Pharmacol 36:225–229

Yoshida Y, Kumaoka H, Sato R (1974) Studies on the microsomal electron-transport system of anaerobically grown yeast. J Biochem 78:1201–1210

Chapter 16

Biosynthesis and Role of Phospholipids in Yeast Membranes

J. E. Hill, C. Chung, P. McGraw, E. Summers, and S. A. Henry [1]

Abbreviations

I-1-P synthase, inositol-1-phosphate synthase; E, ethanolamine; MME, monomethylethanolamine; DME, dimethylethanolamine; C, choline; PA, phosphatidic acid; CDP-DG, cytidine diphosphate-diglyceride; PI, phosphatidylinositol; PS, phosphatidylserine; PE, phosphatidylethanolamine; PMME, phosphatidylmonomethylethanolamine; PDME, phosphatidyldimethylethanolamine; PC, phosphatidylcholine.

1 Regulation of Phospholipid Synthesis in *S. cerevisiae* – A Review

An understanding of the global regulation of phospholipid biosynthesis in *S. cerevisiae* has emerged from studies in a number of laboratories. R. Lester and his colleague, C. Waechter, provided the original biochemical analysis demonstrating that the activities responsible for the three sequential methylations of PE are repressed in cells grown in the presence of choline (Waechter and Lester 1971, 1973). C. Waechter and colleagues later demonstrated a similar pattern of regulation for PS synthase (Carson et al. 1982) and PS decarboxylase (Carson et al. 1984). PS synthase, PS decarboxylase and the *N*-methyltransferases are all membrane-associated enzymes and are part of the methylation pathway for PC biosynthesis (Fig. 1; see also Chapter 17).

    S. Yamashita and colleagues (Yamashita and Oshima 1980; Yamashita et al. 1982) subsequently demonstrated that repression by choline of the *N*-methyltransferases occurs only if inositol is also present in the medium. Inositol is also required for proper regulation of CDP-DG synthase and PS synthase by choline (Homann et al. 1985; Klig et al. 1985). The basic pattern of regulation for all of these enzymes is similar. Full repression of enzymatic activity occurs in cells grown in the presence of inositol plus choline. Partial repression occurs in cells grown in the presence of inositol. Complete derepression occurs when inositol or choline and inositol are absent from the medium. CDP-DG synthase and PS synthase are also repressed in the presence of

[1] Department of Biological Sciences, Carnegie Mellon University, 4400 Fifth Avenue, Pittsburgh, Pennsylvania 15213-3890, USA

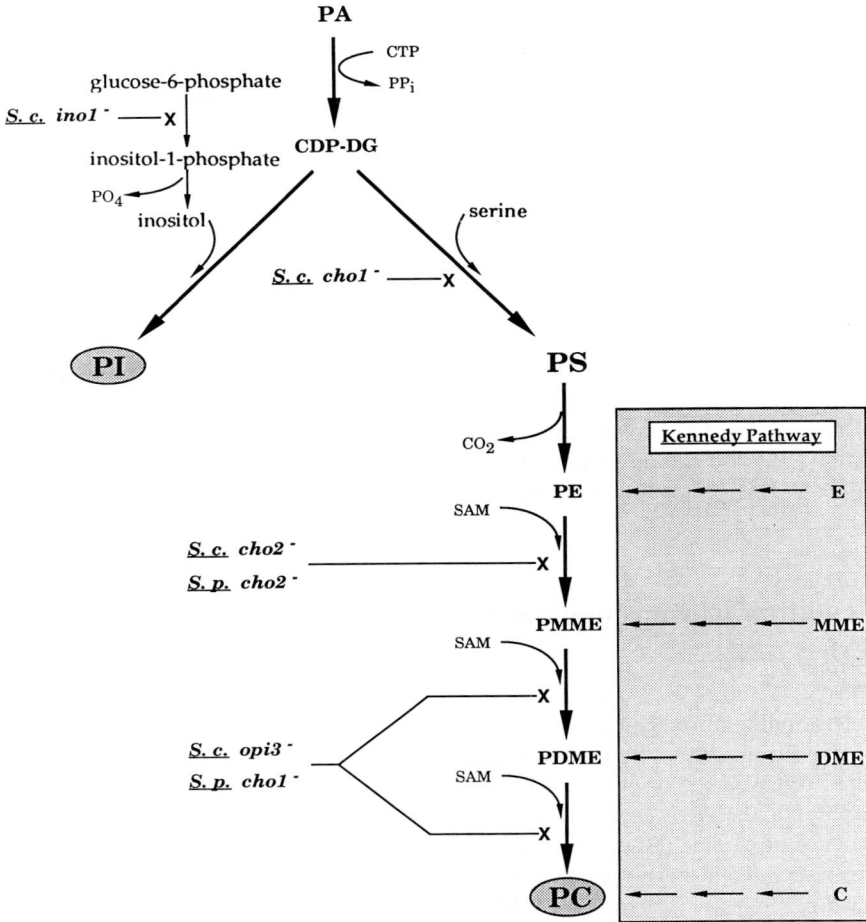

Fig. 1. Biosynthesis of phospholipids in *S. cerevisiae* and *S. pombe*. The pathways in *S. cerevisiae* were determined by Waechter and Lester (1971, 1973) and Steiner and Lester (1972). The "Kennedy pathway" was first described by Kennedy and Weiss (1956). Phospholipid-biosynthetic pathways in *S. pombe* are described in Fernandez et al. (1986). The pathways are similar in both organisms with one exception: the conversion of glucose-6-phosphate to inositol-1-phosphate is absent in *S. pombe*. *S. cerevisiae* mutants are indicated by the designation *S. c.* before a given genotype, and *S. pombe* mutants are indicated by the designation *S. p.* The position of the biochemical lesion in structural gene mutants is indicated by an *X* next to the reaction in question

serine or ethanolamine when inositol is present (Homann et al. 1985, 1987a; Poole et al. 1986).

1.1 Regulation of the *S. cerevisiae* *INO1* and *CHO1* Structural Genes

Originally, the cytoplasmic enzyme I-1-P synthase was shown to be repressed in cells grown in the presence of inositol (Donahue and Henry 1981). The structural gene for this enzyme, *INO1*, was isolated (Klig and Henry 1984) and used to show that the *INO1* transcript is regulated in response to inositol and choline, analogous to PS synthase and the *N*-methyltransferases (Hirsch and Henry 1986). The *INO1* transcript is maximally derepressed in cells grown in the absence of choline and inositol. The abundance of the *INO1* transcript is reduced ten-fold when inositol is added to the growth medium but in the presence of inositol plus choline, the abundance of the transcript is reduced 30-fold compared to the fully derepressed level (Hirsch and Henry 1986).

The structural gene for PS synthase, *CHO1*, was also isolated (Letts et al. 1983) and used to study expression of the *CHO1* transcript (Bailis et al. 1987). The *CHO1* transcript is fully derepressed in cells grown in the absence of any supplement or in the presence of choline. When inositol is present in the medium, the abundance of the *CHO1* transcript is reduced to about 60% of the derepressed level. With both inositol and choline present in the medium the *CHO1* transcript is reduced to approximately 15% of the derepressed level (Bailis et al. 1987).

Although the repression ratio at the level of mRNA abundance is approximately 30-fold for *INO1* and approximately six-fold for *CHO1*, the overall pattern of response to exogenous precursors is similar for both genes. These studies on the expression of *INO1* and *CHO1* mRNA support the hypothesis that regulation of phospholipid metabolism in *S. cerevisiae* is controlled in a coordinated manner and occurs, at least in part, at the transcriptional level.

1.2 Regulatory Genes Involved in Control of Phospholipid Synthesis in *S. cerevisiae*

Genetic studies provide further evidence about the mechanism(s) of regulation of phospholipid biosynthesis. The *ino2* and *ino4* regulatory mutants are unable to derepress I-1-P synthase (Donahue and Henry 1981). Another class of regulatory mutants, *opi1*, are constitutive for I-1-P synthase expression (Greenberg et al. 1982a). Subsequent analysis of the *ino2*, *ino4* and *opi1* mutants revealed that they possessed pleiotropic phenotypes. Not only was regulation of I-1-P synthase expression affected, but regulation of PS synthase, the *N*-methyltransferases and CDP-diglyceride synthase were simultaneously defective (Klig et al. 1985; Loewy and Henry 1984; Homann et al. 1985; Bailis et al. 1987; Hirsch and Henry 1986). Thus, whether or not inositol and choline are present, *ino2* and *ino4* mutants express the basal or repressed levels of the coordinately regulated genes while the *opi1* mutant constitutively overproduces these same enzymes.

All of these regulatory mutations are recessive and appear to be loss of function mutations. It is proposed, therefore, that the wild-type gene products of *INO2* and *INO4* are positive regulatory factors required for the derepression of the structural genes regulated by inositol and choline. Likewise, the wild-type gene product of the *opi1* locus would be a negative regulator required for the repression of the same genes. *INO2* and *INO4* are better candidates for the ultimate regulator than *OPI1* because double mutants (i.e., *ino2, opi1* and *ino4, opi1*) confer an Ino- phenotype (Loewy et al. 1986). Since these regulatory mutations affect *INO1* and *CHO1* mRNA abundance (Hirsch and Henry 1986; Bailis et al. 1987), it is likely that some or all of their gene products function in transcriptional regulation.

Recently, we have reported the isolation of the *INO4* regulatory gene (Klig et al. 1988 b). DNA sequence analysis of *INO4* revealed an open reading frame predicting a highly basic protein of 17 390 molecular weight (Hoshizaki and Henry, unpublished data). The *OPI1* gene has also been isolated, and a molecular analysis is underway.

Deletion analysis of 5'-untranslated sequences of *INO1* suggests that sites located between nucleotides -332 and -259 and between -155 and -120 act negatively to reduce transcription of the gene under repressing conditions (Hirsch 1987). This result contrasts with the genetic studies cited above which suggested that the ultimate regulation would be positive. To elucidate the mode of action of the *INO2*, *INO4* and *OPI1* gene products, it will be necessary to demonstrate which, if any, interact directly with regulatory sequences adjacent to the structural genes they control. Such studies are in progress.

1.3 Studies of Mutants Defective in the Structural Genes Encoding Phospholipid Biosynthetic Enzymes

The structural gene mutants *ino1* (I-1-P synthase deficient) and *cho1* (PS synthase deficient) were identified by screening for inositol and choline auxotrophs, respectively (Culbertson and Henry 1975; Atkinson et al. 1980b; Kovac et al. 1980). The *cho1* auxotrophic requirement is satisfied by ethanolamine, monomethylethanolamine (MME), dimethylethanolamine (DME) or choline (Atkinson et al. 1980a). When grown with one of these supplements, *cho1* cells bypass their lack of PS and synthesize PC by employing the Kennedy pathway (Fig. 1) (Kennedy and Weiss 1956).

Subsequent attempts to isolate mutants defective in the synthesis of PC via methylation of PE by screening for choline auxotrophs were unsuccessful (Letts and Henry 1985). The only choline auxotrophs obtained by our laboratory and by the laboratories of S. Fogel and I. Dawes were *cho1* mutants (Atkinson et al. 1980a; Letts and Dawes 1979). S. Yamashita and colleagues, however, reported the isolation of mutants defective in the methylation of PE which were auxotrophic for choline (Yamashita et al. 1982). These mutants (called *pem-*) were used to clone the structural genes for the *N*-methyltransferases (Kodaki and Yamashita 1987). However, no gene disruption experiments were reported with the cloning experiments, and no extensive genetic characterization of the mutants was included in their original description.

We have isolated mutants completely defective in the N-methyltransferases and find that they are not choline auxotrophs. The first of these mutants, *opi3-3*, was isolated whilst screening for mutants which overproduce and excrete inositol (*Opi-*) (Greenberg et al. 1982b; Greenberg et al. 1983). The original *opi3* mutant (*opi3-3*) has a slightly "leaky" phenotype and displays a membrane-phospholipid composition (expressed as a % of total extractable phospholipid) of 48% PMME and only 2% PC. It has now been demonstrated that the *Opi-* phenotype in *opi3* mutants is conditional and is observed only if cells are grown in the absence of DME or choline. This phenotype was employed in the isolation of additional "tight" *opi3* mutants (McGraw and Henry 1989). These mutants synthesize virtually no PC when grown in the absence of choline. The biochemical defect in all *opi3* mutants appears to lie in the final two methylations of PC biosynthesis (i.e., PMME→PDME→PC) (Greenberg et al. 1983; McGraw and Henry 1989).

The *cho2* mutants which are defective in the first methylation, i.e., PE→PMME (Henry et al. 1984; Summers et al. 1988), like the *opi3* mutants, are not auxotrophic for choline (or MME) and have very abnormal phospholipid compositions, including reduced levels of PC.

We have isolated the *CHO2* and *OPI3* genes and have constructed strains bearing gene disruptions for these loci. The *CHO2* and *OPI3* gene disruptants are not choline auxotrophs (Summers et al. 1988; McGraw and Henry 1988) and have phenotypes similar to the *cho2* and *opi3* mutants isolated following chemical mutagenesis. Comparison of restriction maps for the isolated *CHO2* and *OPI3* genes with the DNA sequences for *PEM1* and *PEM2* (Kodaki and Yamashita 1987) leads us to conclude that *OPI3* and *PEM2* are identical sequences as are *CHO2* and *PEM1*. We believe, as do Kodaki and Yamashita (1987), that these DNA fragments contain the structural genes encoding N-methyltransferases. The evidence for this claim is not yet definitive, however, because the native enzymes have never been purified or sufficiently characterized in yeast.

We do not agree, however, with the identification and characterization of the *Pem*-mutants as structural gene mutants. Null mutants containing disruption of the *CHO2* and *OPI3* genes are clearly not choline auxotrophs. When grown in the absence of DME or choline, the *opi3* mutant produced by gene disruption has a highly abnormal phospholipid composition consisting of 44% PMME, 2% PDME and no detectable PC. In the presence of DME, the same strain has a composition consisting of 36% PDME but no detectable PC (Table 1).

When assayed in vitro, membranes obtained from *opi3* mutants cannot carry out the final two methylations leading to PC (V. Letts, unpublished data; G. Carman, personal communication); in vivo, cells carrying the *OPI3* gene disruption accumulate a small amount of PDME but no PC. This result suggests that the mutant retains some limited capacity in vivo to carry out the second methylation but not the third methylation. The complete absence of the *OPI3* gene product in the strain carrying the gene disruption means that the limited methylation of PMME is not due to the product of the *OPI3* gene. We believe that this residual activity is due to the *CHO2* gene product because the activity is eliminated in strains bearing mutations in both genes (Summers et al. 1988).

The *CHO2* gene disruptant has a phospholipid composition consisting of 55% PE, 7% PC and no detectable PMME (Table 1). In vitro assays of N-methyltransfer-

Table 1. Phospholipid composition of *S. cerevisiae* and *S. pombe* strains

Organism		Relative amount of label incorporated (%)						
Strain	Medium	PI	PS	PE	PMME	PDME	PC	Other
Saccharomyces cerevisiae								
Wild type	I+	21.8	6.6	17.7	0.8	2.9	42.2	8.0
cho2-Δ	I+	27.3	5.2	49.6	–	–	6.8	11.1
opi3-Δ	I+	35.0	3.0	5.6	44.0	2.0	–	10.4
opi3-Δ	I+D+	21.1	4.6	11.6	12.0	35.8	–	14.9
opi3-Δ	I+C+	24.7	5.1	12.2	8.9	1.8	39.4	7.9
Schizosaccharomyces pombe								
Wild type	D+	22.0	7.3	14.6	1.0	3.5	48.3	3.3
Wild type	C+	20.1	8.0	33.8	–	–	34.5	3.6
chol⁻	D+	21.6	7.3	9.0	2.7	56.3	0.7	2.4
chol⁻	C+	19.4	8.8	15.8	–	–	51.4	4.6

The numbers in the body of the table represent percentages of total lipid extractable ^{32}P detected in each phospholipid after steady-state labelling of cells with $H_3{}^{32}PO_4$ as described for *S. cerevisiae* (Atkinson et al. 1980a, b) and for *S. pombe* (Fernandez et al. 1986). The two-dimensional paper chromatographic system of Steiner and Lester (1972) was used to separate the phospholipids. The *cho2-Δ* and *opi3-Δ* strains are the null alleles created by gene disruption and replacement (Summers et al. 1988; McGraw and Henry 1989). *S. cerevisiae*-minimal medium contained 75 µM inositol (I+); *S. pombe*-minimal medium contained 50 µM inositol. For both organisms 1 mM DME (D+) or 1 mM choline (C+) were added where indicated. Abbreviations: Other, phosphatidic acid, cardiolipin, and all other phospholipids including lipids remaining at the origin; –, not detectable; rest as defined in Fig. 1

ase activities show that this strain retains about 10% of wildtype capacity to methylate PE (Summers et al. 1988). Again, the residual synthesis of PC is eliminated in the strains bearing mutations in both genes, suggesting that the *OPI3* gene product is responsible for the limited methylation of PE observed in *cho2* mutants.

1.4 Altered Regulation of Phospholipid Synthesis in Structural Gene Mutants of *S. cerevisiae*

All of the mutants analyzed to date which are defective in PC biosynthesis via methylation of PE (i.e., the series PA→CDP-DG→PS→PE→→→PC) have pleiotropic phenotypes which include altered regulation of inositol biosynthesis. Thus, the *cdg1* mutant, which has a partial defect in CDP-diglyceride synthase activity, has a strong OPI- phenotype and expresses I-1-P synthase constitutively (Klig et al. 1988a). Inositol alone or inositol plus choline has no effect upon expression of I-1-P synthase in *cdg1* cells.

The *chol* mutants regulate I-1-P synthase normally in response to inositol if ethanolamine or choline is also present in the growth medium, but *chol* cells starved for ethanolamine and choline derepress I-1-P synthase whether or not inositol is present (Letts and Henry 1985). The *cho2* mutants excrete inositol and do not regulate I-1-P synthase in response to inositol unless MME, DME or choline is provided in the medium (Summers et al. 1988; Hirsch and Henry 1986). In *opi3* mutants, inositol fails to repress I-1-P synthase unless choline or DME is also present (McGraw and Henry 1989). In each of these cases, regulation of the *INO1* gene and its product I-1-P synthase is restored only if the metabolic defect is bypassed and PC (or PDME) is synthesized.

It is also clear that restoration of *INO1* regulation occurs in *cho1*, *cho2* and *opi3* mutants if PC (or PDME) is synthesized via the methylation pathway (as in *cho1* mutants growing in the presence of ethanolamine or *cho2* mutants growing in the presence of MME) or via the Kennedy pathway (i.e., via incorporation of exogenous choline or DME – Fig. 1). These results suggest that the active synthesis of PC or PDME is a required component of the regulatory circuit which senses the presence of precursors and responds to them by regulating expression of structural genes involved in phospholipid synthesis.

1.5 Regulation of Phospholipid Synthesis in *S. cerevisiae* Includes Several Levels of Control

In vivo metabolic studies of phospholipid metabolism suggest that transfer of wildtype *S. cerevisiae* cells to inositol-containing medium rapidly shifts the pattern of phospholipid synthesis such that CDP-DG is preferentially converted to PI rather than to PS (Kelley et al. 1988). The mechanism of this response is not completely understood, but it is clearly too rapid to be due to the transcriptional mechanism described in the preceding section. It is known that inositol is a noncompetitive inhibitor of PS synthase (Kelley et al. 1988) and that the phospholipid environment directly associated with PI synthase and PS synthase directly influences their activities (Hromy and Carman 1986; Fischl et al. 1986).

Another rapid alteration in the pattern of phospholipid synthesis is observed in *cho1* cells deprived of choline and ethanolamine (Letts and Henry 1985). As expected under these conditions, PC biosynthesis declines because *cho1* cells lacking PS synthase are unable to synthesize PE or PC without the precursors ethanolamine or choline. However, PI synthesis declines simultaneously, whether inositol is present or not. Furthermore, the precursors phosphatidic acid or CDP-diglyceride do not accumulate in this situation. The decline in PI biosynthesis coupled to the decline in PC biosynthesis is too rapid to be accounted for by the repression-mediated control mechanisms, and it occurs under conditions that fully derepress the coordinately regulated enzymes (i.e., absence of inositol and choline).

Phospholipid biosynthetic activities in *S. cerevisiae* are also regulated in response to growth stage. The activities of CDP-DG synthase, PS synthase and the phospholipid *N*-methyltransferases are all maximal in logarithmically growing cells and repressed two- to five-fold in stationary phase cells (Homann et al. 1987b).

Finally, it should be mentioned that the membrane-associated phospholipid-biosynthetic enzymes are not uniformly distributed in all subcellular membranes

(Kuchler et al. 1986). PS decarboxylase, for example, is restricted to the inner mitochondrial membrane, the N-methyltransferases are found in the microsomal fraction and the PI and PS synthases are distributed to both microsomal membranes and the outer mitochondrial membrane. These observations suggest that considerable movement of phospholipids must occur during the course of biosynthesis, and compartmentalization could well play a role in overall regulation.

2 Regulation of Phospholipid Synthesis in *S. pompe* – Introduction

From the above discussion it is clear that inositol plays a major role in the regulation of phospholipid biosynthesis in *S. cerevisiae*. The fission yeast *S. pombe* is a natural inositol auxotroph, suggesting that it might regulate phospholipid biosynthesis differently from *S. cerevisiae*. Two major lines of research with *S. pombe*, the study of the effect of inositol deprivation and the study of the methylation pathway for synthesizing PC, have confirmed that phospholipid metabolism differs in these two yeasts. The work on PC biosynthesis, however, has also identified some important similarities, especially in regards to the physiological importance of the methylated phospholipids in cell growth.

2.1 Effects of Inositol Deprivation on *S. pombe*

In minimal medium lacking inositol, Ino- strains of *S. cerevisiae* undergo inositol-less death, a phenomenon in which cell viability decreases by three to four orders of magnitude within 24 hours (Henry et al. 1977). With *S. pombe*, however, a decrease in viability of less than two orders of magnitude is seen after 120 hours of inositol starvation, closely resembling the effect of starvation for a required amino acid (Fernandez et al. 1986). Several mechanisms facilitate the survival of *S. pombe* in the absence of inositol. First, during inositol starvation the extreme decrease in PI is coupled to a reciprocal increase in PS, the other anionic phospholipid in the membranes. Second, the major excreted turnover product of inositol-containing phospholipids in *S. pombe* is inositol, which can be reutilized, rather than glycerophosphorylinositol as found in *S. cerevisiae*. Finally, it appears that the decrease in PI synthesis during inositol starvation of *S. pombe* is more tightly coupled to a decrease in other cellular metabolic processes than is seen for *S. cerevisiae* (Fernandez et al. 1986).

2.2 PC Biosynthesis via the Methylation Pathway in *S. pombe* – Biochemical Analysis

The inverse relationship between the amounts of PI and PS in the membranes of *S. pombe* is particularly interesting with respect to the biosynthesis of PC. When the phospholipid composition is determined for cells grown in concentrations of inositol

Table 2. [Methyl-^{14}C]-methionine pulse-labelling of *Schizosaccharomyces pombe*

Strain	Medium	Relative amount of label incorporated (%)				Total ^{14}C (cpm) Incorporated
		PMME	PDME	PC	NL	
Wild type	C+	0.74	1.82	2.39	92.23	69485
Wild type	C−	4.03	14.72	41.57	38.05	134640
chol⁻	C+	0.98	1.48	1.39	93.01	54355
chol⁻	C−	19.14	1.71	1.02	74.69	71495
cho2⁻	C+	0.64	1.60	1.80	93.58	50195
cho2⁻	C−	0.58	1.38	1.67	94.06	53925

The numbers in the body of the table represent percentages of total lipid extractable ^{14}C detected in each phospholipid after pulse labelling cells with [methyl-^{14}C]-methionine as indicated in Fig. 3. Minimal media contained 50 µM inositol with (C+) or without (C−) 1 mM choline. Abbreviations: PMME, phosphatidylmonomethylethanolamine; PDME, phosphatidyldimethylethanolamine; PC, phosphatidylcholine; NL, neutral lipids (i.e., methylated sterols)

ranging from standard conditions (50 µM) to limiting concentrations (5 µM), the amount of PI decreases from 23% to 2% of the total membrane phospholipid while the amount of PS increases from 6% to 15% of the total membrane phospholipid. However, at all concentrations of inositol, the amount of PC remains at approximately 50% of the total membrane phospholipid (Fernandez et al. 1986). These results are intriguing because PS is the first committed precursor for PC synthesis via the methylation pathway, and they suggest that *S. pombe* carefully regulates the concentration of PC in its membrane despite being able to accommodate wide variations in the amounts of PI and PS.

One point at which this regulation could occur is in the methylation pathway. We have examined the regulation of the methylation reactions that convert PE→PM-ME→PDME→PC. These reactions can be easily assayed in vivo by pulse-labelling cells with [methyl-^{14}C]-methionine, which rapidly equilibriates into the S-adenosyl-methionine (SAM) pool that donates the methyl groups for the methylation reactions. The results from such an experiment are shown in Table 2.

For the wildtype (Cho$^+$) strain grown in the absence of choline, over 40% of the ^{14}C incorporated into the lipid soluble fraction is present in PC. In medium containing 1 mM choline, the same strain incorporates only about 2% of the ^{14}C into PC. Thus, choline represses the methylation pathway in both yeasts, but based on these in vivo assays, choline has at least a 5- to 10-fold greater effect on *S. pombe* than on *S. cerevisiae*.

The *N*-methyltransferases can also be assayed in vitro using a modification of the procedure described for *S. cerevisiae* (Waechter and Lester 1971, 1973). In these experiments crude membrane preparations are incubated in the presence of [methyl-^{3}H]-SAM, and the synthesis of methylated phospholipids is assayed by recovery of label into the lipid-soluble fraction. When membrane preparations from wildtype cells grown in the presence or the absence of choline are assayed in vitro, the repression of the *N*-methyltransferases by choline is only 2- to 3-fold compared to the 10- to 20 fold repression seen with the in vivo labellings. These in vitro activities closely resemble the in vitro activities of *S. cerevisiae* grown in the presence versus the absence of choline. In *S. cerevisiae*, however, the repression ratios determined by the in vitro and in vivo assays are quite similar. The in vitro studies suggest that the effect of choline on the activities of the *N*-methyltransferases may be similar for both organisms. The discrepancy in *S. pombe* between the results from in vivo and in vitro assays suggests that additional regulatory mechanisms present in this organism permit the careful control of PC levels in the membrane under a wide range of environmental conditions.

2.3 PC Biosynthesis via the Methylation Pathway in *S. pombe* – Genetic Analysis

Phospholipid metabolism can also be compared between *S. pombe* and *S. cerevisiae* by studying the types of mutants that can be isolated. Screening for inositol auxotrophs, one of the most useful types of mutant searches in *S. cerevisiae*, is obviously not possible for a natural inositol auxotroph like *S. pombe*. For this reason we have screened for mutants that are choline auxotrophs (Cho⁻).

Three independent Cho⁻ strains were isolated in our laboratory. An additional six Cho⁻ strains were generously provided to us by Prof. J. Kohli from a collection made by Prof. U. Leupold. All of these mutants have a "leaky" growth phenotype on plates containing minimal medium-lacking choline. In fact, the growth phenotype is extremely difficult to score on plates unless the vital dye Magdala Red (Phloxine B) is present in the medium.

As an initial step in characterizing these mutants, their growth was tested on minimal medium containing ethanolamine, MME or DME, but no choline. None of the mutants could grow with ethanolamine as the supplement. These results clearly distinguish the *S. pombe* Cho⁻ strains from *S. cerevisiae* Cho⁻ strains, which are blocked in the synthesis of PS and can grow with ethanolamine, MME, DME, or choline. The lack of growth on ethanolamnine also indicates that they are not blocked in the synthesis of PE.

Seven of the nine mutants could grow with MME or with DME. These mutants are likely to be defective in the first *N*-methyltransferase, which converts PE to PMME. With only the first methylation reaction blocked, MME or DME in the medium could be converted into PMME or PDME via the Kennedy pathway and then further methylated to PC via the methylation pathway. This phenotype is also seen in *Neurospora crassa* mutants that are defective in the first methyltransferase (Crocken and Nyc 1964; Scarborough and Nyc 1967).

The other two Cho⁻ strains require either DME or choline for growth. This phenotype could indicate that these mutants are blocked only in the second *N*-methyltransferase, which would prevent them from growing on MME. Mutants of *N. crassa* with the same phenotype, however, have been reported to be blocked in both the second and the third *N*-methyltransferase (Crocken and Nyc 1964; Scarborough and Nyc 1967). Likewise, as discussed previously, the *opi3* mutants of *S. cerevisiae* are defective in both *N*-methyltransferase activities.

The *S. pombe* mutants have been placed into two complementation groups by standard genetic techniques: *chol⁻*, which comprises the two mutants that require either DME or choline for growth, and *cho2⁻*, which include the other seven mutants that can also utilize MME for growth. Subsequent crosses have confirmed that the two *chol⁻* mutants are tightly linked, eliminating the possibility that they could define two genes each blocked in one of the final two methylation reactions.

The biochemical lesions in the *chol⁻* and *cho2⁻* strains were further characterized by in vivo pulse labelling studies using [methyl-^{14}C]-methionine. Similar studies with Cho⁺ strains had shown that very little label is incorporated into the methylated phospholipids when the cells are grown in the presence of choline, but the Cho⁻ strains can only grow for about two to three generations in the absence of choline. Consequently, the cells were grown overnight in the presence of choline, washed and suspended in either choline-free or choline-containing medium, and grown for 6 hours (about two generations) before labelling. The results of such an experiment are shown in Table 2. Wildtype, *chol⁻* and *cho2⁻* strains all have a similar pattern of incorporation of ^{14}C into phospholipids when grown in the presence of choline. In the absence of choline, however, *chol⁻* cells shows a tremendous amount of incorporation of radioactivity into PMME, approximately five-fold greater than wildtype in the same medium, but virtually no incorporation into PDME of PC. This result confirms that the second methylation reaction is defective in *chol⁻* strains but does not provide any information about the activity of the third methylation reaction. The *cho2⁻* strain shows essentially the same pattern of incorporation in the absence of choline

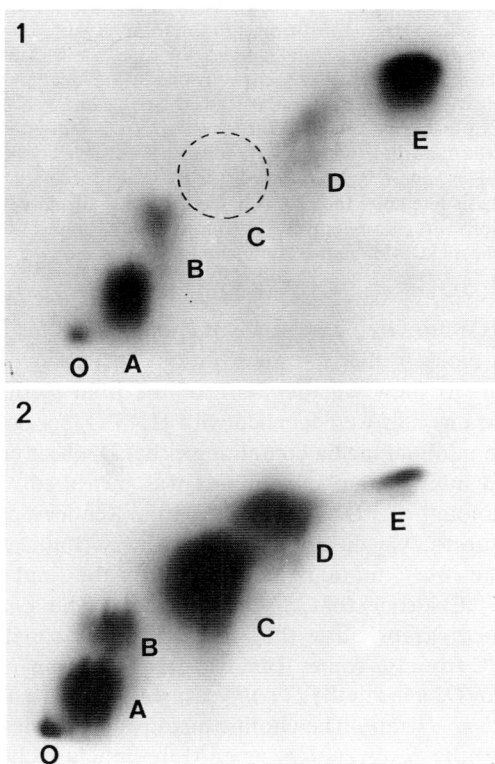

Fig. 2. Phospholipids of the *S. pombe chol⁻* mutant. Autoradiograms of ^{32}P-steady-state-labelled phospholipids separated by two-dimensional paper chromatography according to Steiner and Lester (1972). When grown in the presence of DME, *chol⁻* cells substitute PDME for PC in their lipid composition without adverse effects upon growth. **Panel 1,** *chol⁻* cells were grown for five to six generations in minimal medium supplemented with 1 mM DME and containing 5 μCi/ml carrier-free H$_3$^{32}PO$_4$. Phospholipids were extracted, processed, and quantitated as previously described (Fernandez et al. 1986). **Panel 2,** the same as Panel 1 except that the medium contained 1 mM choline instead of DME. Positions of phospholipids on the chromatogram: *O,* origin; *A,* phosphatidylinositol; *B,* phosphatidylserine; *C,* phosphatidylcholine; *D,* phosphatidylethanolamine; *E,* phosphatidyldimethylethanolamine

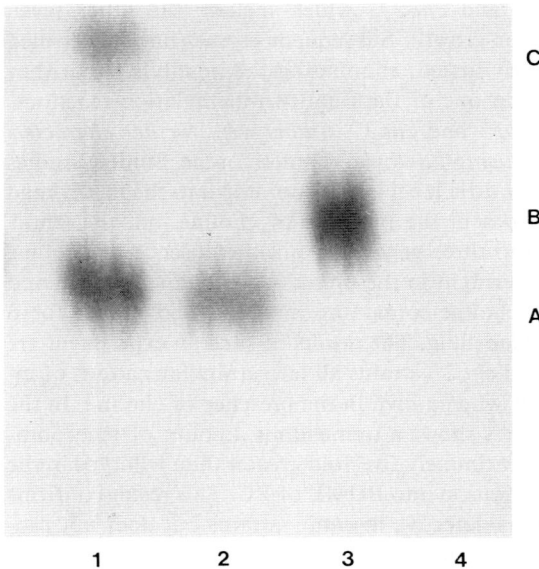

Fig. 3. Complementation of *S. pombe* Cho- strains with the *S. cerevisiae CHO2* gene. Autoradiogram of a one-dimensional chromatographic separation of [methyl-^{14}C]-labelled lipids from wildtype and Cho- *S. pombe* strains transformed with various plasmids. Labelling, extraction and processing of samples were adapted from procedures used with *S. cerevisiae* (Loewy and Henry 1984). *Lane 1, S. pombe* wild type; *Lane 2, S. pombe cho2$^-$* cells transformed with YEp13-*CHO2* plasmid carrying the *S. cerevisiae CHO2* gene provided by E. Summers (Summers et al. 1988) (note restoration of PC biosynthesis to level similar to wild type); *Lane 3, S. pombe cho1$^-$* cells transformed with plasmid YEp13-*CHO2* carrying *S. cerevisiae CHO2* gene (note buildup of PMME characteristic of *cho1$^-$* cells indicating that the *S. cerevisiae CHO2* gene does not restore PC biosynthesis to *S. pombe cho1$^-$* cells); *Lane 4, S. pombe cho2$^-$* cells transformed with the parent vector, YEp13. (This pattern of synthesis resembles the *cho2$^-$* mutant; there is virtually no detectable phospholipid methylation). Positions of phospholipids: *A*, phosphatidylcholine; *B*, phosphatidylmonomethylethanolamine; *C*, phosphatidyldimethylethanolamine

as it does in the presence of choline, and this result is consistent with the presence of a block in the first methylation reaction.

To show whether both of the final methylation steps are defective in a *cho1$^-$* strain, cells were labelled with $H_3{}^{32}PO_4$ for 5 to 6 generations (steady-state labelling) in minimal media containing either DME or choline, and the incorporation of ^{32}P into individual phospholipids was measured. In the presence of DME an excessive amount of PDME is made via the Kennedy pathway, but virtually none is converted into PC via the methylation pathway (Fig. 2 and Table 1). This result confirms that the *cho1$^-$* mutants are defective in the final methylation reactions.

Evidence that *S. pombe cho1$^-$* and *cho2$^-$* are structural genes for the *N*-methyltransferases that catalyze the three methylation reactions in the formation of PC from PE comes from two "plasmid swap" experiments. The wild-type *S. cerevisiae CHO2* gene in the *E. coli*-yeast shuttle vector YEp13 was transformed into a *S. pombe cho2$^-$* strain. The choline auxotrophy of the mutant appeared to be complemented by the *S. cerevisiae* gene. More conclusively, an in vivo assay with [methyl-^{14}C]-

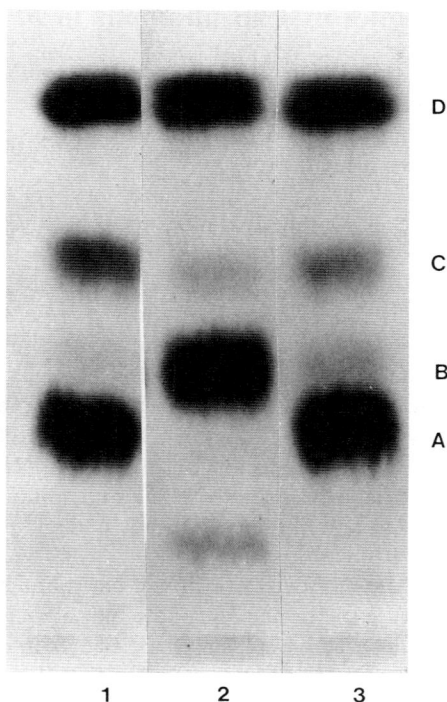

Fig. 4. Complementation of a *S. cerevisiae* *opi3* strain with the *S. pombe chol*[+] gene. Autoradiogram of a one-dimensional chromatographic separation of [methyl-^{14}C]-labelled lipids from *S. cerevisiae* strains, showing complementation of *opi3* cells transformed with plasmid pVX1 (pDB248 containing *S. pombe* genomic DNA complementing the *S. pombe chol*[−] mutant lesion). *Lane 1,* *S. cerevisiae* wild type; *Lane 2, opi3* parent mutant (note buildup of PMME); *Lane 3, opi3* transformed with pVX1 (note restoration of PC biosynthesis showing a pattern similar to wild type). Positions of phospholipids: *A,* phosphatidylcholine; *B,* phosphatidylmonomethylethanolamine; *C,* phosphatidyldimethylethanolamine; *C,* neutral lipids (i.e., methylated sterols)

methionine showed that the biochemical lesion was corrected, i.e., essentially wild-type levels of ^{14}C were incorporated into PC (Fig. 3).

In the other swap, a plasmid had been isolated from a *S. pombe* genomic library that could complement the choline auxotrophy and the biochemical lesion of the *S. pombe chol*[−] mutant. When this plasmid was transformed into a *S. cerevisiae opi3* strain, it was found, in an in vivo assay, to complement the biochemical lesion of the *opi3* strain (Fig. 4). These results provide compelling evidence that the *S. pombe* *chol*[−] and *cho2*[−] strains represent lesions in the structural genes for the *N*-methyltransferases.

Thus, the effects of inositol deprivation, the high degrees of repression of the methylation pathway by choline, and the fact that *S. pombe* mutants in the *N*-methyltransferases are choline auxotrophs confirm that regulation of phospholipid metabolism in *S. pombe* is quite different from that in *S. cerevisiae*. However, several aspects of phospholipid metabolism are similar in the two organisms, notably that PDME seems to be interchangeable with PC in providing normally functioning membranes and that for two cases the *S. pombe* and *S. cerevisiae* structural genes for the *N*-methyltransferases can function in the heterologous system.

3 Roles of Phospholipids in *S. cerevisiae* and *S. pombe*

The mutants of *S. cerevisiae* and *S. pombe* that are defective in phospholipid synthesis have, as discussed above, made possible an analysis of the regulation which controls the phospholipid biosynthetic pathways. Such mutants also provide unique possibilities to study the roles of individual membrane components by modifying the composition of biological membranes in vivo.

The phospholipid mutants of both yeasts have provided evidence of a large degree of tolerance for the substitution of certain phospholipids with others. The types of substitutions which are tolerated involve phospholipids sharing the same net charge. As discussed above, wildtype *S. pombe* cells grown in low levels of inositol substitute PS for much of the PI in the membrane (Fernandez et al. 1986). PI and PS are both anionic phospholipids carrying a net charge of [-1]. Some PI appears to be necessary, however, because *S. pombe* cells stop growing in the total absence of inositol. In *S. cerevisiae chol* cells, no PS at all is detected (Atkinson et al. 1980a; Letts and Henry 1985), but the proportion of PI is elevated compared to wildtype. Thus, PS is not an essential component of cellular membranes in *S. cerevisiae*, although the total proportion of anionic phospholipids is maintained in *S. cerevisiae chol* cells (Atkinson et al. 1980a).

Mutants defective in the phospholipid-methylation pathway in both organisms provide the opportunity to manipulate the relative content of PE, PMME, PDME, and PC. These lipids are zwitterionic and have a net neutral charge. Analysis of *opi3* mutants provides evidence that PDME can substitute entirely for PC in *S. cerevisiae*. *opi3* cells grown in the presence of DME have an almost normal growth rate and viability, despite the absence of PC in their membranes (Table 1). Under these conditions *opi3* cells also regulate I-1-P synthase properly in response to inositol. *opi3* cells grown in the absence of choline or DME have an abnormally high content of PMME (Table 1). Under these conditions I-1-P synthase is not regulated properly. Furthermore, we have found that *opi3* cells grown under these conditions lose viability in stationary phase (McGraw and Henry 1989). Thus, PMME can substitute for PC (or PDME) in supporting growth of *S. cerevisiae*, but poor viability in stationary phase and abnormal regulation of I-1-P synthase suggest that PMME cannot substitute for all of the cellular functions of PC.

In *S. pombe*, PDME also substitutes for PC in *chol*⁻ mutants (Fig. 2). Unlike *opi3* cells of *S. cerevisiae*, however, *chol*⁻ cells of *S. pombe* will not grow unless choline or DME is supplied. In this organism PMME, which accumulates in *chol*⁻ cells in the absence of DME and choline, is not adequate even to support growth. Thus, *S. pombe* appears to be less flexible in the substitutions that it tolerates among the zwitterionic lipids. In addition, the failure to find PS synthase mutants among choline auxotrophs of *S. pombe* suggests that the absence of PS in *S. pombe* may be lethal. In regard to the permissible phospholipid substitutions and growth requirements of phospholipid mutants, *S. pombe* appears to resemble *N. crassa* and *Aspergillus nidulans* more than *S. cerevisiae*. Indeed, it is quite possible that *S. cerevisiae* is unique in its extreme tolerance for modification of its membrane-phospholipid composition.

Mutants of the types described in this report allow the detailed analysis of the effects of membrane-lipid composition on the function of membranes, thereby permitting the roles of and interactions of individual membrane components to be assessed in vivo. Thus far we have only begun to exploit the unique possibilities offered by these mutants for studying membrane function.

References

Atkinson K, Fogel S, Henry SA (1980a) Yeast mutant defective in phosphatidylserine synthesis. J Biol Chem 255:6653–6661

Atkinson KA, Jensen B, Kolat AI, Storm EM, Henry SA, Fogel S (1980b) Yeast mutants auxotrophic for choline or ethanolamine. J Bacteriol 141:558–564

Bailis AM, Poole MA, Carman GA, Henry SA (1987) The membrane associated enzyme phosphatidylserine synthase is regulated at the level of mRNA abundance. Mol Cell Biol 7:67–176

Carson MA, Atkinson KD, Waechter CJ (1982) Properties of particulate and solubilized phosphatidylserine synthase activity from *Saccharomyces cerevisiae*: inhibitory effect of growth medium. J Biol Chem 257:8115–8121

Carson MA, Emala M, Hogsten P, Waechter CJ (1984) Coordinate regulation of phosphatidylserine decarboxylase activity and phospholipid N-methylation in yeast. J Biol Chem 259:6267–6273

Crocken BJ, Nyc JF (1964) Phospholipid variations in mutant strains of *Neurospora crassa*. J Biol Chem 239:1727–1730

Culbertson MR, Henry S (1975) Inositol requiring mutants of *Saccharomyces cerevisiae*. Genetics 80:23–40

Donahue TF, Henry SA (1981) Myo-inositol-1-phosphate synthase: characteristics of the enzyme and identification of its structural gene in yeast. J Biol Chem 256:7077–7085

Fernandez S, Homann MJ, Henry SA, Carman GM (1986) The metabolism of the phospholipid precursor inositol and its relationship to growth and viability in the natural auxotroph *Schizosaccharomyces pombe*. J Bacteriol 166:779–786

Fischl AS, Homann MJ, Poole MA, Carman GM (1986) Phosphatidylinositol synthase from *Saccharomyces cerevisiae*. Reconstitution characterization and regulation of activity. J Biol Chem 261:3178–3183

Greenberg ML, Goldwasser P, Henry S (1982a) Characterization of a yeast regulatory mutant constitutive for inositol-1-phosphate synthase. Mol Gen Genet 186:157–163

Greenberg ML, Reiner B, Henry S (1982b) Regulatory mutations of inositol biosynthesis in yeast: isolation of inositol excreting mutants. Genetics 100:19–33

Greenberg ML, Klig LS, Letts VA, Shicker-Loewy B, Henry SA (1983) Yeast mutant defective in phosphatidylcholine synthesis. J Bacteriol 153:791–799

Henry SA, Atkinson DD, Kolat AI, Culbertson MR (1977) Growth and metabolism of inositol-starved cells of *Saccharomyces cerevisiae*. J Bacteriol 130:472–484

Henry SA, Klig LS, Loewy BS (1984) The genetic regulation of biosynthetic pathways in yeast: amino acid and phospholipid synthesis. Annu Rev Genet 18:207–231

Hirsch JP, Henry SA (1986) Expression of the *Saccharomyces cerevisiae* inositol-1-phosphate synthase (*INO1*) gene is regulated by factors that affect phospholipid synthesis. Mol Cell Biol 6:3320–3328

Hirsch JP (1987) Cis- and trans-acting regulation of the *INO1* gene of *Saccharomyces cerevisiae*. Ph D Thesis, Albert Einstein College of Medicine, Bronx, NY

Homann MJ, Henry SA, Carman G (1985) Regulation of CDP-diacylglycerol synthase activity in *Saccharomyces cerevisiae*. J Bacteriol 163:1265

Homann MJ, Bailis AM, Henry SA, Carman GM (1987a) Coordinate regulation of phospholipid biosynthesis by serine in *Saccharomyces cerevisiae*. J Bacteriol 169:3276–3280

Homann MJ, Poole MA, Gaynor PM, Ho C-T, Carman GM (1987b) Effect of growth phase on phospholipid biosynthesis in *Saccharomyces cerevisiae*. J Bacteriol 169:533–539

Hromy JM, Carman GM (1986) Reconstitution of *Saccharomyces cerevisiae* phosphatidylserine synthase into phospholipid vesicles: Modulation of activity by phospholipids. Jour Biol Chem 261:15572–15576

Kelley MJ, Bailis AM, Henry SA, Carman GM (1988) Regulation of phospholipid biosynthesis in *S. cerevisiae* by inositol. Inositol is an inhibitor of phosphatidylserine synthase activity. J Biol Chem 263:18078–18085

Kennedy EP, Weiss SB (1956) The function of cytidine coenzymes in the biosynthesis of phospholipids. J Biol Chem 222:193–214

Klig LS, Henry SA (1984) Isolation of the yeast *INO1* gene: located on an autonomously replicating plasmid the gene is fully regulated. Proc Natl Acad Sci USA 81:3816–3820

Klig LS, Homann MJ, Carman GM, Henry SA (1985) Coordinate regulation of phospholipid biosynthesis in *S. cerevisiae*: pleiotropically constitutive *opi1* mutant. J Bacteriol 162:1135–1141

Klig LS, Homann MJ, Kohlwein SD, Kelley MJ, Henry SA, Carman GM (1988a) *Saccharomyces cerevisiae* mutant with a partial defect in the synthesis of CDP-diacylglycerol and altered regulation of phospholipid biosynthesis. J Bacteriol 170:1878–1886

Klig LS, Hoshizaki DK, Henry SA (1988b) Isolation of the yeast *INO4* gene a positive regulator of phospholipid synthesis. Curr Genet 13:7–14

Kodaki T, Yamashita S (1987) Yeast phosphatidylethanolamine methylation pathway. J Biol Chem 262:15428–15435

Kovac L, Goelska I, Poliachova V, Subik J, Kovacova N (1980) Membrane mutants: a yeast mutant with a lesion in phosphatidylserine biosynthesis. Eur J Biochem 111:291–301

Kuchler K, Daum G, Paltauf F (1986) Subcellular and submitochondrial localization of phospholipid-synthesizing enzymes in *Saccharomyces cerevisiae*. J Bacteriol 165:901–910

Letts VA, Dawes IW (1979) Mutations affecting lipid biosynthesis of *Saccharomyces cerevisiae*: isolation of ethanolamine auxotrophs. Biochem Soc Trans 7:976–977

Letts VA, Henry SA (1985) Regulation of phospholipid synthesis in phosphatidylserine synthase-deficient (*cho1*) mutants of *Saccharomyces cerevisiae*. J Bacteriol 163:560–567

Letts VA, Klig LS, Bae-Lee M, Carman GM, Henry SA (1983) Isolation of the yeast structural gene for the membrane-associated enzyme phosphatidylserine synthase. Proc Natl Acad Sci USA 80:7279–7283

Loewy BS, Henry S (1984) The *INO2* and *INO4* loci of *Saccharomyces cerevisiae* are pleiotropic regulatory genes. Mol Cell Biol 4:2479–2485

Loewy BS, Hirsch J, Johnson M, Henry SA (1986) Coordinate regulation of phospholipid synthesis in yeast. Proc UCLA Symp Cell Biol 33:551–565

McGraw P, Henry SA (1989) Mutations in the *Saccharomyces cerevisiae OPI3* gene: effects on phospholipid methylation, growth and cross-pathway regulation of inositol synthesis. Genetics 122:317–330

Poole MA, Homann MJ, Bae-Lee MS, Carman GM (1986) Regulation of phosphatidylserine synthase from *Saccharomyces cerevisiae* by phospholipid precursors. J Bacteriol 168:668–672

Scarborough GA, Nyc JF (1967) Methylation of ethanolamine phosphatides by microsomes from normal and mutant strains of *Neurospora crassa*. J Biol Chem 242:238–242

Steiner SM, Lester RL (1972) In vitro studies of phospholipid biosynthesis in *Saccharomyces cerevisiae*. Biochem Biophys Acta 260:222–243

Summers EF, Letts VA, McGraw PM, Henry SA (1988) *Saccharomyces cerevisiae cho2* mutants are deficient in phospholipid methylation and cross-pathway regulation of inositol synthesis. Genetics 120:909–922

Waechter CJ, Lester RJ (1971) Regulation of phosphatidylcholine biosynthesis in *Saccharomyces cerevisiae*. J Bacteriol 105:837–843

Waechter CJ, Lester RJ (1983) Differential regulation of the *N*-methyltransferases responsible for phosphatidylcholine synthesis in *Saccharomyces cerevisiae*. Arch Biochem Biophys 158:401–410

Yamashita S, Oshima A (1980) Regulation of phosphatidylethanolamine methyltransferase level by myo-inositol *Saccharomyces cerevisiae*. Eur J Biochem 104:611–616

Yamashita S, Oshima A, Nikawa J, Hosaka K (1982) Regulation of the phosphatidylethanolamine methylation pathway in *Saccharomyces cerevisiae*. Eur J Biochem 128:589–595

Chapter 17

Inhibitors of Phospholipid Biosynthesis

G. D. Robson[1], M. Wiebe[1], P. J. Kuhn[2], and A. P. J. Trinci[1]

1 Introduction

Antifungal compounds used to prevent or cure fungal diseases should display selective toxicity, i.e. they should inhibit growth of the fungus but not of the host. Selective toxicity may result from a compound affecting a site present in the fungus but not the host, e.g. polyoxins inhibit the biosynthesis of chitin, a polymer present in fungi but not in plants (Hori et al. 1974). Alternatively, a site in the fungus may be more sensitive to a compound than an equivalent site in the host, e.g. the antimitotic compound, carbendazim has a greater affinity for fungal β-tubulin than plant β-tubulin (Davidse 1982). Finally, an antifungal compound may display selective toxicity if its uptake, detoxification or activation in the fungus differs from that in the host (Baldwin 1984).

Phospholipids, predominantly as glycerophospholipids, are integral structural components of all membranes and the phospholipid composition of fungal membranes (Brennan and Lösel 1978; see also Chapter 9) is very similar to that of membranes in plants (Kates and Marshall 1975) and animals (Dittmer 1962). Furthermore, phospholipids in fungi appear to be synthesized by pathways identical to those present in other eukaryotes (Brennan and Lösel 1978; Weete 1980). On this basis, it seems unlikely that inhibitors of phospholipid biosynthesis would exhibit selective toxicity. However, the phosphatidylcholine inhibitors, edifenphos (HINOSAN) and iprobenfos (KITAZIN P) have been used successfully as fungicides to control rice sheath blast caused by *Pyricularia oryzae*, even though they exhibit some phytotoxicity. Furthermore, the antibiotic, validamycin (VALIDACIN), a putative inhibitor of inositol biosynthesis, has been used to control several diseases caused by *Rhizoctonia solani* (syn. *Pellicularia sasakii*), particularly rice sheath blight. Despite its apparent mode of action, validamycin has a very low acute toxicity to mammals and, of 150 species of plants tested, only mulberry was sensitive to the antibiotic (Wakae and Matsuura 1975). The basis of the selectivity of these antifungal compounds is not known but their efficacy suggests that phospholipid biosynthesis is a potential target for selective toxicity.

[1] Microbiology Group, Department of Cell & Structural Biology, Stopford Building, University of Manchester, Manchester M13 9Pt, UK
[2] Shell Research Ltd, Sittingbourne Research Centre, Sittingbourne, Kent ME9 8AG, UK

2 Phospholipid Structure and Biosynthesis

2.1 Phospholipid Structure

Glycerophospholipids are by far the largest class of phospholipids and are composed of a glycerol 'backbone' with fatty acid chains esterified onto both C1 and C2 of the glycerol molecule, together with a polar headgroup esterified via the phosphate group to C3. Six main types of glycerophospholipids (differing from one another in the nature of the polar head group) are usually present in eukaryotic membranes (Fig. 1). In fungi, as in other eukaryotes, phosphatidychloline (PC) and phosphatidylethanolamine (PE) are usually the most abundant phospholipids, typically making up 16% to 59% and 14% to 50% respectively of the total glycerophospholipid present (Brennan and Lösel 1978; Weete 1980). Phosphatidylserine (PS), phosphatidylinositol (PI) and diphosphatidylglycerol (cardiolipin, DPG) are found in variable amounts making up between 0% – 19%, 0% – 23% and 0 – 14%, respectively, of the total glycerophospholipid present in fungi. Phosphatidylglycerol (PG) is typically present in much lower concentrations than the other glycerophospholipids.

2.2 Phospholipid Biosynthesis

The biosynthesis of phospholipids in fungi has been reviewed by Brennan and Lösel (1978), Weete (1980), Choprar and Khuller (1983) and Henry et al. (1984). Although there is still some confusion over the existence of certain pathways for phospholipid biosynthesis in fungi, it is generally accepted that, as with most other eukaryotes, three main pathways are present (Fig. 2). In recent years, the regulation of phospholipid biosynthesis in *Saccharomyces cerevisiae* has been the subject of various studies (Greenberg et al. 1982; Henry et al. 1984; Klig et al. 1985; Letts and Henry 1985; Homann et al. 1985; see also Chapter 16) but few comparable studies have been made with filamentous fungi.

2.2.1 The CDP-Diglyceride Pathway

The CDP-diglyceride pathway leads to the formation of PS, PI and PG from CDP-diglyceride (formed from phosphatidic acid and CTP) and serine, inositol or glycerol-3-phosphate respectively; the reactions are catalyzed by membrane-bound PS, PI or PG synthases (Steiner and Lester 1972; Carman and Matas 1981; Fischl and Carman 1983; Kutchler et al. 1986) (Fig. 2). The minor phospholipid, DPG is formed by the addition of a second diacylglycerol group to PG from CDP-diacylglycerol (Steiner and Lester 1972).

$$H_2C-OR$$
$$|$$
$$R'O-CH \qquad O$$
$$| \qquad \parallel$$
$$H_2C-O-P-O-X$$
$$|$$
$$O^{\ominus}$$

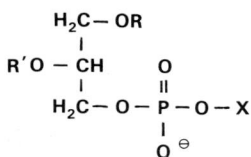

Glycerophospholipid

R,R' = fatty acid acyl groups

X = polar head group

X

$-H$	Phosphatidic acid	PA
$-CH_2CH_2\overset{\oplus}{N}H_3$	Phosphatidylethanolamine	PE
$-CH_2CH_2\overset{\oplus}{N}H_2CH_3$	Phosphatidylmonomethylethanolamine	PMME
$-CH_2CH_2\overset{\oplus}{N}H(CH_3)_2$	Phosphatidyldimethylethanolamine	PDME
$-CH_2CH_2\overset{\oplus}{N}(CH_3)_3$	Phosphalidylcholine	PC
$-CH_2\overset{\oplus}{C}HNH_3$ $\quad\ \ |$ $\quad COO^{\ominus}$	Phosphatidylserine	PS
(inositol ring)	Phosphatidylinositol	PI
$\begin{array}{c} OH \\ | \\ -CH_2CHCH_2OH \end{array}$	Phosphatidylglycerol	PG
(diphosphatidylglycerol structure)	Diphosphatidylglycerol	DPG

For Phosphatidylinositol:

OH OH

HO

OH

OH

For Diphosphatidylglycerol:

$$RO-CH_2$$
$$|$$
$$OH \qquad O \qquad HC-OR'$$
$$| \qquad \parallel \qquad |$$
$$-CH_2CHCH_2-O-P-O-CH_2$$
$$|$$
$$O^{\ominus}$$

Fig. 1. Structures of glycerophospholipids

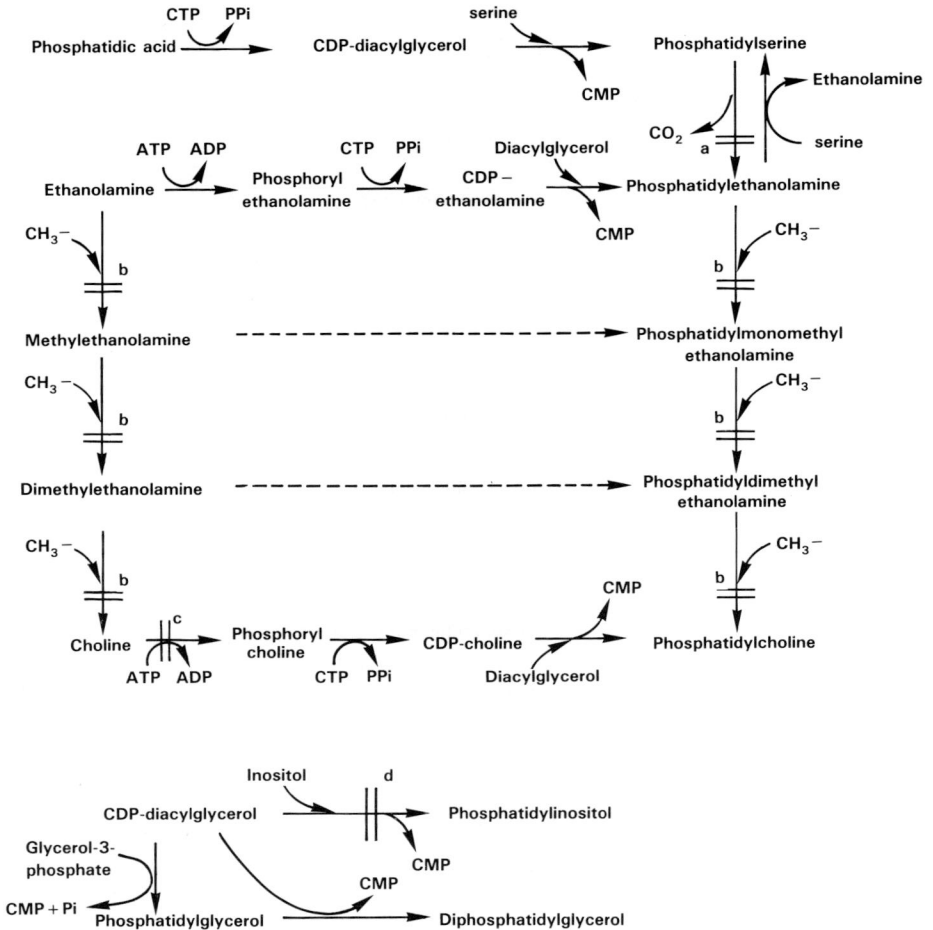

Fig. 2. Pathways of phospholipid biosynthesis. The sites at which inhibitors (**a – d**) are thought to act are indicated. **a** Serine hydroxymate, hydroxylamine. **b** Edifenphos, iprobenfos, 3-deazaadenosine, *S*-adenosylhomocysteine, *L*-homocysteine thiolactone, 2-hydroxyethylhydrazine, ethionine, **c** Hemichol-ninium 3, **d** Validamycin

2.2.2 The Kennedy Cytidine Nucleotide Pathway

A pathway starting with choline and leading to PC by way of CDP-choline was discovered by Kennedy and Weiss (1956) (Fig. 2). Phosphorylcholine, formed by the action of choline kinase, forms CDP-choline which is then esterified to the diacylglycerol molecule with the release of CMP. Similarly, PE can also be formed from ethanolamine by the Kennedy pathway (Waechter et al. 1969; Steiner and Lester 1972). Choline appears to be derived from ethanolamine by successive methylations, with ethanolamine being formed by the decarboxylation of serine (Matysiak et al. 1974).

2.2.3 The Bremner-Greenberg Methylation Pathway

In the alternative pathway for PC biosynthesis, PC arises by three successive transfers of methyl groups from S-adenosylmethionine to PE (Bremner and Greenberg 1961) (Fig. 2). PE is formed by the decarboxylation of PS catalyzed by the enzyme PS decarboxylase (Sherr and Byk 1971; Steiner and Lester 1972). Although this particular reaction is irreversible, it appears that PS may be formed from PE by base exchange (Kasinathan and Khuller 1983). PE then undergoes three methylations which successively form phosphatidylmonomethylethanolamine (PMME), phosphatidyldimethylethanolamine (PDME) and PC; S-adenosylmethionine acts as the methyl donor in each step, the reaction being catalyzed by S-adenosylmethionine transferase (Scarborough and Nyc 1967a, b; Steiner and Lester 1970; Sherr and Byk 1971; A. C. Wilson and Barran 1980). In *Neurospora crassa* (Scarborough and Nyc 1967a, b) and *S. cerevisiae* (Kodaki and Yamashita 1987) it has been proposed that two membrane-associated *N*-methyltransferases are involved in the three methylations that convert PE to PC. One catalyzes the first step in PC formation (the formation of PMME from PE), whilst the second either catalyzes the last two methylations (PMME→PDME→PC) (in *N. crassa*) or all three methylations involved in the formation of PC from PE (in *S. cerevisiae*).

In most eukaryotic cells the Kennedy pathway usually predominates (Pelech and Vance 1984), but in fungi PC is mostly formed by the methylation pathway (Waechter et al. 1969; Steiner and Lester 1970; Sherr and Byk 1971; Chin and Bloch 1988). Selective toxicity may be based on this difference.

3 Inhibitors of Phospholipid Biosynthesis

3.1 Inhibitors Other Than Validamycin, Iprobenfos and Edifenphos

Several compounds have been reported to be inhibitors of phospholipid biosynthesis in bacteria, fungi or mammalian cells (Table 1). Hydroxylamine has been used to change the phospholipid composition of *S. cerevisiae* and *Candida albicans* (Trivedi et al. 1983), and the methionine analogue, ethionine (Chin and Bloch 1988) and 2-hydroxyethyl hydrazine (Nikawa and Yamashita 1983) have been used to inhibit the Bremner-Greenberg methylation pathway in *S. cerevisiae*.

3.2 Validamycin

Validamycin is an aminoglycosidic antibiotic (Fig. 3) produced by *Streptomyces hygroscopicus* var *limoneus* and was developed in Japan as an antifungal agent to control *R. solani*, the causal agent of rice sheath blight (Iwasa et al. 1970, 1971). Studies of the effects of the antibiotic on growth and morphology of *Rhizoctonia* spp. revealed that, unlike most other antifungal compounds, validamycin was neither fun-

Table 1. Inhibitors of phospholipid biosynthesis other than validamycin, iprobenfos and edifenphos

Compound	Site of Action	Organism	Reference
Hydroxylamine	PS decarboxylase	*E. coli*	Raetz and Kennedy (1972)
			Cronan and Vagelos (1972)
			Ohta et al. (1977)
		S. cerevisiae	Trivedi et al. (1983)
		C. albicans	
Serine hydroxymate	PS decarboxylase	*E. coli*	Pizer and Merlie (1973)
3-deazaadenosine	PE methylation	Rat erythrocytes	Randon et al. (1981)
		Rat hepatocytes	Pritchard et al. (1982)
Ethionine	PE methylation	*S. cerevisiae*	Chin and Bloch (1988)
2-hydroxyethyl-hydrazine	PE methylation	*S. cerevisiae*	Nikawa and Yamashita (1983)
L-homocysteine-thiolactone	PE methylation	Rat erythrocytes	Randon et al. (1981)
Hemicholinium 3	Choline kinase	Castor bean endosperm	Kinney and Moore (1987)

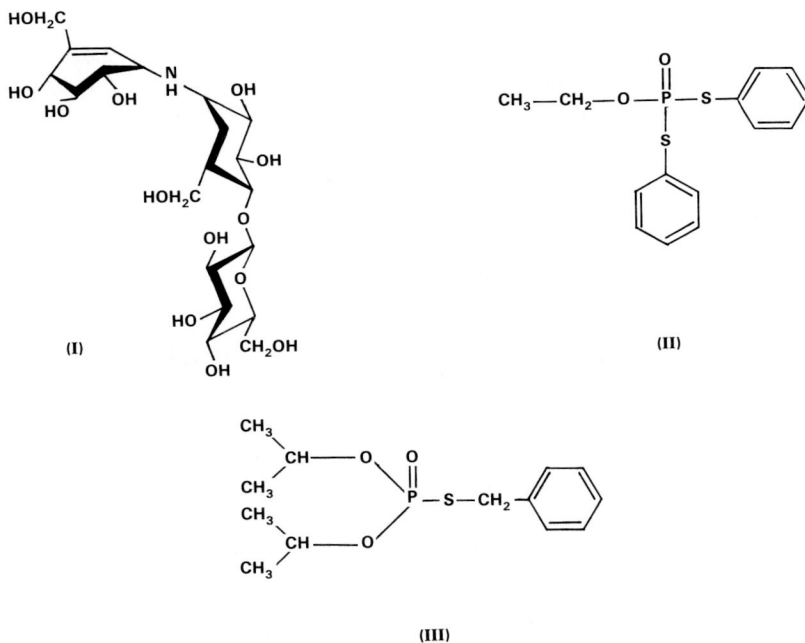

Fig. 3. Structures of (*I*) validamycin, (*II*) edifenphos and (*III*) iprobenfos

gicidal nor fungistatic; restriction of the rate of spread of the pathogen, in vivo and in vitro was caused by a decrease in the rate of hyphal extension coupled with a stimulation of hyphal branching (Nioh and Mizushima 1974; Trinci 1985). Compounds which alter the morphology (spatial distribution) of a mycelium but not its rate of production (specific growth rate) are known as paramorphogens (Tatum et al. 1949),

Table 2. Effects of validamycin on the growth and morphology of *Rhizoctonia cerealis*

Culture condition	Parameter measured	Medium lacking validamycin		Medium containing 1 µM validamycin	
Semi-solid culture	Colony*, radial growth rate (Kr, µm h^{-1})	278	± 5*	96	± <1[e]
	Internode length (I, µm)[a]	188	±11	94	± 6[e]
	Colony surface area (cm^{-2})[b]	26.4	± 0.4	2.3	± 1[e]
	Colony density (mg, cm^{-2})[b]	0.22	± 0.02	0.78	± 0.08[e]
Shake-flask, liquid culture	Specific growth rate (µ, h^{-1})[c]	0.059	± 0.005	0.062	± 0.004
	Internode length (I, µm)[d]	129	±14	131	±10
Stationary liquid culture	Internode length (I, µm)[d]	139	±10	120	±12

R. cerealis was grown at 25°C on Vogel's medium (containing 5 mM glucose as the carbon source) in the presence and absence of validamycin in semi-solid, shake-flask and stationary liquid cultures.
* Mean ± standard error.
[a] Mean of 45 internode lengths taken from the first 3 internodes in each of 3 leading hyphae from the margins of 5 replicate colonies in the linear growth phase.
[b] Measurements were made on colonies 120 h after inoculation and represent the means of 5 replicate colonies.
[c] Growth was followed by measuring dry weight.
[d] Mean of at least 21 internode lengths measured randomly in cultures 48 h after inoculation.
[e] Significantly different (using the t-test, P < 0.05) from the control.

and they include L-sorbose (Trinci and Collinge 1973), cellobiose (R. W. Wilson and Niederpruem 1967), glucosamine (Jejelowo and Trinci 1988), 3-*O*-methylglucose (Galpin et al. 1977; Jejelowo and Trinci 1988) and the diterpenes sclareol and episclareol (Bailey et al. 1974).

Validamycin has no effect on the specific growth rate of *R. cerealis* (Table 2) but causes a decrease in internode length (= the distance between two adjacent branches on a hypha and is thus a measure of branch frequency) and colony radial growth rate. An unusual-dose-response relationship was observed at low antibiotic concentrations (Fig. 4); at concentrations up to 1 µM, a period of normal growth occurred before the colony became inhibited, and the duration of this period of normal growth decreased as the concentration of validamycin was increased until, at a concentration of 1 µM, no period of normal growth was observed (Trinci 1985). Furthermore, all validamycin-'inhibited' colonies had approximately the same colony radial growth rate. The basis of this apparent "all or none" response is not known but may be dependent on the ratio between glucose concentration and antibiotic concentration at the colony margin (G. D. Robson, P. J. Kuhn and A. P. J. Trinci, unpublished observations). By contrast, at high antibiotic concentrations (up to 5 mM), internode length and colony radial growth rate both decreased with increase in validamycin concentration (Fig. 5).

The similarity between the validoxylamine moiety of validamycin and inositol led to an investigation of the effect of the antibiotic on inositol biosynthesis. It was found

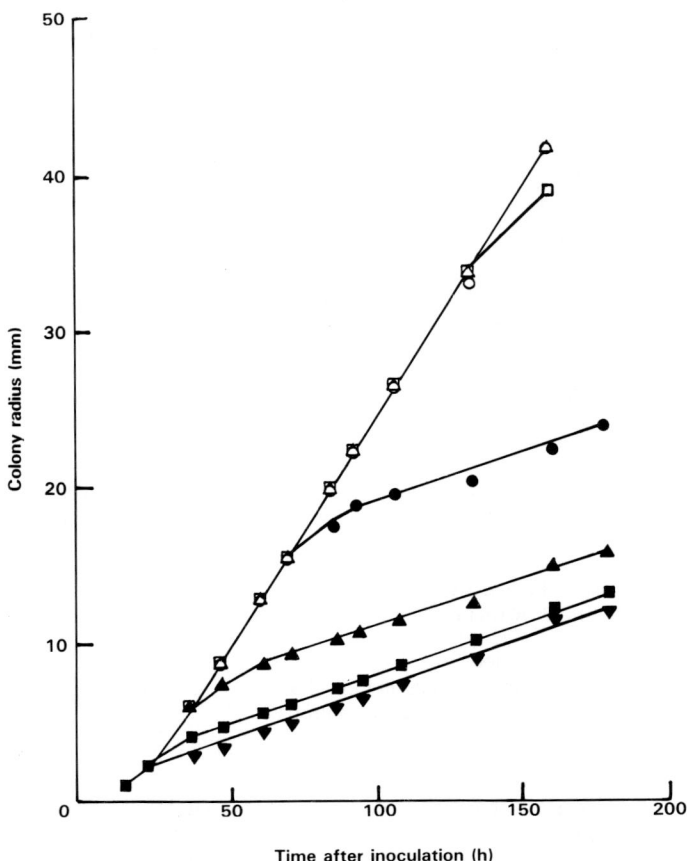

Fig. 4. Effect of validamycin on the growth of *R. cerealis* colonies cultured on Vogel's medium containing 5 mM glucose as carbon source. Validamycin concentrations: ○, 0; △, 0.05; □, 0.10; ●, 0.20; ▲, 0.40; ■, 0.80; ▼, 1.00 µM

that the germination of cucumber seeds infected with *R. solani* was increased following treatment with validamycin, but when the application of antibiotic was followed by treatment with inositol there was a decrease in seed germination although inositol alone had no effect. Nonetheless, addition of inositol to the growth medium failed to reverse the morphological effects induced in *R. cerealis* by the antibiotic (Shibata et al. 1980; Trinci 1984b, 1985).

The inositol content of the supernatant of cultures of *R. cerealis* was increased if it was hydrolysed with 6M HCl prior to analysis, indicating that the fungus secretes water-soluble inositol-containing compounds into the medium, as well as free inositol (Table 3). Validamycin reduced the total inositol content of *R. cerealis* cultures by 62% (Table 3), but addition of the antibiotic to cultures after they had reached stationary phase had no effect on their (endogenous or exogenous) inositol content, indicating that the antibiotic did not stimulate inositol degradation. Over the concentration

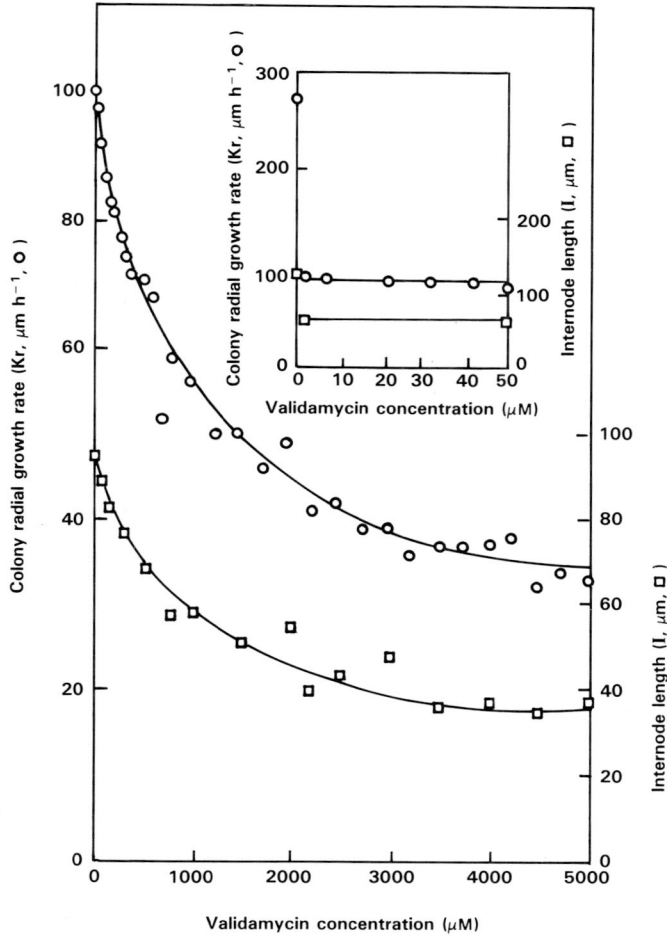

Fig. 5. Effect of validamycin on the colony radial growth rate and internode length of *R. cerealis*. *R. cerealis* colonies were grown at 25 °C on Vogel's medium containing 5 mM glucose as carbon source in the presence and absence of various concentrations of validamycin. All colony radial growth rates were measured during the linear growth phase and are the mean of 10 replicates. Each internode length was the mean of 45 internode lengths taken from the first 3 consecutive internodes in each of 3 leading hyphae from the margins of 5 colonies in the linear phase of growth. The inset represents the effect of validamycin up to a concentration of 50 μM

range 1 to 5000 μM, there was an inverse relationship between antibiotic concentration and inositol content of *R. cerealis* colonies (Fig. 6). Positive correlations were found between inositol content and internode length and between inositol content and colony radial growth rate (Fig. 7). Similarly, inositol auxotrophs of *N. crassa* grown on media containing sub-optimal concentrations of inositol formed highly branched mycelia, and colonies which expanded at much slower rates than control colonies (Fuller and Tatum 1956; Shatkin and Tatum 1961; Hanson and Brody 1979).

Table 3. The effects of validamycin on the distribution of inositol in cultures of *Rhizoctonia cerealis*

Material analysed	Inositol content[a]		
	Medium lacking validamycin	Medium containing 1 µM validamycin added 6 days after inoculation	Medium with 1 µM validamycin
Mycelium[b]	2.51 ± 0.32*	0.96 ± 0.19*,[d]	2.71 ± 0.36*
Unhydrolysed culture supernatant[c]	0.49 ± 0.08	0.12 ± 0.01[d]	0.42 ± 0.05
Hydrolysed culture	1.71 ± 0.22	0.86 ± 0.10[d]	1.57 ± 0.39

R. cerealis was grown at 25 °C in Vogel's medium (containing 5 mM glucose as the carbon source) in shake-flask culture in the presence and absence of validamycin. Validamycin was added at 0 or 6 days after inoculation and the mycelium and culture supernatant were hydrolysed in 6 M HCl for 40 – 44 h and 20 – 25 h respectively and the inositol liberated measured by bioassay (Paranjapye et al. 1964). The inositol content of unhydrolysed culture supernatant was also determined by bioassay.
* Mean ± standard error.
[a] Results are the means of 5 replicate cultures.
[b] Results are expressed as µg inositol mg^{-1} mycelial dry weight.
[c] Results are expressed as µg inositol mg^{-1} culture supernatant.
[d] Significantly different (using the t-test, P < 0.05) from control).

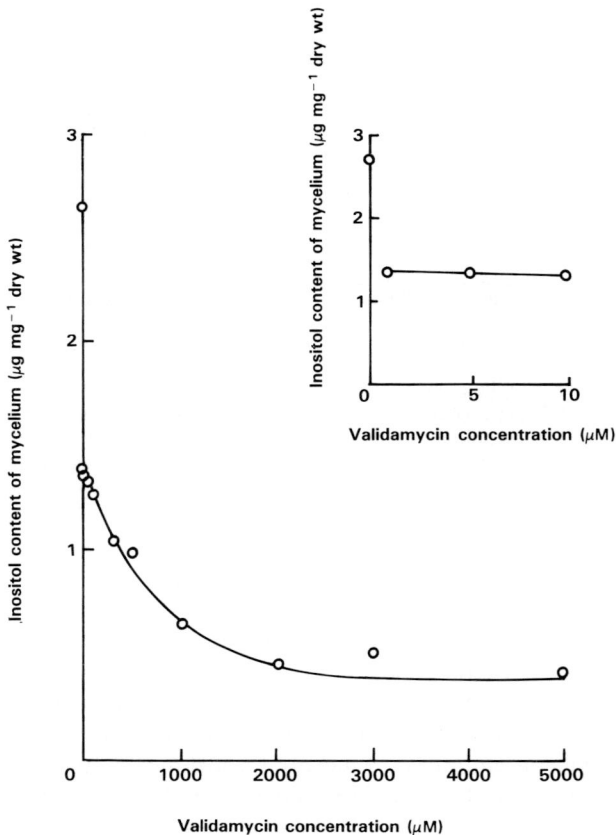

Fig. 6. Effect of validamycin on the inositol content of *R. cerealis*. *R. cerealis* colonies were grown at 25 °C on Vogel's agar medium containing 5 mM glucose as carbon source overlaid with Cellophane in the presence and absence of various concentrations of validamycin. Colonies were harvested 5 days after inoculation. Mycelium was hydrolyzed in 6 M HCl for 40 – 44 h and the inositol liberated was determined by bioassay (Paranjapye et al. 1964). Results are expressed as µg inositol mg^{-1} mycelial dry weight and represent the means of 2 determinations made on each of 5 replicate colonies. The inset represents the effect on inositol content of validamycin up to a concentration of 10 µM

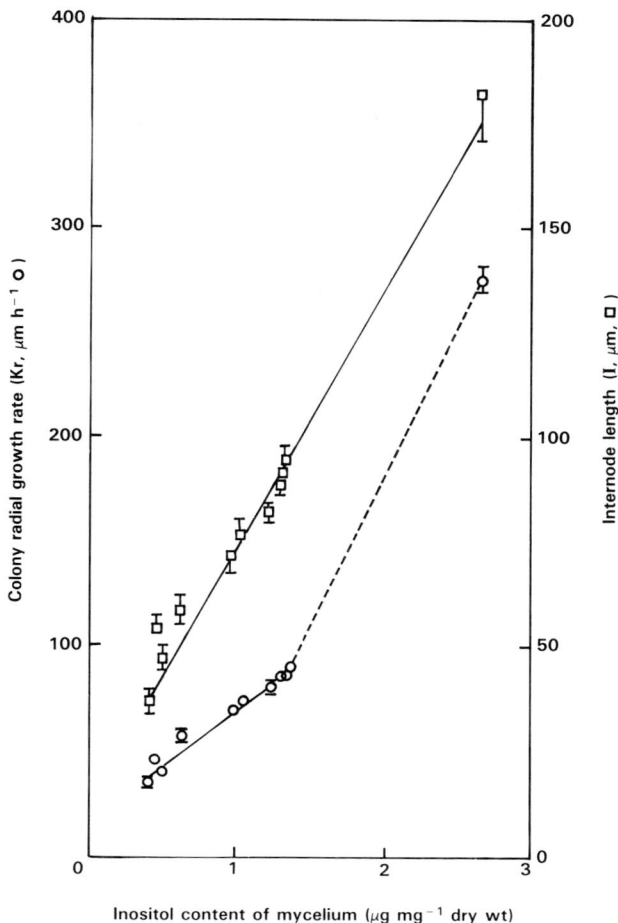

Fig. 7. Relationships between inositol content, colony radial growth rate and internode length of colonies of *R. cerealis*. Inositol content of *R. cerealis* mycelium from Fig. 6 was plotted against the corresponding colony radial growth rates and internode lengths from Fig. 5. Lines were plotted using least squares-regression analysis

In control cultures of *R. cerealis*, 66% of the total inositol was extracted in the lipid fraction (Robson et al. 1989); most of the remaining inositol was probably present as inositol-containing phospholipid associated with the cell wall (Hanson and Brody 1979). PI was found to be the major inositol-containing phospholipid in *R. cerealis*. Validamycin caused a 47% reduction in PI which was correlated with a rise in the levels of PS and an unidentified phospholipid (designated, B). Other phospholipids in *R. cerealis* were not significantly affected by validamycin treatment (Table 4) and the total phospholipid content of the fungus was not significantly altered by the antibiotic. The rise in PS in validamycin-treated cultures may compensate for the decrease in PI. Certainly, this kind of compensation was observed when inositol auxotrophs of *N. crassa* were grown on sub-optimal concentrations of inositol (Hubbard and Brody 1975); despite wide variations in the relative proportions of individual phospholipids in the auxotroph, the cardiolipin content, the total phospholipid content, the total content of the zwitteronic species, the total content of the anionic

Table 4. The effect of validamycin on the phospholipid composition of *Rhizoctonia cerealis*

Phospholipid	Phospholipid composition[a]	
	Medium-lacking validamycin	Medium containing 1 μM validamycin
Phosphatidic acid	0.7 ± 0.2*	0.7 ± 0.2*
Phosphatidylinositol	4.9 ± 0.3	2.4 ± 0.3[b]
Phosphatidylserine	9.0 ± 1.4	11.7 ± 0.7[b]
Phosphatidylcholine	46.4 ± 3.1	46.1 ± 3.9
Phosphatidylethanolamine	25.6 ± 2.0	24.5 ± 1.7
Phosphatidylglycerol	0.8 ± 0.3	1.0 ± 0.4
Diphosphatidylglycerol	4.2 ± 0.2	3.7 ± 0.1
Unknown A	0.3 ± 0.2	0.3 ± 0.1
Unknown B	0.3 ± 0.1	0.8 ± 0.1[b]

R. cerealis was grown at 25 °C in shake-flask culture in Vogel's medium (containing 5 mM glucose as the carbon source) in the presence and absence of validamycin. Cultures were harvested after 4 days, and phospholipids were extracted (Angus and Lester 1972), separated by 2-D TLC (Steiner and Lester 1972) and quantified by phosphorus analysis (Rouser et al. 1970).

* Mean ± standard error.

[a] Results are expressed as percentage of total phospholipid and represent the mean of 8 replicate cultures pooled from 2 independent experiments.

[b] Significantly different (using the t-test, $P < 0.05$) from control.

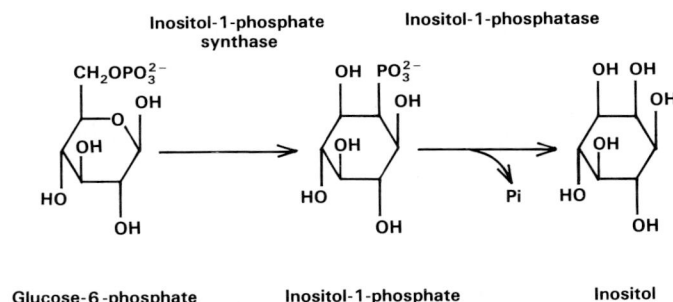

Fig. 8. Biosynthesis of inositol

species, and the ratio of the zwitterionic to ionic totals, all remained approximately constant. Thus, it is believed that an internal compensation mechanism operates in *N. crassa* to maintain the contribution of the phospholipid components to the overall membrane charge constant.

The inhibition of inositol biosynthesis caused by validamycin suggests that the antibiotic might act by inhibiting inositol-1-phosphate synthase or inositol-1-phosphatase (Fig. 8). However, addition of up to 5 mM inositol to a medium containing mannitol as the carbon source (unlike glucose, mannitol does not repress inositol uptake; G. D. Robson, P. J. Kuhn and A. P. J. Trinci, unpublished observations) caused only a

Table 5. Partial reversal of the effect of validamycin by inositol

Inositol concentration in medium (mM)	Colony radial growth rate (Kr, μm h^{-1})	
	Medium lacking validamycin	Medium containing 1 μM validamycin [a]
0	247 ± 3*	61 ± 1*
0.01	252 ± 6	61 ± <1
0.1	245 ± 4	66 ± 2
1.0	249 ± 7	73 ± 3 [c]
5.0	250 ± 11	79 ± 3 [c]
10.0	127 ± 2 [b]	33 ± <1 [c]

R. cerealis colonies were grown at 25 °C on Vogel's medium (containing 50 mM mannitol as the carbon source) in the presence and absence of validamycin and inositol.
* Mean ± standard error.
[a] Each result significantly different (using the t-test, $P < 0.05$) from control lacking validamycin.
[b] Significantly different ($P < 0.05$) from control lacking inositol.
[c] Significantly different ($P < 0.05$) from control lacking inositol.

Table 6. The effect of inositol on the phosphatidylinositol content of untreated and validamycin-treated mycelium of *Rhizoctonia cerealis*

Concentration of inositol in medium	Phosphatidylinositol content (μmol P g^{-1})[a]	
	Medium lacking validamycin	Medium containing 1 μM validamycin
0	2.72 ± 0.22*	1.20 ± 0.17 [b],*
5	3.05 ± 0.18 [c]	1.16 ± 0.10 [b,c]

R. cerealis was grown at 25 °C in shake-flask culture in Vogel's medium (containing 50 mM mannitol as the carbon source) in the presence and absence of validamycin and inositol. Mycelium was harvested after 5 days during late exponential phase; the phospholipids were extracted and then separated by 2-D TLC. Phosphatidylinositol was quantified by phosphorus analysis.
* Mean ± standard error.
[a] Results are expressed as μmol phosphorus g^{-1} mycelial dry weight mycelium and represent the means of 5 replicates.
[b] Significantly different (using the t-test, $P < 0.05$) from control-lacking validamycin.
[c] No significant difference ($P > 0.05$) from control-lacking inositol.

small increase in colony radial growth rate (Table 5), and, at higher concentrations, inositol caused a decrease in colony radial growth rate. The limited ability of exogenous inositol to reverse the morphological effects caused by validamycin and its inability to restore PI levels (Table 6), suggest that the antibiotic does not inhibit inositol biosynthesis directly. However, it is possible that validamycin inhibits the incorporation of inositol into PI by membrane-bound phosphatidylinositol synthase. Inhibition of this enzyme would lead to a rise in the free inositol pool which would inhibit inositol biosynthesis by negative feedback inhibition. Strong negative feedback inhibition of inositol biosynthesis by inositol has been demonstrated in other fungi (Culbertson et al. 1976; Greenberg et al. 1982; Zsindely et al. 1983).

3.3 Iprobenfos and Edifenphos

The organophosphorus fungicides, iprobenfos (S-benzyl-O-O'-diisopropyl-phospho-rothioate, IBP) and edifenphos (O-ethyl-S,S-diphenyl-phosphorothioate) were developed in the 1960s to control rice blast caused by *Pyricularia oryzae*. They have also been used to control *Fusarium solani* (Sisler 1986) and *Colletotrichum gloeospories* (Dickman et al. 1983; Sisler 1986).

Low concentrations of edifenphos did not reduce the specific growth rate of *Fusarium graminearum*, but caused a decrease in hyphal growth unit length and a decrease in colony radial growth rate (Table 7). Thus, at these concentrations, edifenphos behaved as a paramorphogen (Shatkin and Tatum 1961), altering the morphology of *F. graminearum* but not its specific growth rate. Choline auxotrophs of *N. crassa* also formed densely branched mycelia when grown on medium containing sub-optimal concentrations of choline (Horowitz et al. 1945; Shatkin and Tatum 1961). The inhibitory effect of edifenphos on colony radial growth rate of *F. graminearum* (Table 7) results from the decrease in hyphal growth unit length which leads to a reduction in peripheral growth zone width and hence in colony radial growth rate (Trinci 1984a, b; Jejelowo and Trinci 1988). At high concentrations (1 mM and above), edifenphos inhibited the germination of spores of *Aspergillus nidulans* (Craig and Peberdy 1983a) and *F. graminearum* (G. D. Robson, M. Wiebe and A. P. J. Trinci, unpublished observations) and dry weight increase of cultures of *P. oryzae* (Kodama et al. 1980).

The primary mode of action of edifenphos was first thought to be chitin synthesis, possibly at the site where UDP-*N*-acetylglucosamine penetrates the cytoplasmic membrane before polymerization into chitin (Maeda et al. 1970). However, it was subsequently shown that iprobenfos and edifenphos cause a significant decrease in the PC fraction of membrane phospholipids of *P. oryzae* (Kodama et al. 1979, 1980) and *F. graminearum* (Tables 7, 8). Craig and Peberdy (1983a,b), however, found that in *A. nidulans*, iprobenfos caused a 70% increase in PE concentration and a 90% inhibition of chitin-synthase activity. These observations have led to the suggestion that inhibition of lipid biosynthesis is the primary site of action of these fungicides and that

Table 7. Effect of edifenphos on the growth and morphology of *Fusarium graminearum*

Edifenphos concentration in the medium (μM)	Colony radial growth rate (Kr, μm h^{-1})	Hyphal growth unit length (G, μm)[a]	Specific growth rate (μ, h^{-1})[b]
0	140 ± 2 *	249 ± 10 *	0.31 ± 0.02 *
44	132 ± 2	142 ± 4	0.28 ± 0.03
88	100 ± 1	90 ± 3	0.30 ± 0.01
265	64 ± 1	61 ± 1	0.27 ± 0.01

F. graminearum was grown in semi-solid culture at 25 °C on Vogel's medium (containing 50 mM glucose as the carbon source) containing various concentrations of edifenphos.
* Mean ± standard error.
[a] Hyphal growth unit lengths were measured on young mycelia with at least 10 hyphal tips. Results are the means of at least 15 replicates.
[b] Determined by time-lapse photomicroscopy. Results are the mean of 5 replicates. Edifenphos-treated cultures were not significantly different (using F-test, P < 0.05) from the control.

Table 8. The effect of edifenphos on the phospholipid composition of *Fusarium graminearum*

Phospholipid	Phospholipid composition[a]	
	Medium lacking edifenphos	Medium containing 132 μM edifenphos
Phosphatidic acid	1.5 ± 0.03 *	4.0 ± 0.7 *
Phosphatidylinositol	7.5 ± 0.6	6.1 ± 0.6
Phosphatidylserine	5.9 ± 0.6	3.9 ± 0.8
Phosphatidylcholine	48.6 ± 2.7	33.4 ± 1.8[b]
Phosphatidylethanolamine	19.0 ± 1.7	17.0 ± 1.1
Phosphatidylglycerol	1.0 ± 0.3	3.3 ± 0.5
Diphosphatidylglycerol	2.1 ± 0.2	1.6 ± 0.3
Unknown 5	Trace	18.2 ± 2.7[b]
Unknown 1 − 4	Trace	Trace

F. graminearum was grown at 25 °C in shake-flask culture in Vogel's medium (50 mM glucose as the carbon source) in the presence and absence of 132 μM edifenphos. Cultures were harvested during the exponential phase of growth. Lipids were extracted, separated by 2-D TLC, and quantified by phosphorus analysis.
* Mean ± standard error.
[a] Results are expressed as percentage of total phospholipid and represent the mean of at least 5 replicate cultures pooled from 2 independent experiments.
[b] Significantly different (using the F-test, $P < 0.05$) from control.

inhibition of chitin synthesis is a secondary effect resulting from the altered phospholipid composition of the membrane (Yoshida et al. 1984).

Although *A. nidulans* is less sensitive than *P. oryzae* to iprobenfos, the fungicide induced the formation of abnormal hyphae in *A. nidulans* (Craig and Peberdy 1983a), but not, apparently, in *P. oryzae* (Kodama et al. 1979). Although it is not known if edifenphos or iprobenfos affect the morphology of *P. oryzae*, balloons (abnormal hyphal swellings) are formed when choline-deficient mutants of *A. nidulans* are grown on a choline-deficient medium (Markham and Bainbridge 1978).

Edifenphos and iprobenfos affect PC biosynthesis by inhibiting the Bremner-Greenberg methylation pathway (Akatsuka et al. 1977; Kodama et al. 1980). Incorporation of choline into PC via the Kennedy cytidine-nucleotide pathway is unaffected but the methylation of ethanolamine to choline is also inhibited (Yoshida et al. 1984). In *F. graminearum*, the decrease in the concentration of PC caused by edifenphos was accompanied by an increase in the concentration of an unknown phospholipid (unknown 5), presumed to be phosphatidyldimethylethanolamine (Table 8). Thus, the ratio between anionic and zwitterionic phospholipids in the membrane of *F. graminearum* remained almost constant following treatment with edifenphos. No decrease in total phospholipid was observed.

4 Validamycin and Edifenphos as Paramorphogens

Validamycin and certain concentrations of edifenphos act as paramorphogens. Validamycin apparently has no effect on the specific growth rate of *R. cerealis* at concentrations up to 5 mM, whereas edifenphos acts as a paramorphogen at low concentrations (<0.5 mM) but becomes inhibitory to germination and growth at high concentrations (1 mM and above). Validamycin and edifenphos both alter the phospholipid composition of sensitive fungi, validamycin by decreasing the PI content, and edifenphos by decreasing the PC content.

It is known that differences exist in the protein composition of the plasmalemma, endoplasmic reticulum, vacuolar membranes and mitochondrial membranes of *N. crassa* (Bowman et al. 1987), and in the phospholipid composition of various membranes of *Agaricus bisporus* (Weete et al. 1985) and *S. cerevisiae* (Daum et al. 1986). Thus, since lipid analyses give only the mean lipid composition of a fungus, it is possible that the effect of these antifungal compounds on some membranes in a mycelium may be greater than on others.

The activities of at least some membrane-bound enzymes display a requirement for a particular phospholipid; in fungi, for example, maximal in vitro activities of chitin synthase (Duran and Cabib 1978; Vermeulen and Wessels 1983; Montgomery and Gooday 1985) and chitinase (Humphreys and Gooday 1984) are dependent on the presence of PC. In addition, low concentrations of PI and PS stimulate chitin synthase activity in *Schizophyllum commune*, although high concentrations are inhibitory (Vermeulen and Wessels 1983). The nature of the fatty acids esterified to the phospholipid can also have a major effect on enzyme activity. For both chitin synthase from *Coprinus cinereus* (Montgomery and Gooday 1985) and chitinase from *Mucor mucedo* (Humphreys and Gooday 1984), dimyristoyl PC stimulated enzyme activity in vitro, whereas dioleoyl PC strongly inhibited enzyme activity.

The reduction in the PC content of *P. oryzae* by edifenphos or iprobenfos is thought to be the indirect cause of the reduction in the chitin content of the cell walls of sensitive fungi (Yoshida et al. 1984). This hypothesis is supported by the observation that in an inositol auxotroph of *S. cerevisiae*, inositol deprivation leads to a decrease in the mannan fraction of the cell wall. The in vitro activity of the membrane-associated enzyme, UDP-*N*-acetylglucosamine-dolichol phosphate *N*-acetylglucosamine-1-phosphotransferase which is involved in mannan synthesis, requires PI for maximal activity (Hanson and Lester 1982; Hanson 1984). Similarly, when an inositol auxotroph of *N. crassa* was grown under inositol-deficient conditions, there was a 50% decrease in the glucosamine content of the wall, suggesting an inhibition of chitin biosynthesis (Hanson and Brody 1979). A decrease in the PI content of *R. cerealis* resulting from validamycin treatment, may, therefore, influence the maximum activity of one or more enzymes involved in the biosynthesis of wall polymers. Support for this hypothesis is provided by the observation that validamycin causes a reduction in the (1→3)-*β*-linked glucomannan fraction of the cell wall of *R. cerealis* (Kido et al. 1986). In addition, alterations in membrane composition affect membrane fluidity which in turn influences enzyme activity. For example, decreased levels of PC in a choline auxotroph of *N. crassa*, grown on sub-optimal concentrations of choline, have been shown to increase membrane fluidity in vivo (Juretic 1977). Finally, it has been demonstrated that PI acts as a hydrophobic anchor for many membrane proteins in mammalian cells (Low et al. 1986) and that glycophospholipids act as an anchor for temperature-specific surface antigens in *Paramecium aurelia* (Capdeville et al. 1987). Glycerophospholipids may have a similar role in fungi, and therefore a reduction in PI or PC caused by validamycin or edifenphos may lead to a decrease in the activities of membrane-bound enzymes.

In fungal mycelia, there is a relationship between branch frequency (hyphal growth unit) and hyphal extension, and the mean rate of hyphal extension is determined by specific growth rate and hyphal growth unit length (Trinci 1974, 1984a):

$$G = E/\mu ,$$

where G is the hyphal growth unit length (mean length of hypha associated with each hyphal tip), μ is the specific growth rate and E is the mean hyphal extension rate. Tip extension involves the incorporation of new wall material at the hyphal apex, followed by the formation of linkages between newly formed polymers causing wall rigidification (Wessels et al. 1983; Sonnenberg et al. 1985). A decrease in the maximum activity of an enzyme(s) involved in wall biosynthesis or assembly at the hyphal tip may lead to a decrease in hyphal extension rate, but, providing that overall synthesis of the polymer in the mycelium is not affected, this would not necessarily have any effect on specific growth rate. Further, any factor which reduces the mean rate of hyphal extension rate without affecting μ will cause a reduction in hyphal growth unit length (i.e. an increase in branching) (Trinci 1984a, b).

The manner in which paramorphogen fungicides control fungal pathogens without inhibiting their growth is not known, but the observation that they reduce the rate of spread of the pathogen in vitro and in vivo may be significant since the natural defence system of a plant is likely to be more effective against a localised lesion than against one that is rapidly spreading. Alternatively, paramorphogens may interfere with penetration of the host by the pathogen or affect the ecological competitiveness of the pathogen at the plant surface.

References

Akatsuka T, Kodama O, Yameda H (1977) A novel mode of action of Kitazin P in *Pyricularia oryzae*. Agric Biol Chem 41:2111–2112

Angus WW, Lester RL (1972) Turnover of inositol and phosphorus-containing lipids in *Saccharomyces cerevisiae*; extracellular accumulation of glycerophosphorylinositol derived from phosphatidylinositol. Arch Biochem Biophys 151:483–495

Bailey JA, Vincent GG, Burden RS (1974) Diterpenes from *Nicotiana glutinosa* and their effect on fungal growth. J Gen Microbiol 85:57–64

Baldwin BC (1984) Potential targets for the selective inhibition of fungal growth. In: Trinci APJ, Ryley JF (eds) Mode of action of antifungal agents. Cambridge Univ Press, Cambridge, pp 43–62

Bowman BJ, Borgeson CE, Bowman EJ (1987) Composition of *Neurospora crassa* vacuolar membranes and comparison to endoplasmic reticulum, plasma membranes and mitochondrial membranes. Exp Mycol 11:197–205

Bremner J, Greenberg DM (1961) Methyl transferring enzyme system of microsomes in the biosynthesis of lecithin. Biochim Biophys Acta 46:205–216

Brennan PJ, Lösel DM (1978) Physiology of fungal lipids: selected topics. Adv Microb Physiol 17:47–179

Capdeville Y, Cardosa de Almeida ML, Deregnaucourt C (1987) The membrane-anchor of *Parame-cium* temperature-specific antigens in a glycosylinositol phospholipid. Biochem Biophys Res Commun 147:1219–1225

Carman GM, Matas J (1981) Solubilization of microsomal-associated phosphatidyl serine and phosphatidylinositol synthase from *Saccharomyces cerevisiae*. Can J Microbiol 27:1140–1149

Chin J, Bloch K (1988) Phosphatidylcholine synthesis in yeast. J Lipid Res 29:9–14

Choprar A, Khuller GK (1983) Lipid metabolism in fungi. CRC Crit Rev Microbiol 11:209–271

Craig GD, Peberdy JF (1983a) The mode of action of *S*-benzyl *O,O*-diisopropyl phosphorothioate and dicloran on *Aspergillus nidulans*. Pestic Sci 14:17–24

Craig GD, Peberdy JF (1983b) The effect of *S*-benzyl *O,O*-diisopropylphosphorothioate (Kitazin) and dicloran on the total lipid, sterol and phospholipids in *Aspergillus nidulans*. FEMS Microbiol Lett 18:11–14

Cronan JE, Vagelos PR (1972) Metabolism and function of the membrane phospholipids of *Escherichia coli*. Biochim Biophys Acta 265:25–60

Culbertson MR, Donahue TF, Henry SA (1976) Control of inositol biosynthesis in *Saccharomyces cerevisiae*: Properties of a repressible enzyme system in extracts of wild type (Ino$^+$) cells. J Bacteriol 126:232–242

Daum G, Heidorn E, Paltauf F (1986) Intracellular transfer of phospholipids in the yeast, *Saccharomyces cerevisiae*. Biochim Biophys Acta 878:93–101

Davidse LC (1982) Benzimidazole compounds: selectivity and resistance. In: Dekker J, Georgopolous SG (eds) Fungicide resistance in crop protection. Pudoc, Wagenigen, pp 60–70

Dickman MB, Patil SS, Kolattukudy PE (1983) Effects of organophosphorus pesticides on cutinase activity and infection of papayas by *Colletotrichum gloeosporoides*. Phytopathology 73:1209–1214

Dittmer JC (1962) Distribution of phospholipids. In: Flortun M, Mason HS (eds) Comparative biochemistry, vol III. Academic Press, London New York, pp 231–264

Duran A, Cabib E (1978) Solubilization and partial purification of yeast chitin synthetase. J Biol Chem 253:4419–4425

Fischl AS, Carman GM (1983) Phosphatidylinositol biosynthesis in *Saccharomyces cerevisiae*: Purification and properties of microsome-associated phosphatidylinositol synthesis. J Bacteriol 154:304–311

Fuller RC, Tatum EL (1956) Inositol phospholipid in *Neurospora* and its relationship to morphology. Am J Bot 43:361–365

Galpin MF, Jennings DH, Thornton JD (1977) Hyphal branching in *Dendryphiella salina*: effect of various compounds and further elucidation of the effect of sorbose and the rôle of cAMP. Trans Br Mycol Soc 69:175–182

Greenberg ML, Reiner B, Henry SA (1982) Regulatory mutants of inositol biosynthesis in yeast: Isolation of inositol-excreting mutants. Genetics 100:19–33

Hanson BA (1984) Rôle of inositol-containing sphingolipids in *Saccharomyces cerevisiae* during inositol starvation. J Bacteriol 159:837–842

Hanson BA, Brody S (1979) Lipid and cell wall changes in an inositol-requiring mutant of *Neurospora crassa*. J Bacteriol 138:461–466

Hanson BA, Lester RL (1982) Effect of inositol starvation on the in vitro synthesis of mannan and N-acetylglucosaminylpyrophosphorydolichol in *Saccharomyces cerevisiae*. J Bacteriol 151:334–342

Henry SA, Klig LS, Loewy BS (1984) The genetic regulation and coordination of biosynthetic pathways in yeast: amino acid and phospholipid synthesis. Annu Rev Genet 18:2017–2031

Homann MJ, Henry SA, Carman GM (1985) Regulation of CDP-diacylglycerol synthase activity in *Saccharomyces cerevisiae*. J Bacteriol 163:1265–1266

Hori M, Kakiki K, Misato T (1974) Studies on the mode of action of polyoxins. Part IV. Further studies on the relation of polyoxin structure to chitin synthetase inhibition. Agric Biol Chem 38:691–698

Horowitz NH, Bonner D, Houlahan T (1945) The utilization of choline analogues by cholineless mutants of *Neurospora*. J Biol Chem 159:145–151

Hubbard SC, Brody S (1975) Glycerophospholipid variation in choline and inositol auxotrophs of *Neurospora crassa*. J Biol Chem 250:7173–7181

Humphreys AM, Gooday GW (1984) Phospholipid requirement of microsomal chitinase from *Mucor mucedo*. Curr Microbiol 11:187–190

Iwasa T, Yamamoto H, Shibata M (1970) Studies on validamycins, new antibiotics I: *Streptomyces hygroscopicus* var *limoneus*, valdidamycin A producing organism. J Antibiot 23:595–602

Iwasa T, Higashide E, Yamamoto H, Shibata M (1971) Studies on validamycins, new antibiotics. II. Production and biological properties of validamycins A and B. J Antibiot 24:102–113

Jejelowo OA, Trinci APJ (1988) Effects of the paramorphogens, 3-*O*-methylglucose, glucosamine and L-sorbose on the growth and morphology of *Botrytis fabae*. Trans Br Mycol Soc 91:653–660

Juretic D (1977) The effect of phosphatidylcholine depletion on biochemical and physical properties of a *Neurospora* membrane mutant. Biochim Biophys Acta 469:137–150

Kasinathan C, Khuller GK (1983) Biosynthesis of major phospholipids of *Microsporum gypseum*. Biochim Biophys Acta 752:187–190

Kates M, Marshall O (1975) Biosynthesis of phosphoglycerides in plants. In: Galliard T, Mercer EI (eds) Recent advances in the chemistry and biochemistry of plant lipids. Academic Press, London New York, pp 115–159

Kennedy EP, Weiss SB (1956) The function of cytidine coenzymes in the biosynthesis of phospholipids. J Biol Chem 222:193–214

Kido Y, Nagasato T, Ono K, Fujimoto Y, Uyeda M, Shibata M (1986) Change in a cell-wall component of *Rhizoctonia solani* inhibited by validamycin. J Antibiot 50:1519–1525

Kinney AJ, Moore TS (1987) Phosphatidylcholine synthesis in castor bean endosperm. I. Metabolism of L-serine. Plant Physiol 84:78–81

Klig LS, Homann MJ, Carman GM, Henry SA (1985) Coordinate regulation of phospholipid biosynthesis in *Saccharomyces cerevisiae*: Pleiotropically constitutive Opi 1 mutant. J Bacteriol 162:1135–1141

Kodaki T, Yamashita S (1987) Yeast phosphatidylethanolamine methylation pathway and characterization of two distinct methyltransferase genes. J Biol Chem 262:15428–15435

Kodama O, Yamada H, Akatsuka T (1979) Kitazin P, inhibitor of phosphatidylcholine biosynthesis in *Pyricularia oryzae*. Agric Biol Chem 43:1719–1725

Kodama O, Yamashita K, Akatsuka T (1980) Hinosan inhibition of phosphatidylcholine biosynthesis in *Pyricularia oryzae*. Agric Biol Chem 44:1015–1021

Kutchler K, Daum G, Paltauf F (1986) Subcellular and submitochondrial localization of phospholipid-synthesizing enzymes in *Saccharomyces cerevisiae*. J Bacteriol 165:901–910

Letts VA, Henry SA (1985) Regulation of phospholipid synthesis in phosphatidylserine synthase-deficient (cho 1) mutants of *Saccharomyces cerevisiae*. J Bacteriol 163:560–567

Low MG, Ferguson MAJ, Futerman AH, Silman I (1986) Covalently attached phosphatidylinositol as a hydrophobic anchor for membrane proteins. TIPS 11:212–215

Maeda T, Abe H, Kakiki K, Misato T (1970) Studies on the mode of action of organophosphorus fungicide, Kitazin. Part II. Accumulation of an amino sugar derivative from Kitazin-treated mycelia of *Pyricularia oryzae*. Agric Biol Chem 34:700–709

Markham P, Bainbridge BW (1978) A morphological lesion (ballooning) related to a requirement for choline in mutants of *Aspergillus nidulans*. Proc Soc Gen Microbiol 5:65

Matysiak Z, Radominska-Pyrek A, Chojnacki T (1974) The cytidine mechanism and methylation pathway in the formation of *N*-methylated ethanolamine phosphoglycerides in *Neurospora crassa*. J Mol Cell Biochem 3:143–151

Montgomery GWG, Goodway GW (1985) Phospholipid-enzyme interactions of chitin synthase of *Coprinus cinereus*. FEMS Microbiol Lett 27:29–33

Nikawa J, Yamashita S (1983) 2-Hydroxyethylhydrazine as a potent inhibitor of phospholipid methylation in yeast. Biochim Biophys Acta 751:201–209

Nioh T, Mizushima S (1974) Effect of validamycin on the growth and morphology of *Pellicularia sasakii*. J Gen Appl Microbiol 20:373–383

Ohta T, Okuda S, Takahashi H (1977) The relationship between phospholid composition and transport activities of amino acids in *Escherichia coli* membrane vesicles. Biochim Biophys Acta 466:44–56

Paranjapye VN, Deshusses J, Posternack TH (1964) Sur une méthode simplifiée de dosage microbiologique du MS-inositol. Anal Chim Acta 31:480–488

Pelech SL, Vance DE (1984) Regulation phosphatidylcholine biosynthesis. Biochim Biophys Acta 779:217–251

Pizer LI, Merlie JP (1973) Effect of serine hydroxymate on phospholipid synthesis in *E. coli*. J Bacteriol 114:980–987

Pritchard PH, Chiang PK, Cantoni GL, Vance DE (1982) Inhibition of phosphatidylethanolamine-*N*-methylation by 3-deazaadenosine stimulates the synthesis of phosphatidylcholine via the CDP-choline pathway. J Biol Chem 257:6362–6367

Raetz CRH, Kennedy EP (1972) The association of phosphatidylserine synthetase with ribosomes in extracts of *Escherichia coli*. J Biol Chem 247:2008–2014

Randon J, Lecompte T, Chignard M, Siess W, Marlas G, Dray F, Vargaftig B (1981) Dissociation of platelet activation from transmethylation of their membrane phospholipids. Nature (London) 293:660–662

Robson GD, Kuhn PJ, Trinci APJ (1989) Effect of validamycin A on the inositol content and branching of *Rhizoctonia cerealis* and other fungi. J Gen Microbiol 135:739–750

Rouser G, Fleischer S, Yamamoto A (1970) Two-dimensional thin-layer chromatographic separation of polar lipids and determination of phospholipids by phosphorus analysis of spots. Lipids 5:494–496

Scarborough GA, Nyc JF (1967a) Properties of a phosphatidylethanolamine-methyltransferase from *Neurospora crassa*. Biochim Biophys Acta 146:111–115

Scarborough GA, Nyc JF (1967b) Methylation of ethanolamine phosphatides by microsomes from normal and mutant strains of *Neurospora crassa*. J Biol Chem 242:238–242

Shatkin AJ, Tatum EL (1961) The relationship of *m*-inositol to morphology in *Neurospora crassa*. Am J Bot 48:760–771

Sherr S, Byk KC (1971) Choline and serine incorporation into the phospholipids of *Neurospora crassa*. Biochim Biophys Acta 239:243–247

Shibata M, Uyeda M, Mori K (1980) Reversal of validamycin inhibition by the hyphal extract of *Rhizoctonia solani*. J Antibiot 34:447–451

Sisler HD (1986) Control of fungal diseases by compounds acting as antipenetrants. Crop Protect 5:306–313

Sonnenberg ASM, Sietsma JH, Wessels JGH (1985) Spatial and temporal differences in the synthesis of $(1\rightarrow3)$-$\beta(1\rightarrow6)$-β linkages in a wall glucan of *Schizophyllum commune*. Exp Mycol 9:141–148

Steiner MR, Lester RL (1970) In vitro study of the methylation pathway of phosphatidylcholine synthesis and the regulation of this pathway in *Saccharomyces cerevisiae*. Biochemistry 9:63–69

Steiner MR, Lester RL (1972) In vitro studies of phospholipid biosynthesis in *Saccharomyces cerevisiae*. Biochim Biophys Acta 260:222–243

Tatum EL, Barratt RW, Cutter VM (1949) Chemical induction of colonial paramorphs in *Neurospora* and *Syncephalastrum*. Science 109:509–511

Trinci APJ (1974) A study of the kinetics of hyphal extension and branch initiation of hyphal extension and branch initiation of fungal mycelia. J Gen Microbiol 81:225–236

Trinci APJ (1984a) Regulation of hyphal branching and hyphal orientation. In: Jennings DH, Rayner ADM (eds) The ecology and physiology of the fungal mycelium. Cambridge Univ Press, Cambridge, pp 22–52

Trinci APJ (1984b) Antifungal agents which affect hyphal extension and hyphal branching. In: Trinci APJ, Ryley JF (eds) Mode of action of antifungal agents. Cambridge Univ Press, Cambridge, pp 113–134

Trinci APJ (1985) Effect of validamycin A and *L*-sorbose on the growth and morphology of *Rhizoctonia cerealis* and *Rhizoctonia solani*. Exp Mycol 9:20–27

Trinci APJ, Collinge A (1973) Influence of *L*-sorbose on the growth and morphology of *Neurospora crassa*. J Gen Microbiol 78:179–192

Trivedi A, Singhal GS, Prasad R (1983) Effect of phosphatidylserine enrichment on amino acid transport in yeast. Biochim Biophys Acta 729:85–89

Vermeulen CA, Wessels JGH (1983) Evidence for a phospholipid requirement of chitin synthase in *Schizophyllum commune*. Curr Microbiol 8:67–71

Waechter CJ, Steiner MR, Lester RL (1969) Regulation of phosphatidylcholine biosynthesis by the methylation pathway in *Saccharomyces cerevisiae*. J Biol Chem 244:3419–3422

Wakae O, Matsuura K (1975) Characteristics of validamycin as a fungicide for *Rhizoctonia* disease control. Rev Plant Protect Res 8:81–92

Weete JD (1980) Lipid biochemistry of fungi and other organisms. Academic Press, London New York

Weete JD, Furter R, Hansler E, Rast DM (1985) Cellular and chitosomal lipids of *Agaricus bisporus* and *Mucor rouxii*. Can J Microbiol 31:1120–1126

Wessels JGH, Sietsma JH, Sonnenberg ASM (1983) Wall synthesis and assembly during hyphal morphogenesis in *Schizophyllum commune*. J Gen Microbiol 129:1607–1616

Wilson AC, Barran LR (1980) The methylation system for 3-Sn-phosphatidylcholine biosynthesis in *Fusarium oxysporum*. Can J Microbiol 26:774–777

Wilson RW, Niederpruem DJ (1967) Cellobiose as a paramorphogen in *Schizophyllum commune*. Can J Microbiol 16:629–634

Yoshida M, Moriya S, Uesugi Y (1984) Observation of transmethylation from methionine into choline in the intact mycelia of *Pyricularia oryzae* by ^{13}C NMR under the influence of fungicides. J Pestic Sci 9:703–708

Zsindely A, Kiss A, Shablik M, Szabolics M, Szabó G (1983) Possible rôle of a regulatory gene product upon the myo-inositol-1-phosphate synthase production in *Neurospora crassa*. Biochim Biophys Acta 741:273–278

Transduction of the Calcium Signal with Special Reference to Ca^{2+}-Induced Conidiation in *Penicillium notatum*

D. PITT and A. KAILE [1]

1 Introduction

Calcium acts as a second messenger within eukaryotic cells and transduces cell surface primary stimuli into intracellular events. The primary stimulus may be a hormone binding to its receptor, an electrical stimulus which induces a change in membrane potential, such as an action potential, or, perhaps, a physical stimulus such as a sperm entering an egg. Thus, neither the external stimulus, nor the manifested response is necessarily due to calcium.

Calcium has many properties that make it better suited to act as a carrier of information than other more abundant cations (e.g. Mg^{2+}), such as its ionic radius (large), energy of hydration (low) and the presence of d orbitals, allowing it to form a variety of complexes with varying co-ordination numbers and bond lengths (Hepler and Wayne 1985).

Since calcium may be cytotoxic, a low intracellular concentration is a necessary condition for the phosphate-driven metabolism characteristic of higher organisms (Carafoli and Penniston 1985; Williamson 1981). If the Ca^{2+} concentration ($[Ca^{2+}]$) was high, it would bind to the phosphate groups and precipitate as hydroxyapatite leading ultimately to cellular dysfunction and death. To control cellular processes effectively therefore, calcium itself must be regulated and to do this cells have evolved an elaborate system of proteins that interact with the calcium ion and govern the transmission of the intracellular message (Carafoli and Penniston 1985).

2 Regulation of Intracellular Calcium

It is the transient rise in free Ca^{2+} in the cytoplasm that is frequently the trigger for intracellular regulatory events. This implies that low levels of Ca^{2+} need to be maintained such that a small rise in calcium concentration will be sufficient to produce the stimulus (Rasmussen and Barrett 1984). The maintenance of a low concentration of

[1] Department of Biological Sciences, University of Exeter, Washington Singer Laboratories, Perry Road, Exeter, EX4 4QG, UK

Ca^{2+} in the cytoplasm is also important in plants (Williamson 1981) and probably in fungi (Pitt and Ugalde 1984).

Regulation of intracellular Ca^{2+} is mediated through various calcium-binding proteins. Carafoli and Penniston (1985) recognised two main classes:

(1) membrane-bound proteins and
(2) soluble proteins.

It is the membrane-bound proteins, in particular those involved in ion pumping that are major transporters and regulators governing the passage of Ca^{2+} into and out of the cytoplasm and between organelles. Soluble proteins located in the cytoplasm and the organelles serve to control the concentration of calcium in the cell, but the amount of calcium that such proteins can bind is limited by the number of protein molecules. Their role is primarily to mediate intracellular events rather than to act as major Ca^{2+} regulators. Of the various calcium regulators currently known, calmodulin (CaM), which is present in all eukaryotic cells examined to date, is probably the most important. This calcium-binding protein is termed CaM since it modulates enzyme activities in a calcium-dependent manner (Means and Dedman 1980).

Meyer et al. (1964) were the first to suggest the possibility that calcium might not act in its free ionic form, but rather require the presence of a binding protein. Cheung (1970) discovered CaM as a protein activator of mammalian brain cyclic nucleotide phosphodiesterase, an enzyme which hydrolyses cyclic AMP.

CaM, which is a single polypeptide chain of 148 amino acid residues, contains four Ca-binding sites and has been highly conserved during evolution (Babu et al. 1985). Although CaM exhibits significant a-helicity in the absence of Ca^{2+}, binding of the ion results in an increase in the a-helical content. It is the Ca-induced conformational change that enables CaM to interact with enzymes, peptides and various pharmacological agents. It is possible that the regulation of different enzymes by CaM is partially a function of the number of Ca ions bound, though controversy exists regarding the order in which the sites are filled and the discrete structural changes that accompany binding (Babu et al. 1985).

There are two main ways in which Ca^{2+} modulates a response via CaM which are referred to as amplitude modulation and sensitivity modulation (Rasmussen et al. 1985). In amplitude modulation the magnitude of the response is directly related to an increase in the concentration of Ca^{2+} in the cell and, hence, to the concentration of the Ca^{2+}-CaM complex. Sensitivity modulation is where the concentration of the Ca^{2+}-CaM response element usually increases without a prior increase in intracellular Ca^{2+} (Hepler and Wayne 1985). These regulatory mechanisms probably function in a co-operative manner rather than individually.

It is now known that CaM activates the Ca^{2+} pump of red blood cells and controls many other enzyme-mediated processes in animal cells (Means and Dedman 1980). A number of physiological events in plants also appear to be regulated by CaM (Dieter 1984).

3 Calcium as an Intracellular Regulator

3.1 In Animal Cells

Four major attributes of messenger Ca^{2+} are recognised (Rasmussen and Barrett 1984).

1. Calcium is an almost *universal* messenger in many cells.
2. It may be a *minatory* messenger in which excess cellular Ca^{2+} leads to dysfunction and cell death.
3. It is a *mercurial* messenger whereby the rise in cytosolic calcium is transient even in those cells displaying a sustained response.
4. It is a *synarchic* messenger in that it generally regulates in concert with other intracellular messengers.

It was suggested by Michell (1975) that turnover of inositol phospholipids in response to hormonal stimulation may be responsible for the regulation of Ca^{2+} gating. Presently, two major second messengers are known (Berridge 1985), one of which is cyclic AMP. The other comprises a combination of second messengers including calcium ions, *myo*-inositol-1,4,5-trisphosphate (IP_3) and diacylglycerol (DG) (Berridge and Irvine 1984; Nishikuza 1984; Hirata et al. 1985).

The operational possibilities for cellular regulation stemming from these various second messenger pathways are enormous and provide the basis for the flexible set of responses to multiple extra-cellular stimuli (Rasmussen et al. 1985). The paths have aspects in common, with the final stages in both inducing changes in the structure of cellular proteins. Alterations in protein structure are brought about by two main mechanisms. In one, the second messenger acts directly by binding to the protein at its regulatory site triggering conformational change. In the alternative and more common mechanism, the second messenger acts indirectly by activating a protein kinase, e.g. calcium-activated, phospholipid-dependent protein kinase (C-kinase) and phosphorylase kinase, which then phosphorylates another protein. This induces conformational changes in the protein enabling it to mediate an intracellular event (Berridge 1985).

Adenylate cyclase, which is itself subject to regulation via CaM, is a membrane-bound protein composed of a regulatory and a catalytic subunit, and which when activated by the binding of a hormone to a cell-surface receptor, catalyses the formation of cyclic AMP from ATP. The level of cyclic AMP is also regulated by cytoplasmic phosphodiesterase which breaks down the nucleotide. Cyclic AMP can activate protein kinases, which phosphorylate appropriate proteins to produce intracellular events (Fig. 1).

In the Ca^{2+} second messenger system, the flow of information from the cell surface to the cell interior proceeds by two operationally distinct branches; the CaM branch and the C-kinase branch (Rasmussen and Barrett 1984) (Fig. 1). When a stimulus (S), e.g. a hormone, interacts with its receptor (R), a phospholipase C is activated, which catalyses the hydrolysis of phosphatidylinositol-4,5-bisphosphate (PIP_2) in the plasma membrane, yielding DG and IP_3. The latter induces a release of Ca^{2+} from an intracellular store, probably the endoplasmic reticulum. In this way the cytosolic free Ca^{2+} concentration ($[Ca^{2+}]_c$) increases giving rise to the amplitude

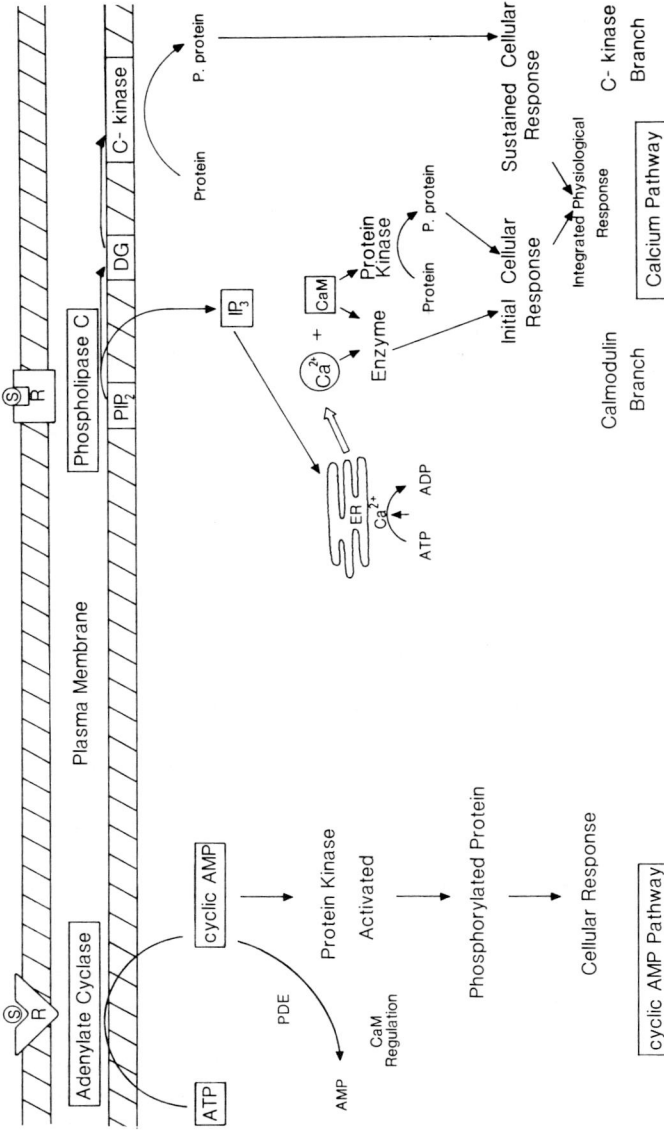

Fig. 1. Summary of the general models of major signal transduction in eukaryotic cells. Information flow in most types of animal cells is via both the cyclic AMP (*left*) and Ca^{2+} (*right*) messenger systems, with interaction occurring between each system and the two limbs of the Ca^{2+} pathway. The concerted operation of the systems in animal cells is explained in the text with discussion of the incomplete evidence for the existence of the routes in plants, fungi and some other eukaryotic organisms. Stimulus, (*S*); receptor, (*R*); phosphatidylinositol-4,5-bisphosphate, (*PIP₂*); diacylglycerol, (*DG*); *myo*-inositol 1,4,5-trisphosphate (*IP₃*); calcium ions, (Ca^{2+}); endoplasmic reticulum, (*ER*); calmodulin, (CaM); phosphodiesterase, (PDE); phosphorylated protein, (P. protein)

modulation of a number of Ca-dependent enzymes; some are activated directly, others via CaM. The activity of response elements can be altered by CaM through a direct allosteric activation of the enzyme or via the action of a protein kinase. The flow of information through this CaM branch leads to the initial cellular response. The increase in the DG content of the plasma membrane leads to the binding of C-kinase to the membrane. This now activated C-kinase catalyses the phosphorylation of a separate subset of cellular proteins; the flow of information through this C-kinase branch is responsible for the maintenance of the sustained cellular response.

3.2 In Plant Cells

Whilst the concept of a second messenger has been successful in explaining many processes in animal cells, the evidence for its operation in plants is so far less convincing, particularly for the cyclic AMP-messenger system. The presence in higher plants of cyclic AMP, adenylate cyclase, and phosphodiesterases does indicate, however, that plant tissues do at least possess the biochemical framework for cyclic AMP-regulated systems. Plants have also been found to contain protein kinases, though none have shown any sensitivity to cyclic AMP (Trewavas 1976). Therefore, although the potential is there, it is unclear whether or not the elements compose a physiologically active system. One particular problem is that cyclic AMP has never been shown as a requirement for any physiological response in plants (Brown and Newton 1981).

In plants, therefore, calcium ions alone may contribute to the coupling of stimulus to response (Hepler and Wayne 1985). Membrane-bound and soluble Ca-dependent protein kinases have been identified in plants (Hetherington et al. 1986). Veluthambi and Poovaiah (1984) discovered that phosphorylation of several polypeptides in corn coleoptiles was promoted by adding calcium. Furthermore, addition of the CaM inhibitor chlorpromazine, reduced the Ca^{2+}-promoted phosphorylation, suggesting that the phosphorylation was modulated by CaM. To date, at least three different classes of protein kinases have been found in plants (Hetherington and Trewavas 1984). The one described by Veluthambi and Poovaiah (1984) is classed as Ca^{2+}-CaM-dependent, while the other two classes are Ca^{2+}-dependent and Ca^{2+}-independent protein kinases; all three have been found in membranes from pea shoots (Hetherington and Trewavas 1984).

It would also seem likely that those pathways involving hydrolysis of inositol phosphates occur in plants, and these could constitute a Ca^{2+}-dependent regulation (Boss and Massel 1985; Poovaiah et al. 1987; Rincon and Boss 1987).

In a review of calcium as a regulator of plant function Williamson (1981) concluded that the case for cytoplasmic calcium as a plant cell regulator relied too heavily on analogies developed from work on animal systems, and that experimental confirmation for the existence of comparable mechanisms in plants was to a great extent lacking. However, more recent work by Williamson and Ashley (1982), who were the first workers to make direct measurements of free cytosolic $[Ca^{2+}]$ in plants using the alga *Chara*, and observations on *Fucus* (Brownlee and Wood 1986) and *Haemanthus* (Keith et al. 1985), suggest that changes in the intracellular $[Ca^{2+}]$ over the range $0.1\ \mu M$ to $1-10\ \mu M$, which activate response elements in animals, may also promote physiological responses in plants.

4 The Role of Calcium in Fungi

Calcium has diverse functions in fungi as reviewed by Pitt and Ugalde (1984). Whilst Ca^{2+} occupies a central position as an intracellular regulator in animal cells (Campbell 1983), and evidence mounts for a similar role in plants (Hepler and Wayne 1985; Kauss 1987), the position in fungi is less clear. Impetus for further investigation in fungi was provided by the discovery of CaM in *Blastocladiella* (Gomes et al. 1979), *Neurospora crassa* (Cox et al. 1982) and Basidiomycete fungi (Grand et al. 1980). Progress has continued with radioimmunoassay determinations of CaM levels in a range of yeasts and some filamentous fungi, including *Penicillium notatum* (Muthukumar et al. 1987).

In addition to the demonstration of a CaM-like protein in *Saccharomyces cerevisiae* (Hubbard et al. 1982) and the identification of a gene for CaM (Davis et al. 1986), a Ca^{2+}-/CaM-dependent protein kinase has been described in this yeast (Londesborough and Nuutinen 1987) strengthening the view that Ca^{2+} homeostasis and Ca^{2+}-/CaM-mediated processes may be involved in the regulation of cell division. Furthermore, the presence of CaM and the identification of a calcium- and phospholipid-dependent protein kinase C in *N. crassa* have led Favre and Turian (1987) to suggest that Ca^{2+} may serve a regulatory function in the development of this filamentous fungus.

Since the occurrence and distribution of polyphosphoinositides in fungi are well documented (Michell 1975; Smith and Berry 1976; Dahl and Dahl 1985; Bowman et al. (1987), as well as phospholipases of suitable specificities (Weete 1974) for the generation of DG and IP_3 from the parent lipids via established schemes (Lester and Steiner 1968; Brennan and Lösel 1978), it appears that fungi may possess the main components of the model proposed for signal transduction in higher eukaryotic cells. Notwithstanding these observations there is, as yet, no evidence for the existence of the complete framework of this transduction route in any particular fungus and, as with studies in plants, an additional primary difficulty relates to problems in the determination of the $[Ca^{2+}]_c$ by existing methods. Although average Ca^{2+} concentrations of $5-25\ \mu m$ have been determined in *S. cerevisiae*, and free $[Ca^{2+}]$ estimated at an order of magnitude lower (Eilam et al. 1985), which would greatly exceed that in the cytosol of animal cells, no accurate values of the latter have been reported in fungi.

5 Ca^{2+}, Signal Transduction and Morphogenesis in Fungi

Whilst a growth requirement for calcium has been found in only a few fungi (Pitt and Ugalde 1984), it is likely that this cation is as essential to fungi as to higher eukaryotes. Calcium is not recognised as a general requirement for reproductive development in fungi, though it is effective in water moulds and in some Penicillia when the latter are grown in submerged, aerated culture.

5.1 In Lower Fungi

The literature relating to Ca^{2+}-induced morphogenesis in water moulds has been reviewed by Fletcher (1979) with particular reference to *Saprolegnia*. Although it was not possible to localise regulatory sites for Ca^{2+} in reproduction in *S. diclina*, a possible rôle linked to phospholipid metabolism was considered. The discovery of calcium-binding protein in *Blastocladiella* suggests a mechanism whereby the Ca^{2+} signal may be transduced, whilst the inhibition of calcium-stimulated reproduction in *Phytophthora* by trifluoperazine and ophiobolin A indicates that calmodulin may be involved in promoting reproduction (Elliott 1986).

5.2 In the Penicillia

Several species of *Penicillium* and *Aspergillus*, among other genera, sporulate in submerged culture on depletion of nutrient sources, according to Klebs' general principles (Klebs 1898). However, certain Penicillia are induced to conidiate by Ca^{2+} (10 mM) in submerged, aerated cultures which are not undergoing nutrient depletion (Pitt and Ugalde 1984). These symptoms have been characterised in *P. notatum* (Pitt and Poole 1981) and *P. cyclopium* (Ugalde and Pitt 1983). In *P. notatum* addition of Ca^{2+} to 36h vegetative cultures promoted a sharp growth check for a 12 h-period, during which spores were formed, followed by resumption of rapid reproductive growth. Although the depletion of carbohydrate is markedly reduced in cultures to which Ca^{2+} is added, the capacity for glucose transport is increased by the cation (Pitt and Barnes 1987). The growth check has been correlated with metabolic changes in the routes of carbohydrate catabolism based on gluconate (Pitt and Mosley 1985 a, b, 1986) following substantial biphasic uptake of Ca^{2+} by the mycelium (Ugalde and Pitt 1986), and accumulation at subcellular sites, including mitochondria (Ugalde and Pitt 1984). Several of the key Ca^{2+}-sensitive sites identified in conidiation, including certain EMP enzymes, pyruvate dehydrogenase and oxoglutarate dehydrogenase (lipoamide) complexes, are known to be influenced (Campbell 1983) well below the average cell calcium content of approximately 2.4 mmol (kg cell water)$^{-1}$ associated with induced mycelium. Whilst Ca^{2+} at low (approximately 30 µM) concentration activates certain mitochondrial matrix enzymes from mammalian sources by decreasing their K_m values (McCormack and Denton 1981; Denton and McCormack 1985), oxoglutarate dehydrogenase and pyruvate dehydrogenase were inactivated in calcium-induced *P. notatum* preparations (Pitt and Mosley 1986). However, it is not known in any system whether activity of these Ca^{2+}-sensitive enzymes is related to calcium-binding proteins and regulated via the $[Ca^{2+}]_c$.

6 Ca^{2+} and Signal Transduction in *Penicillium notatum*

Although the above observations identified Ca^{2+}-sensitive sites in *P. notatum*, which when affected could lead to a cascade effect on ATP depletion and influence biosynthesis, it is not apparent if these events are a cause or a consequence of the conidiation event.

Recent work in our laboratory has been directed towards elucidating the molecular mechanisms whereby the external stimulus, in this case Ca^{2+} itself, is transduced. Some preliminary findings are discussed here in relation to the routes known in higher eukaryotes and those now emerging in fungi.

6.1 Cyclic AMP and Adenylate Cyclase

Diverse functions have been assigned to cyclic AMP in fungi and the vast literature relating to its putative control of physiological processes, ranging from nutrition to morphogenesis, has been reviewed (Pall 1981). Adenylate cyclase, the enzyme that synthesizes cyclic AMP and controls its concentration in conjunction with phosphodiesterase in most organisms, occurs in several fungi. Recent observations in *Neurospora, Saccharomyces* and *Phycomyces* (Janssens 1987) suggest that the adenylate cyclases are activated by guanine nucleotides and experiments in vivo with *Neurospora* indicate regulation by GTP-binding *G* proteins as in vertebrates.

Although Cooper et al. (1985) found that cyclic AMP may be involved in the regulation of dimorphism in *Aureobasidium pullulans*, our observations in *P. notatum* do not directly implicate the cyclic nucleotide in calcium-induced conidiation. Radioimmunoassay of cyclic AMP levels in *P. notatum* during the 12 h period of calcium-induced conidiation revealed no significant changes in the cyclic AMP pool from that of vegetative controls $[5.2 \pm 1.6 \ (8) \ \text{pmol (mg dry wt)}^{-1}]$. In addition, whilst cyclic AMP levels were increased up to 8-fold by administration of the phosphodiesterase inhibitors theophylline (0.1 – 1.0 mM) and caffeine (0.1 mM), conidiation was unaffected. Likewise the addition to cultures of cyclic AMP (1 – 100 μM) or dibutyryl cyclic AMP (0.5 – 200 μM), or the presence of forskolin $(5 \times 10^{-6} – 5 \times 10^{-9} \ \text{M})$, which stimulates activity of adenylate cyclase and elevated cellular cyclic AMP levels approximately 3-fold failed to influence sporulation or affect growth of either induced or vegetative cultures.

6.2 Calmodulin Branch

CaM is considered of general occurrence in eukaryotes and its presence has been confirmed in *P. notatum* (Muthukumar et al. 1987).

Since Roufogalis (1982) counsels against concluding Ca^{2+}/CaM interactions based on trifluoperazine (TFP) effects alone, experiments investigated the inhibition of calcium-induced conidiation in *P. notatum* using TFP and a second CaM-interacting compound, ophiobolin A (Leung et al. 1985). Whilst these antagonists did not substantially affect Ca^{2+} uptake by the mycelium, nor influence growth, Fig. 2 shows the inhibitory effects of TFP and ophiobolin A on conidiation in *P. notatum* over the physiological ranges of concentrations. The data imply that a Ca^{2+}/CaM interaction is involved in conidium development, as found with reproduction in the Oomycetes

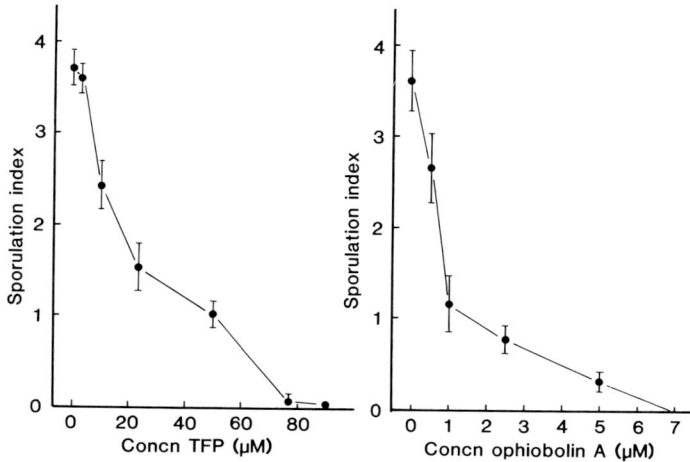

Fig. 2. Effects of trifluoperazine (TFP) and ophiobolin A on calcium-induced conidiation in *P. notatum*. Vegetative cultures were grown for 36 h in shaken flask culture and antagonists were added followed by addition of Ca^{2+} (10 mM) after 30 min. Cultures were harvested after further growth for 48 h and the sporulation index determined. There were six replicates for each treatment and the means $\pm$ SEM are shown

(Elliott 1986). These observations suggest that a CaM-protein kinase link may be involved at a stage(s) in morphogenesis, but the mechanisms await elucidation.

6.3 Calcium-activated, Phospholipid-dependent Protein Kinase Branch

Although, as indicated above, fungi may possess the main features of the phosphatidylinositol cycle found in higher eukaryotes, there is limited evidence for the functional significance of this signal transduction route in eukaryotic microbes. However, recent observations in *Saccharomyces* show that glucose addition to cells arrested at the G_0G_1 phase results in ^{32}P incorporation into phosphatidic acid and various phosphoinositides. Also, such cells show ^{32}P labelling of a 105 kDa protein and a protein of similar weight is phosphorylated by incubation with phorbol esters (Janssens 1987).

Aspects of a putative phosphatidylinositol cycle have been investigated in *P. notatum* using the calcium ionophore, A23187, which facilitates entry of Ca^{2+} into animal cells, alters fluxes of Ca^{2+} between cell compartments and elevates $[Ca^{2+}]_c$, and phorbol esters, which mimic the action of DG through direct activation of C-kinase. The data in Table 1 show that certain biologically active phorbol esters, at a concentration near the upper end of the physiological range and which did not influence Ca^{2+} uptake by the fungus, did, however, augment conidiation at sub-maximal levels of Ca^{2+} (5 mM) in cultures. A biologically inactive phorbol ester (4α-PDD) was ineffective in enhancing morphogenesis. Additional experiments showed that whilst 5 μM A23187, but not ionomycin, promoted uptake of Ca^{2+} into

Table 1. Effects of phorbol esters (0.75 µM) on conidiation of *P. notatum* in the presence of 5 mM Ca^{2+} in shaken flask culture

Phorbol ester	Dry wt ($mg\ ml^{-1}$)	Sporulation index (10^6 spores·mg dry wt^{-1})
4α-Phorbol-12,13-didecanoate (4α-PDD)	8.77 ± 0.41	6.6 ± 0.8
Phorbol-12,13-didecanoate (PDD)	7.77 ± 0.14	10.1 ± 1.2
12-*O*-Tetradecanoylphorbol-13-acetate (TPA)	7.44 ± 0.32	10.9 ± 1.4
Phorbol 12,13-dibutyrate (PDBu)	7.69 ± 0.16	11.4 ± 1.9
Solvent (12 µl EtOH), control	8.78 ± 0.46	7.5 ± 0.9

Phorbol esters added to 36 h vegetative cultures; sporulation index determined after 96 h; six replicates per treatment; data are means ± SEM

mycelium by 30% – 50%, and also significantly enhanced conidiation at low calcium levels, the combined administration of these agents, whilst additive was not synergistic. In the absence of added Ca^{2+}, phorbol esters and A23187, either singly or in combination, failed to promote conidiation.

Although these observations differ substantially from those of Tyers and Harley (1986) who showed that exposure of human HL-60 cells to sub-threshold concentrations of TPA synergistically promoted monocytic differentiation when cells were co-treated with A23187, our results nevertheless suggest that a Ca^{2+} component of phosphoinositide-based signaling may be involved in morphogenesis in *P. notatum*. These initial observations, combined with the CaM requirement indicate possible involvement of both the Ca^{2+}/CaM kinase and C-kinase limbs of the signal transduction route, though currently there is no evidence of an IP_3 link between these in *P. notatum*.

6.4 Cytoplasmic Free Calcium During Conidiation in *P. notatum*

Cytoplasmic Ca^{2+} has emerged as a regulator of processes in many types of animal cells where it has frequently been shown that a response is accompanied by a change in free intracellular $[Ca^{2+}]$. Additionally, experimental manipulation of the natural $[Ca^{2+}]$ accompanying the stimulus can modify the Ca^{2+}-mediated response. The few measurements of $[Ca^{2+}]_c$ so far achieved in plants suggest similarities with animal systems, but much of the evidence for the rôle of Ca^{2+} arises from experiments presumed to adjust $[Ca^{2+}]$ (Hepler and Wayne 1985). Although the evidence for a second messenger function for Ca^{2+} in fungi is mounting, and a rôle for Ca^{2+} in morphogenesis established (Pitt and Ugalde 1984; Wessels 1986; Harold and Harold 1986), measurements of $[Ca^{2+}]_c$ have not yet been made in these organisms. Experiments with fungi, as with plant cells, encounter difficulties relating

(a) to the small size and vacuolate nature of the cells, which preclude the use of microinjection techniques and of currently available Ca^{2+} electrodes, and

(b) to permeability towards and hydrolysis of Ca^{2+} chelators which were essentially designed for use with animal cells (Gilroy et al. 1986).

Initial attempts to monitor $[Ca^{2+}]_c$ in mycelial suspensions of *P. notatum* by the use of quin-2 encountered problems from aggregation of the biomass and from wall-bound calcium. Subsequent experiments used protoplasts of *P. notatum*, obtained by modifications of the methods of Hamlyn et al. (1981), as a model system for the study of Ca^{2+} transport and $[Ca^{2+}]_c$. Whilst the general procedure for the use of quin-2 for measuring $[Ca^{2+}]_c$ in lymphocytes was used (Tsien et al. 1982), problems were encountered beyond those discussed by Rink and Pozzan (1985), and were only partly resolved.

Protoplasts were loaded with quin-2 by incubation with the acetoxymethyl ester (50 µM) for 90 min at 25 °C, which permeated the membrane and accumulated within

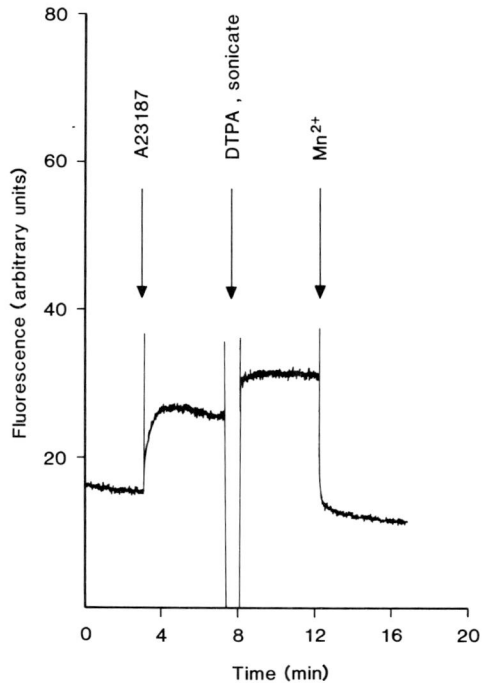

Fig. 3. Estimation of $[Ca^{2+}]_c$ in protoplasts derived from 36-h vegetative cultures of *P. notatum*. Traces show fluorescence, above autofluorescence, of 4×10^8 protoplasts ml^{-1} in 50 mM-potassium-hydrogen phthalate buffer adjusted to pH 5.3 with tetramethylammonium hydroxide containing 1.2 M-sorbitol and 50 mM-glucose. Protoplasts loaded, with aeration, for 90 min in quin-2 acetoxymethyl ester (50 µM) and washed in fresh medium, had a final cell concentration of about 0.5 mM quin 2.

Fluorescence (*F*) was continuously recorded, with stirring at 25 °C, using a Perkin-Elmer LS-5B luminescence spectrophotometer with excitation and emission wavelengths of 339 nM and 492 nm, respectively, and slit widths of 2.5 nm. Calcium ($1-2.5$ mM, as appropriate) was added to the suspension and fluorescence (*F*) monitored for 3 min before addition of 5 µM A23187 (in dimethyl sulphoxide) followed by sonication in the presence of 250 µM Ca-DTPA (diethylene-triaminepentaacetic acid), to prevent quenching by ions released in lysis, and F_{max} was monitored. Addition of 500 µM Mn^{2+} displaced Ca^{2+} from the dye and F_{min} was recorded. From these fluorescence values cytoplasmic free calcium concentration was estimated from the equation $[Ca^{2+}]_c = (115 \text{ nM}) (F-F_{min})/(F_{max}-F)$ according to Tsien et al. (1982)

the cells. However, low uptake of the ester occurred at pH $5.8-7.0$ and its hydrolysis within the cells was low over this pH range. Consequently, average cell pH was adjusted to pH 5.0, by modifications to the incubation buffer, at which uptake of ester was enhanced to 1.5 mM concentration and appreciable hydrolysis of the ester occurred to give an intracellular quin-2 concentration of 0.5 mM. The necessity to maintain low intracellular pH precluded the use of EGTA for the determination of F_{min}, and this was achieved by Mn^{2+} (Fig. 3). Although only a sluggish F_{max} was obtained by the use of Triton X-100, a satisfactory response occurred with 5 µM A23187. Since the cytoplasmic free $[Mg^{2+}]$, which markedly influences the K_d for quin-2 (Rink and Pozzan 1985), in unknown for *P. notatum*, but falls in the range $1-10$ mM, and because $[Ca^{2+}]_c$ determinations were necessarily made at pH 5.0 and 25 °C, as opposed to 1 mM Mg^{2+}, pH 7.0 and 37 °C used with animal systems, the effective K_d of 115 nM, as used with the latter (Tsien et al. 1982) may not be entirely appropriate. Furthermore, since Ca^{2+}-EGTA buffers cannot be used to set and calibrate $[Ca^{2+}]$ under the pH conditions of our incubation buffer, further work is required to determine the apparent K_d for quin-2 in the *P. notatum* system. Therefore, as with plant investigations (Gilroy et al. 1986), a K_d value of 115 nM has been used presently for estimating $[Ca^{2+}]_c$. Although this is probably too low for determinations of absolute values for $[Ca^{2+}]_c$, it serves to compare relative changes in its level.

Figure 4 shows the capacity of protoplasts isolated from 36 h vegetative mycelium, for Ca^{2+} accumulation from medium containing 2.5 mM Ca^{2+} and the relationship to $[Ca^{2+}]_c$. Whilst Ca^{2+} uptake was biphasic and involved an initial rapid binding to the membrane $[K_d, 3.07$ mM-Ca^{2+}; Y_t, 1.92 nmol-Ca^{2+} $(10^8$ cells$)^{-1}]$, followed by an

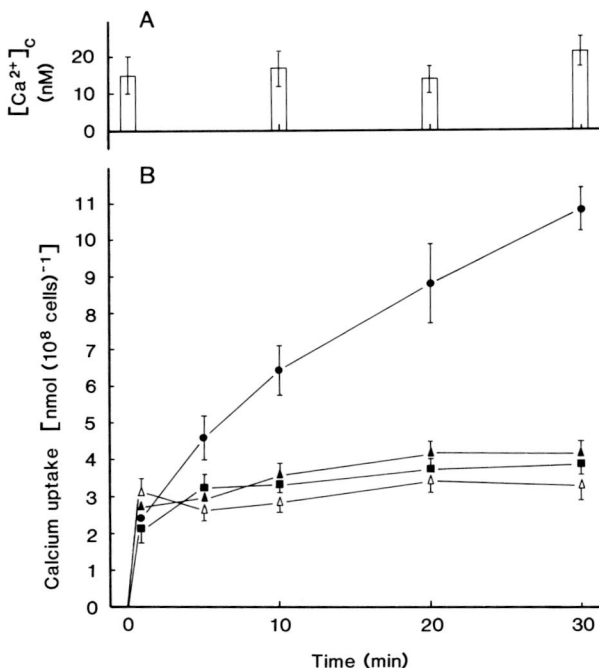

Fig. 4A–B. Calcium accumulation from 2.5 mM-Ca^{2+} by protoplasts of *P. notatum* and the relationship to cytosolic free calcium concentration, $[Ca^{2+}]_c$. **A** records $[Ca^{2+}]_c$ of protoplasts over the uptake period. **B** compares Ca^{2+} uptake at 25 °C in the presence of 50 mM-glucose $(-\bullet-)$ and the analogue 50 mM-deoxy-*D*-glucose $(-\blacksquare-)$ and shows the effects of respiratory inhibitors in the presence of 50 mM-glucose; $(-\blacktriangle-)$, 160 µM-antimycin A and $(-\triangle-)$, 25 µM-carbonyl cyanide *m*-chlorophenyl hydrazone (CCCP)

energy-dependent transport displaying Michaelis-Menten kinetics [K_m, 0.28 mM-Ca^{2+}; V_{max}, 0.6 nmol (mg dry wt)$^{-1}$], and is accompanied by sequestration at mitochondrial and other subcellular sites (Ugalde and Pitt 1984), the level of cytoplasmic free Ca^{2+} remained unchanged over 30 min [15 nM ±7 (8)]. Elevated values for $[Ca^{2+}]_c$ were only recorded for protoplasts derived from mycelium previously exposed to Ca^{2+} (10 mM) for periods greater than 2 h [81 nM ±9 (10)] or for protoplasts obtained from vegetative cells and subsequently exposed to ⩾2.5 mM Ca^{2+} for 2–4 h [37 nM ±4 (9)]. Thus, although the average content of cell calcium in protoplasts increased from an initial resting level of approximately 2 µmol to 46 µmol (kg cell water)$^{-1}$, with the latter value being approximately 2000-fold that of $[Ca^{2+}]_c$ at 30 min exposure to calcium, the constant value of $[Ca^{2+}]_c$ implies that calcium entering the cell is bound or sequestered, presumably in the mitochondria and elsewhere (Ugalde and Pitt 1984) without accumulating in the cytosol as Ca^{2+}.

These puzzling observations imply that transient changes in $[Ca^{2+}]_c$ may not be involved in the initial morphogenetic response to the cation but could be significant at a later stage(s). Alternatively, the data may be interpreted, in conjunction with our earlier findings which have a bearing on the signal transduction route, as an initial phase of sensitivity modulation where the concentration of a Ca^{2+}-receptor response element may increase without an accompanying increase in $[Ca^{2+}]_c$. The belated rise in $[Ca^{2+}]_c$ after 2 h exposure of mycelium and protoplasts to Ca^{2+} may reflect either an amplitude response or saturation of subcellular stores and subsequent discharge to the cytoplasm (Hepler and Wayne 1985). Further work is in progress to characterise more fully the Ca^{2+}-receptor response elements of this system, including any relationships between calcium binding at the cell surface and the changes in $[Ca^{2+}]_c$ during the various phases of conidial morphogenesis.

7 Concluding Comments

Although ideas on the mechanisms of transmembrane-signal transduction and the rôle of calcium in eukaryotic microbes have hitherto depended heavily on analogies with animal cells, and to a lesser extent with plants, recent data suggest some striking similarities. Whilst important progress has been forthcoming in *Neurospora* and *Saccharomyces*, with observations being presented on *Penicillium*, the complete chain of events involving cell surface receptors to intracellular effectors has not yet been reported in a single fungus. However, sufficient progress has been made to indicate that fungi, and other eukaryotic microbes, may provide useful experimental material for future studies on the general mechanisms of signal transduction in eukaryotes as they have in other fundamental areas of biology in the past.

Acknowledgement. AK received financial support from the Science and Engineering Research Council and Shell Research Ltd.

References

Babu YS, Sack JS, Greenhough TJ, Bugg CE, Means AR, Cook WJ (1985) Three-dimensional structure of calmodulin. Nature (London) 315:37−40

Berridge MJ (1985) The molecular basis of communication within the cell. Sci Am 253:124−134

Berridge MJ, Irvine RF (1984) Inositol trisphosphate, a novel second messenger in cellular signal transduction. Nature (London) 312:315−321

Bild GS, Bhat SG, Ramados CS, Axelrod B (1978) Biosynthesis of a prostaglandin by a plant enzyme. J Biol Chem 253:21−23

Boss WF, Massel M (1985) Polyphosphoinositides present in plant tissue culture cells. Biochem Biophys Res Commun 132:1018−1023

Bowman BJ, Borgeson CE, Bowman EJ (1987) Composition of *Neurospora crassa* vacuolar membranes and comparison to endoplasmic reticulum, plasma membranes, and mitochondrial membranes. Exp Mycol 11:197−205

Brennan PJ, Lösel DM (1978) Physiology of fungal lipids: selected topics. Adv Microb Physiol 17:47−179

Brown EG, Newton RP (1981) Cyclic AMP and higher plants. Phytochemistry 20:2453−2456

Brownlee C, Wood JW (1986) A gradient of cytoplasmic free calcium in growing rhizoid cells of *Fucus serratus*. Nature (London) 320:624−626

Campbell AK (1983) Intracellular calcium: its universal rôle as regulator. Wiley, New York

Carafoli E, Penniston JT (1985) The calcium signal. Sci Am 253:70−116

Cheung WY (1970) Cyclic 3′,5′-nucleotide phosphodiesterase: demonstration of an activator. Biochem Biophys Res Commun 33:533−538

Cooper LA, Edwards SW, Gadd GM (1985) Involvement of adenosine 3′:5-cyclic monophosphate in the yeast-mycelium transition of *Aureobasidium pullulans*. J Gen Microbiol 131:1589−1593

Cox JA, Ferrax C, Demaille JG, Perez RO, Van Tuinen D, Marmé D (1982) Calmodulin from *Neurospora crassa*, general properties and conformational changes. J Biol Chem 257:10694−10700

Dahl JS, Dahl CE (1985) Stimlulation of cell proliferation and polyphosphoinositide metabolism in *Saccharomyces cerevisiae* GL7 by ergosterol. Biochem Biophys Res Commun 133:844−850

Davis TN, Urdea MS, Masiarz FR, Thorner J (1986) Isolation of the yeast calmodulin gene: calmodulin is an essential protein. Cell 47:423−431

Denton RM, McCormack JG (1985) Physiological rôle of Ca^{2+} transport by mitochondria. Nature (London) 315:635

Dieter P (1984) Calmodulin and calmodulin-mediated processes in plants. Plant Cell Environ 7:371−380

Eilam Y, Lavi H, Grossowicz N (1985) Cytoplasmic Ca^{2+} homeostasis maintained by a vacuolar Ca^{2+} transport system in the yeast *Saccharomyces cerevisiae*. J Gen Microbiol 131:623−629

Elliott CG (1986) Inhibition of reproduction by *Phytophthora* by calmodulin-interacting compounds trifluoperazine and ophiobolin A. J Gen Microbiol 132:2781−1785

Favre B, Turian G (1987) Identification of a calcium- and phospholipid-dependent protein kinase (protein kinase C) in *Neurospora crassa*. Plant Sci 49:15−21

Fletcher J (1979) Effect of calcium chloride concentration on growth and sporulation of *Saprolegnia terrestris*. Ann Bot (London) 44:589−594

Gilroy S, Hughes WA, Trewavas AJ (1986) The measurement of intracellular calcium levels in protoplasts from higher plant cells. FEBS Lett 199:217−222

Gomes SL, Mennucci L, Maia JCC (1979) A calcium-dependent protein activator of mammalian cyclic nucleotide phosphodiesterase from *Blastocladiella emersonii*. FEBS Lett 99:39−42

Grand RJA, Nairn AC, Perry SV (1980) The preparation of calmodulins from barley (*Hordeum* sp.) and basidiomycete fungi. Biochem J 181:755−760

Hamlyn PF, Bradshaw RE, Mellon FM, Santiago CM, Wilson JM, Peberdy JF (1981) Efficient protoplast isolation from fungi using commercial enzymes. Enzyme Microb Technol 3:321−325

Harold RL, Harold FM (1986) Ionophores and cytochalasins moderate branching in *Achlya bisexualis*. J Gen Microbiol 132:213−219

Hepler PK, Wayne RO (1985) Calcium and plant development. Annu Rev Plant Physiol 36:397−439

Hetherington AM, Trewavas A (1984) Activation of pea membrane protein kinase by calcium ions. Planta 161:409−417

Hetherington AM, Blowers D, Trewavas A (1986) Calcium/calmodulin dependent membrane bound protein kinase. In: Trewavas AJ (ed) Molecular and cellular aspects of calcium in plant development. NATO ASI Ser. Series A, Life Sci, vol 104. Plenum, New York London, pp 123–131

Hirata M, Sasaguri T, Hamachi T, Hashimoto KM, Kukita M, Koga T (1985) Irreversible inhibition of Ca^{2+} release in saponin-treated macrophages by the photoaffinity derivative of inositol-1,4,5-trisphosphate. Nature (London) 317:723–725

Hubbard M, Bradley M, Sullivan P, Shepherd M, Forrester I (1982) Evidence of the occurrence of calmodulin in the yeasts Candida albicans and Saccharomyces cerevisiae. FEBS Lett 317:85–88

Janssens PMW (1987) Did vertebrate signal transduction mechanisms originate in eukaryotic microbes? Trends Biochem Sci 12:456–459

Kauss H (1987) Some aspects of calcium-dependent regulation in plant metabolism. Annu Rev Pl Physiol 78:47–72

Keith CH, Ratan R, Maxfield FR, Bajer A, Shelanski ML (1985) Local cytoplasmic calcium gradients in living mitotic cells. Nature (London) 316:848–850

Klebs G (1898) Zur Physiologie der Fortpflanzung einiger Pilze. Jahrb Wiss Bot 32:1–70

Lester RL, Steiner MR (1968) The occurrence of diphosphoinositide and triphosphoinositide in Saccharomyces cerevisiae. J Biol Chem 243:4889–4893

Leung PC, Taylor WA, Wang JH, Tipton CL (1985) Rôle of calmodulin inhibition in the mode of action of ophiobolin A. Plant Physiol 77:303–308

Londesborough J, Nuutinen M (1987) Ca^{2+}/calmodulin-dependent protein kinase in Saccharomyces cerevisiae. FEBS Lett 219:249–253

McCormack JG, Denton RM (1981) A comparative study of the regulation by Ca^{2+} of the activation of the 2-oxoglutarate dehydrogenase complex and NAD^+-isocitrate dehydrogenase from a variety of sources. Biochem J 196:619–624

Means AR, Dedman JR (1980) Calmodulin – an intracellular calcium receptor. Nature (London) 285:73–77

Meyer WL, Fischer WH, Krebs EG (1964) Activation of skeletal muscle phosphorylase β-kinase by Ca^{2+}. Biochemistry 3:1033–1039

Michell RH (1975) Inositol phospholipids and cell surface receptor function. Biochim Biophys Acta 415:81–147

Muthukumar G, Nickerson AW, Nickerson KW (1987) Calmodulin levels in yeasts and filamentous fungi. FEMS Microbiol Lett 41:253–255

Nishikuza Y (1984) Turnover of inositol phospholipids and signal transduction. Science 225:1365–1370

Pall ML (1981) Adenosine 3′,5′-phosphate in fungi. Microbiol Rev 45:462–480

Pitt D, Barnes JC (1987) Hexose transport during calcium induced conidiation in Penicillium notatum. Trans Br Mycol Soc 89:859–865

Pitt D, Mosley MJ (1985a) Enzymes of gluconate metabolism and glycolysis in Penicillium notatum. Antonie Leeuwenhoek Microbiol 51:353–364

Pitt D, Mosley MJ (1985b) Pathways of glucose catabolism and the origin and metabolism of pyruvate during calcium-induced conidiation of Penicillium notatum. Antonie Leeuwenhoek Microbiol 51:365–384

Pitt D, Mosley MJ (1986) Oxidation of carbon sources via the tricarboxylic acid cycle during calcium-induced conidiation of Penicillium notatum. Antonie Leeuwenhoek Microbiol 52:467–482

Pitt D, Poole PC (1981) Calcium-induced conidiation in Penicillium notatum in submerged culture. Trans Br Mycol Soc 76:219–230

Pitt D, Ugalde UO (1984) Calcium in fungi. Plant Cell Environ 7:467–475

Poovaiah BW, Reddy ASN, McFadden JJ (1987) Calcium messenger system: Rôle of protein phosphorylation and inositol biphospholipids. Physiol Plant 69:569–573

Rasmussen H, Barrett PQ (1984) Calcium messenger system: An integrated view. Physiol Rev 64:938–984

Rasmussen H, Zawalich W, Kojima J (1985) Ca^{2+} and cAMP in the regulation of cell function. In: Marmé D (ed) Calcium and cell physiology. Springer, Berlin Heidelberg New York Tokyo, pp 3–14

Rincon M, Boss WF (1987) myo-inositol trisphosphate mobilises calcium from fusogenic carrot (Daucus carota L.) protoplasts. Plant Physiol 83:395–398

Rink J, Pozzan T (1985) Using Quin 2 in cell suspensions. Cell Calcium 6:133–144

Roufogalis BD (1982) Specificity of trifluoperazine and related phenothiazines for calcium-binding proteins. In: Cheung WY (ed) Calcium and cell function, vol 3. Academic Press, London New York, pp 130–159

Smith JE, Berry DR (1976) The filamentous fungi. Biosynthesis and metabolism, vol 2. Arnold, London

Trewavas A (1976) Post-translational modification of proteins by phosphorylation. Annu Rev Plant Physiol 27:349–374

Tsien RY, Pozzan T, Rink TJ (1982) Calcium homeostasis in intact lymphocytes: Cytoplasmic free calcium monitored with a new intracellularly trapped fluorescent indicator. J Cell Biol 94:325–334

Tyers M, Harley CB (1986) Ca^{2+} and phorbol ester synergistically induce HL-60 differentiation. FEBS Lett 206:99–105

Ugalde UO, Pitt D (1983) Morphology and calcium-induced conidiation of *Penicillium cyclopium* in submerged culture. Trans Br Mycol Soc 80:319–325

Ugalde UO, Pitt D (1984) Subcellular sites of calcium accumulation and relationships with conidiation in *Penicillium cyclopium*. Trans Br Mycol Soc 83:547–555

Ugalde UO, Pitt D (1986) Calcium uptake kinetics in relation to conidiation in submerged cultures of *Penicillium cyclopium*. Trans Br Mycol Soc 87:199–203

Veluthambi K, Poovaiah BW (1984) Calcium-promoted protein phosphorylation in plants. Science 223:167–169

Weete JD (1974) Fungal lipid biochemistry. Plenum, New York London

Wessels JGH (1986) Cell wall synthesis in apical growth. Int Rev Cytol 104:37–79

Williamson RE (1981) Free Ca^{2+} concentration in the cytoplasm: A regulator of plant cell function. Physiol Plant 12:45–48

Williamson RE, Ashley CC (1982) Free Ca^{2+} and cytoplasmic streaming in the alga *Chara*. Nature (London) 296:647–651

Chapter 19

Structure and Function of Fungal Plasma-Membrane ATPases

C. L. SLAYMAN[1], P. KAMINSKI[1,2], and D. STETSON[1,3]

1 Introduction

The characteristic marker enzyme for the plasma membranes (PM) of almost all living cells is a cation-dependent ATPase of $M_r \sim 100000$ which has at least three crucial functions: maintaining ionic balance between the cell interior and the environment; creating background conditions for water balance (i.e. volume control in animal cells; turgor regulation in plants, fungi, and bacteria), and mediating energy transfer from redox and dehydration reactions to a wide variety of so-called active transport processes carried out at the plasma membrane. The cationic substrate for this enzyme in fungal membranes is protons, which usually must be exported in order to offset excess metabolic production. With one or two important exceptions, the entire class of enzymes can transfer net electric charges through membranes, so that they function as "biochemical fuel cells" (Mitchell 1967) or "electroenzymes" (Chapman et al. 1983).

The purposes of this chapter are a) to describe some remarkable physiological properties of this enzyme, along with certain biochemical correlates; b) to summarize recent structural work on the enzyme, particularly as brought out by molecular-biological studies; and c) to consider briefly several rather different ways to view the enzyme and to model its behavior.

2 Membrane Appearance of the Plasma Membrane (PM) ATPase

The plasma membranes of fungi, like those of most other cells, are often split tangentially during freeze-fracturing to reveal surfaces heavily studded with particles. This is illustrated in the electron micrographs of Fig. 1, for mycelia of the ascomycete fungus, *Neurospora crassa*. The distribution of total particles has been found to be thoroughly inhomogeneous: from cell to cell in a population, tangentially within the membrane of a single cell, and perpendicularly through the membrane. Frames A and

---

[1] Department of Cellular and Molecular Physiology, Yale School of Medicine, 333 Cedar Street, New Haven, CT 06510, USA
[2] Current address: BIOKAM Inc., Lewistown, PA 17044, USA
[3] Current address: Department of Zoology, Ohio State University, 1735 Neil Avenue, Columbus, OH 43210, USA

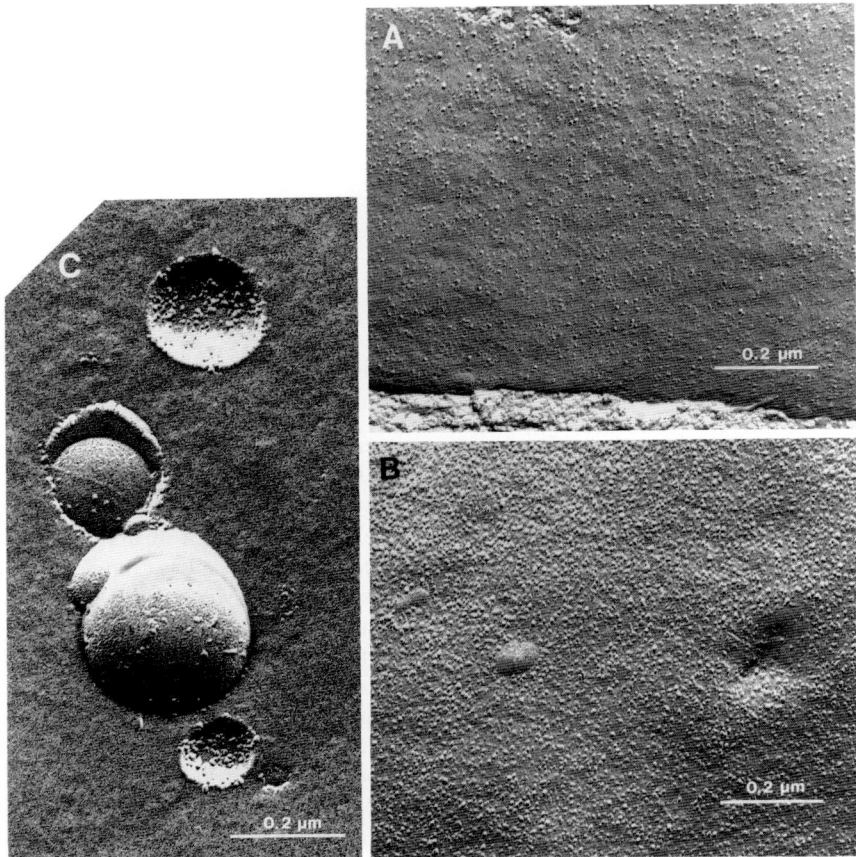

Fig. 1A–C. Freeze-fracture views of the *Neurospora crassa* plasma membrane. **A, B** Random samples from intact mycelium of wild-type RL21a, showing P-face of the fracture; **C** everted plasma-membrane vesicles, showing concave P-faces (at top and bottom) and convex E-faces (middle). Exponentially growing hyphae (average length, ~300 µm) were fixed in 2.5% glutaraldehyde/0.1 M sodium cacodylate (pH 5.8), filter-harvested, frozen in Freon cooled by liquid N_2, fractured on a Balsers BAF apparatus, shadowed with platinum, and replicated with carbon. The replicas were picked onto Formvar grids and examined with a Zeiss EM 10B microscope. Everted membrane vesicles (**C**) were prepared and examined by Dr. R.J. Brooker, as described previously (Perlin et al. 1984). All three specimens were shadowed from bottom to top as mounted. Particle frequencies: **A** 1940/µm²; **B** 3490/µm²

B show two fracture samples from a single pellet of a wild-type *N. crassa*, revealing P (cytoplasmic) faces with ~1940 and ~3490 particles/µm², respectively; individual fracture patches can also display a similar spread of particle frequencies. Frame C shows fracture faces from isolated plasma-membrane vesicles, to illustrate the difference in observable particles between the P-face (concave) and the E-face (convex). Perlin et al. (1984), using the same wild-type strain of *N. crassa*, found the ratio of particle frequencies P : E to be approximately 8 : 1. [The membrane vesicles in frame C are morphologically everted, with the physiological interior facing outward, which is a fortunate consequence of the fractionation procedures.] Measured particle diameters scatter broadly between 25 and 140 Å, which may reflect heterogeneity both in particle identity and in depth of penetration of particles into the membrane.

Proof of the chemical identity of these membrane-bound particles is lacking for fungi, but indirect evidence indicates a major fraction to be the plasma-membrane ATPase. This enzyme is the most abundant membrane protein in fungi: e.g. $5\% - 10\%$ of total membrane protein in *N. crassa* (Bowman et al. 1981). Electric current driven by the enzyme through mature hyphal membranes of *N. crassa* can be $10 - 20 \,\mu A/cm^2$, which would require several thousand ATPase molecules/μm^2 for reasonable turnover numbers. The median size of particles in the P faces, $\sim 80 \,\mu m$ diameter, is consistent with the expected size of the ATPase (~ 60 Å dia.) allowing for an increase due to shadowing. In *N. crassa*, mutant strains with depressed ATPase activity (e.g. *poky* f^-) have proportionately depressed particle frequencies. Artificial membrane vesicles made by adding partially purified native ATPase to liposomes display particles similar to those in plasma membranes, but of more uniform size. Finally, quantitative analyses of closely related enzymes, especially the Ca^{2+}-ATPase from sarcoplasmic reticulum — during progressive reconstitution and during trypsin degradation (Inesi and Scales 1974) — have demonstrated those enzymes to exist as particles very similar to the particles in fungal membranes.

3 Functional Properties of the Proton Pump

The large hyphal size of *N. crassa* (up to 20 μm dia. and many millimeters long) has made that species particularly useful for functional studies of the fungal plasma-membrane ATPase. Intracellular micropipette electrodes have yielded measurements of membrane potential, membrane resistance, pumped ion current through the

Table 1. Steady-state voltages and resistances of *N. crassa* plasmalemma

Membrane potential (mV)[a]	Membrane resistivity (Kohms·cm²)	Conditions of measurement
-180 to -220	5 to 15	Normal, mature hyphae, aerated
-10 to -40	5 to 15	Hyphae, H^+ pump fully blocked[b]
-220 to -250	20 to 50	Hyphae, carbon-starved $1-3$ h
-230 to -320	NM[c]	Hyphae nitrogen-starved >5 h
-220 to -300	40 to 100	Spherocytes[d]
-250 to -350	40 to 120	Spherocytes, carbon-starved >5 h

[a] Units are millivolts, measured as cell interior minus the bath potential.
[b] Complete pump block is not easily obtained; values given are extrapolated from "nearly complete" block by cyanide (ATP withdrawal) and vanadate (direct inhibition of the pump).
[c] Not measured.
[d] Conidia incubated at $25° - 30°C$ in the presence of glucose and $15\% - 18\%$ ethylene glycol enlarge $2-4$ diameters without germinating. Measurements made after the glycol and sugar have been removed.

ATPase, and the steady-state relationship between pump current and membrane potential, as well as between membrane potential and cellular ATP levels. A summary of voltage and resistance measurements is given in Table 1.

3.1 Energy and Stoichiometry

Under normal conditions, current through the proton pump is responsible for 80% or more of the resting membrane potential, which − at −180 to −200 mV − would be nearly at equilibrium if the pump were to export *two* protons for each ATP-molecule split. But conditions are easily found which elevate membrane resistance 5- to 10-fold and push the resting membrane potential to the neighborhood of −300 mV. An important elementary consequence of this fact is to rule out any integral stoichiometry (ions transported per molecule of ATP split) greater than unity. This follows because the free energy available to the pump, calculated as the energy from ATP hydrolysis less that conserved in the proton-chemical gradient, is only −400 to −440 mV (Warncke and Slayman 1980). A normal stoichiometry of 1 : 1 has been confirmed by experiments on plasma-membrane vesicles, where both ATP consumption and H^+ transfer could be measured (Perlin et al. 1984); it has also been confirmed in kinetic activation experiments with varied cytoplasmic pH (Sanders et al. 1981). [Special conditions of energy starvation, however, *may* increase the stoichiometry to 2 : 1, which opens the possibility of actually observing fractional stoichiometries (i.e. between 1 : 1 and 2 : 1; Warncke and Slayman 1980).]

3.2 Current-Voltage Relationship

Although it seems paradoxical for membrane potential to rise during energy restriction (by carbon starvation), the effect is an indirect consequence of other metabolic adjustments, chiefly the large increase of membrane resistance, which overcompensates for diminished ion pumping. Such behavior very simply demonstrates coordinate regulation − during metabolic stress − of different classes of transport systems in a single cell membrane: proton pumping by the H^+-ATPase is diminished, but "leaks" (including bona fide channels and electrophoretically coupled nutrient transporters) are also diminished, thus tightening the membrane and reducing ionic turnover but with minimal effects upon steady-state poise. The mechanisms for such transport regulation are unknown at present.

The overall behavior of the proton pump in relation to the "leaks" is illustrated in Fig. 2, for the case of *N. crassa* spherocytes. In this example of so-called current-voltage (I−V) analysis, the membrane potential was briefly (100 ms each) clamped at 36 different values away from the resting potential (−224 mV), and the required displacement currents are shown as the plotted points, representing the behavior of all functional transporters in the intact membrane. Comparable measurements made during specific inhibition of the pump, plus a simple kinetic model for active transport (Hansen et al. 1981), make possible dissection of the experimental I−V plot into

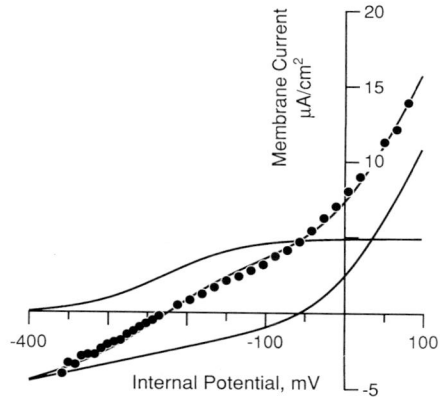

Fig. 2. Current-voltage relationships for the plasma membrane of *N. crassa*. Spherocytes of strain RL21a grown 3 days in a minimal medium, plus 1% glucose and 15% ethylene glycol (Blatt et al. 1987). Plotted points: corresponding values of membrane current and clamped voltage obtained in a 10-sec multi-pulse scan. Smooth curves: *Top*, characteristic for the proton pump; *Bottom*, ensemble of membrane leaks; *Middle* (through points), sum of the other two. Curves obtained by using a least-squares algorithm to fit the plotted points with a two-state reaction equation for the pump (see text section 6.2), plus a constant-field expression for the leak. Data from Dr. M. R. Blatt

components representing the proton pump (upper smooth curve, sigmoid) and the ensemble leak (lower smooth curve). Saturating pump velocity in this example, which is representative for spherocytes, was ~5 $\mu A/cm^2$, or about 50 pmol/cm^2·sec., and the calculated reversal potential was in the neighborhood of −400 mV, i.e. near the theoretical limit. The values of other parameters implied by the pump curve will be discussed later (see Sect. 6.2).

3.3 Proton Extrusion

A major function of the H^+-ATPase in fungi is maintenance of a macroscopic proton-motive force (PMF), which drives uptake of sugars, amino acids, and other critical nutrients via electrophoretic transport systems coupled to protons. Another function is maintenance of background conditions for regulation of cytoplasmic pH, which is not energetically trivial. Fungi are well known — and commercially useful — for their acid-secreting ability, an example of which is illustrated in Fig. 3, again for *N. crassa*. At the normal pH of the culture medium (pH 5.5 – 6.0) apparent steady-state acid secretion (straight line limits in Fig. 3) is not large (about 2 mmol/kg cell water · min), but at high pH the rate approaches 30 mmol/kg cell water · min (nearly 2 M/hour!). The latter rate, rescaled to the surface area of *N. crassa* hyphae, equals the total current-pumping capacity of the plasma-membrane ATPase. A simple interpretation of the pH sensitivity of acid secretion, then, is that the actual proton-pumping rate is insensitive to extracellular pH — a point which has been confirmed independently (Slayman and Sanders 1984) — but that high pH blocks H^+ reentry,

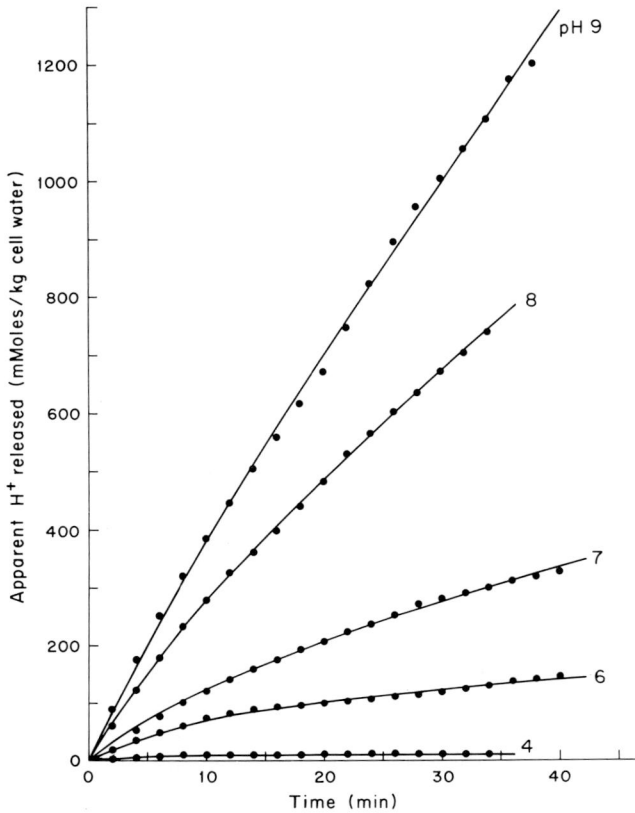

Fig. 3. pH-Dependence of acid secretion by *N. crassa*. Shaking-culture cells averaging ~ 300 μm long, were harvested, rinsed, resuspended in 10 mM KCl + 1% glucose at 25 °C, tritrated quickly to the test pH, and held by means of an electronic pH-stat. H⁺ release was calculated from the amount of base required for stating. The stable rate, attained after 10 – 15 min, is closely related to the proton pump

thus revealing a large net H^+ efflux. [Gas chromatographic analyses of cell filtrates from high-pH experiments have shown the H^+ flux to be almost matched by release of organic acid anions including malate, α-ketoglutarate, glutamate, citrate, succinate, and fumarate (P. Kaminski, unpublished results)]. At low pH, 80% or more of the pumped H^+ efflux would be recycled; but failure of the ATPase to shut off at high pH — when proton *return* becomes difficult — should be energetically expensive, thus accounting qualitatively for the known poor growth of *N. crassa* at high extracellular pH.

4 Chemistry of the Enzyme

The simplest observation clearly identifying the fungal plasma-membrane ATPase with the general family of cation-pumping ATPases is its inhibitor spectrum, shown

Table 2. Some inhibitors tested on the fungal plasma-membrane ATPase[a]

Inhibitor	IC_{50}[b] (μM) in *N. crassa*	Family/known function
Oligomycin	NI[c]	Blocks F_0F_1 ATPases
Azide	NI	Blocks F_0F_1 ATPases
DCCD	3	Blocks H^+ "channels"
Octylguanidinium	5	Lipophilic cation
Quercetin	200	Flavanoid
Ouabain	NI	Blocks Na^+,K^+-ATPase
Diethylstilbesterol	9	Blocks plant H^+-ATPase
Orthovanadate	1	Phosphate analogue; blocks all known phosphorylating transport ATPases

[a] Modified from Bowman et al. 1978.
[b] Approximate concentration for 50% inhibition of ATPase activity measured on broken membranes.
[c] Not inhibitory.

in abbreviated form in Table 2. Typical inhibitors of non-phosphorylating (mitochondrial type) H^+-ATPase/ATP-synthase, such as oligomycin and azide, are ineffective, but the one unequivocal class inhibitor of phosphorylating (P-type) cation-ATPases, orthovanadate, blocks all species of the fungal ATPase at micromolar concentrations. Several other, less specific, inhibitors of P-type ATPases also affect the fungal enzyme, as do carbodiimides (including the dicyclohexyl derivative, DCCD), which block a variety of H^+ "channel" proteins. Further evidence of familial relationships is the fact that all of the P-type enzymes − fungal enzymes included − are deeply embedded in the membrane bilayer, and can be removed only by extensive detergent treatment. Also, when analyzed by SDS-polyacrylamide gel electrophoresis, the ATPases display a major catalytic subunit of $M_r = 95 - 105$ kDa. For the fungal enzymes this is apparently the *sole* subunit (Goffeau and Slayman 1981). Finally, crucial evidence of the fungal enzyme's mechanistic affinity was the demonstration (Dame and Scarborough 1981) of transient covalent bonding of the γ-phosphate from ATP to an aspartic acid residue in the protein.

Purely on the basis of functional properties, the closest relative of the fungal PM ATPase is likely to be the plant-cell enzyme, which is strongly electrogenic and extrudes protons, probably unaccompanied by other ions (this is a much disputed point; see Leonard 1982; Villalobo 1984). Plasma membranes of animal cells contain at least three members of the family: the sodium-pump enzyme or Na^+,K^+-ATPase, which extrudes 3 Na^+ in exchange for 2 K^+; the gastric proton pump or H^+, K^+-ATPase, which carries out an electroneutral exchange to make gastric acid (pH < 1!); and a Ca^{2+}-ATPase which scavenges and extrudes cytoplasmic free calcium in partial exchange for H^+. A separate enzyme in sarcoplasmic reticulum sequesters Ca^{2+} during muscle relaxation; and a truncated relative in bacterial membranes takes up K^+, probably in exchange for H^+.

5 Primary Structure and Predictions

5.1 Sequences and Hydropathy

The most important recent development in our knowledge of fungal plasma-membrane ATPases has been the cloning and sequencing of structural genes for the enzyme from at least three species: *Saccharomyces cerevisiae* (Serrano et al. 1986), *N. crassa* (Hager et al. 1986; Addison 1986), and *Schizosaccharomyces pombe* (Ghislain et al. 1987). The amino acid sequence deduced from the cDNA sequence in *N. crassa* is shown in Fig. 4, with groupings and superscripts to designate special features. Hydropathy analysis of the sequence (Engelman et al. 1986) has yielded the plot in Fig. 5 A, identifying eight segments which are sufficiently hydrophobic (>20 kcal/mol for transfer from lipid to water) to be deeply embedded in (or spanning) the membrane bilayer. These segments are presumed to be coiled into α-helices and to be folded in pairs, forming hydrophobic hairpins plugged into the membrane as depicted in Fig. 5 B. Other ways of folding this polypeptide have been proposed (see Serrano

```
        c -                     - +    --t -       + t ---   ---- -    -   - - - -
   1  MADHSASGAP  ALSTNIESGK  FDEKAAEAAA  YQPKPKVEDD  EDEDIDALIE  DLESHDGHDA

        ------              t  --          - +        --  ++++     +--+-         +
  61  EEEEEEATPG  GGRVVPEDML  QTDTRVGLTS  EEVVQRRRKY  GLNQMKEEKE  NHFLK

                      d                          --    -
 116  FLGFFVGPIQFVMEGAAVLAAGL   ED   WVDFGVICGLLLLNAVVGFV

        -         - - ++   +     +-   +-           -              --         - +    --
 161  QEFQAGSIVD  ELKKTLALKA  VVLRDGTLKE  [IEAPEVVPGD  ILQVEEGTII  PADGRIVTDD

        -              -    -+   - + -        ++ -         -           +
 221  AFLQVDQSAL  TGESLAVDKH]  KGDQVFASSA  VKRGEAFVVI  TATGDNTFVG  RAAALVNAAS  GGS

        h  -                                          +                   -
 284  GHFTEVLNGIGTILLILVIFTLLIVWVSSFY  RSNPIV    QILEFT[LAITIIGVPVGLPAVVTTTMAVGAAYLA

        +++        +       -        -   i p+     + +    -          -      --      i     ++++  -
 355  KKKAIV  QKLSAIESLA  GVEI[LCSDKT  GTLT)KN]KLSL  HDPYTVAGVD  PEDLMLTACL  AASRKKKGID

         -+   +       +   + +       + +        -   ++        -     -+    i   f
 421  AIDKAFLKSL  KYYPRAKSVL  SKYKVLQFHP  FDPVSKKVVA  VVESPQGERI  TC[VKGAPLFV

        +  ---            +  -      -   +         +    +++           -    n - + -
 481  LKTVEEDHPI  PEEVDQAYKN  KVA]EFATRGF  RSLGVARKRG  EGSWEILGIM  PCMDPPRHDT

        +  n- +          +        -          +- +             -+          -    - - +
 541  YKTVCEAKTL  GLSIKMLTGD  AVGIARETSR  QLGLGTNIYN  AERLGLGGGG  DMPGSEVYDF

        -    -   -       +     -        +           -    -       +a -       -   - +
 601  VEA[ADGFAEV  FPQHKYNVVE  ILQQRGYLVA  MTGDGVNDAP  SLKKADTGIA  VEGSSDAARS

        -                    -   +     +   +           +             -                  +
 661  AAD]IVFLAPG  LGAIIDALKT  SRQIFHRMYA  YVVYRIALSI  HLEIFLGLWI  AILNR

        -                    -              -           +   +
 716  SLNIELVVFIAIFADVATLAIAY   DNAPYSQTPVKWNLPKLWGM   SVLLGVVLAVGTWITVTTMYA

                             --           -              +
 780  Q   GENGGIVQNF  GNMDEVLFLQ  ISLTENWLIF  ITRANGPFWS  S

                         - -       -   +                            -
 822  IPSWQLSGAIFLVDILATCFTIWGWF   EHSD    TSIVAVVRIWIFSFGIPCIMGGVYYILQD

        -               + + + +t   --          t      -+
 881  SVGFDNLMHG  KSPKGNQKQR  SLEDFVVSLQ  RVSTQHEKSQ
```

Fig. 4

1988), and the actual arrangement must await crystallographic data. A close evolutionary relationship among the PM ATPases of different fungi is indicated by the fact that the predicted sequences for the *S. cerevisiae* and *S. pombe* enzymes are 75% – 80% identical to each other and to the *N. crassa* enzyme.

5.2 Homologies

Sequence homology between the fungal enzymes and the cation ATPases of animal cells is much lower, however, in the range 20% – 25% over the whole polypeptide chain. Several regions of substantially higher homology between the *N. crassa* enzyme

---

Fig. 4. Primary amino acid sequence of the *N. crassa* PM ATPase, deduced from the DNA sequence. N-terminus at #1, C-terminus at #920. Both termini located intracellularly. Bold-face segments designate predicted hydrophobic helices (see Fig. 5), and each pair represents a "hairpin". Superscripts designate charged amino acids (+ and −), or specific reactive residues as follows: a, probable FSBA-binding site; c, methionine which is clipped during maturation; d, DCCD-binding site; f, FITC-binding site; n, cysteines labeled by NEM in native enzyme; i, extra cysteines labelled by IAEDANS in denatured enzyme; p, phosphorylation site; t, trypsin-cleavage sites. Chemical labelling demonstrated in laboratory of C. W. Slayman by Drs. R. J. Brooker, J. W. Davenport, S. M. Mandala, J. P. Pardo, and M. R. Sussman. Primary sequence from Hager et al. (1986). Reagent abbreviations: DCCD = *N,N'*-dicyclohexylcarbodiimide; FITC = fluorescein isothiocyanate; FSBA = fluorosulfonylbenzoyl-5'-adenosine; IAEDANS = 5-(iodoacetamido-ethyl)-aminonaphthalene-1-sulfonic acid; NEM = *N*-ethylmaleimide

Several regions of high homology with Na^+,K^+-ATPase

Sequence	Begin	End	Length	Na, K enz
1	Isoleucine 191	Histidine 240	50	17
2	Leucine 327	Asparagine 396	70	26
3	Valine 473	Alanine 503	31	11
4	Alanine 604	Aspartate 663	60	26

Putative membrane-spanning helices

Helix #	Begin	End	Length
1	Phenylalanine 116	Leucine 138	23
2	Tryptophane 141	Valine 160	20
3	Glycine 284 (Valine 289)	Tyrosine 314	31 (26)
4	Glutamine 321 (Phenylalanine 325)	Alanine 354	34 (30)
5	Serine 716	Tyrosine 738	23
6	Serine 759	Alanine 779	21
7	Isoleucine 822	Phenylalanine 847	26
8	Threonine 852	Aspartate 880 (Glutamine 879)	29 (28)

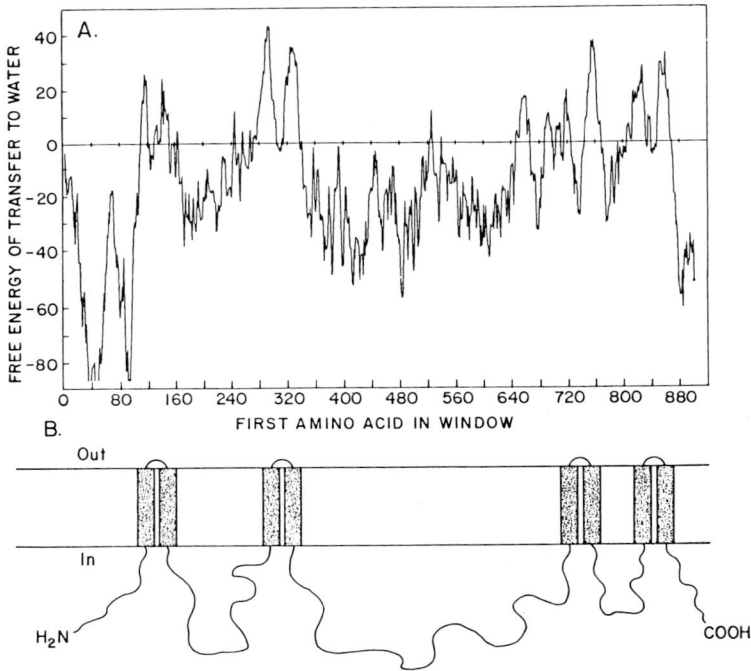

Fig. 5. Hydropathy diagram and expected transmembrane folding of *N. crassa* PM ATPase. Computation with a window of 20 amino acids. Free energy in kcal/mol. *In* and *out* refer to the normal physiological orientation of the membrane, with *in* = cytoplasm. Reproduced by permission from Hager et al. (1986)

and the α-subunit (main subunit) of the Na$^+$,K$^+$-ATPase from *Torpedo* electroplax are marked off by square brackets in Fig. 4; as evidenced in the legend table, these regions contain between 34% and 43% coincident amino acids. Three of the four regions shown lie in the major cytoplasmic loop (Lys^{355} to Arg^{715}), which contains the principal nucleotide-binding site and the phosphorylated residue (Asp^{378}). Indeed, the nucleotide- and phosphate-binding regions have the greatest sequence conservation throughout the P-type membrane ATPases. Specifically, 9 amino acids (Cys^{376} to Thr^{384}) bracketing the phosphorylation site are identical in all of the cation ATPases (except for the bacterial K$^+$-ATPase), and 23 residues (Lys^{628} to Ala^{650}) thought to lie along one side of bound ATP are 80% – 90% conserved. Offsetting such regions of high sequence conservation between the fungal PM ATPases and the other cation ATPases, however, is the long stretch including the last four presumptive membrane-spanning helices plus the entire C-terminus: 205 residues (Ser^{716} through Gln^{920}) in which amino acid coincidences with, e.g. the Na$^+$,K$^+$-ATPase, are barely greater than random.

The most obvious *physical* features of the sequence designated in Fig. 4 are the expected hydrophobic α-helices, whose origins and termini are summarized in the table legend. The most probable helix to span a 30-Å bilayer would contain 20 amino acids (Engelman et al. 1986), so it is reasonable to expect the shorter of these presumptive helices (No.'s 1, 2, 5, and 6) to be essentially fully embedded. That is not true, however, for the longer ones, particularly when near-terminal amino acids are charged, like Glu^{288}, Glu^{324}, and Asp^{880}. These helices would be expected to protrude from the bilayer, and Brandl et al. (1986) have drawn an explicit model for such an arrangement in the Ca^{2+}-ATPase of sarcoplasmic reticulum; there the extra lengths of helix are supposed to extend into the cytoplasm.

5.3 Charged Residues

Another salient feature of the amino acid sequence of the *N. crassa* ATPase is its charge distribution. A total of only 9 frankly charged residues are found in the presumed hydrophobic helices, and — as discussed above — that probably means only 6 charged residues actually embedded in the membrane. This compares, with 14 frank charges in the membrane portions of the Ca^{2+}-ATPase, and more than 30 in putative helical regions adjacent to the membrane (Brandl et al. 1986). Still more remarkable is the high density of charged residues at the N-terminus of the protein: among the 115 amino acids preceding the first hydrophobic helix, there are 20 glutamates, 12 aspartates, 8 lysines, and 5 arginines, for a net excess of *19 negative charges*. This feature is unique to the fungal plasma-membrane ATPases, and seems to be preserved in all of them, even though detailed sequence homology is lower in the N-terminal segment than anywhere else in the fungal enzymes.

5.4 Trypsin Hydrolysis

Protein chemical studies have recently elucidated several other important features in the amino-acid sequence of the *N. crassa* enzyme, of which the most intriguing concerns the effects of trypsin. As was first described by Addison and Scarborough (1982), controlled trypsin degradation creates large, well-defined enzyme fragments whose distribution and stability are strongly influenced by the presence or absence of nucleotides and orthovanadate. Mandala and Slayman (1988) refined these experiments by blotting the large fragments with antibodies specific for the N-terminus (amino acids 3−46) or C-terminus (amino acids 886−920) and by sequencing the fragments via Edman degradation. Three trypsin cleavage sites were thereby identified near the N-terminus: Lys^{24}, Lys^{36}, and Arg^{73}; and two sites were found near the C-terminus: Arg^{900} and Arg^{911}. Removal of the first 36 residues had little effect on hydrolytic activity of the enzyme. Removal of the next 37 residues (cleavage at Arg^{73}) inactivated the enzyme. The reaction was greatly slowed, however, by the presence of micromolar orthovanadate; simultaneously, the Arg^{900} site was revealed.

The unliganded enzyme is assumed to be in a "pre-reactive" state (M_1), and that bound to orthovanadate in a "post-reactive" state (M_2). The major implication of the trypsin results, then, is that during the overall reaction cycle both the N- and C-termini of the protein undergo significant conformation changes. An important spin-off from these studies is confirmation of the cytoplasmic locations of both termini, which follows from the fact that the N- and C-specific antibodies bind only to inverted plasma-membrane vesicles, not to intact cellular protoplasts. This result in turn strengthens the argument, based on the hydropathy plot (Fig. 5 A), for eight (an even number of) trans-membrane helices.

6 Models for Proton Pumping

6.1 A Morphological Model

Different kinds of data on plasma-membrane ATPases evoke different kinds of models, i.e. ways of viewing, describing, or analyzing the enzymes' behavior. The pictures in Fig. 1, for example, lead to a *morphological* model, in which the largest mass of the protein is supposed to protrude from the cytoplasmic face of the plasma membrane. A variety of antibody experiments and tests with side-specific reagents have been carried out in confirmation of this picture for the Na^+,K^+-ATPase and for the Ca^{2+}-ATPase of sarcoplasmic reticulum; Nicholas (1984) assigned relative volumes for the α-subunit of Na^+,K^+-ATPase that correspond to roughly 73% of the molecule protruding into the cytoplasm, 12% spanning the membrane, and 15% exposed on the outside. These numbers compare with 74%, 22.5%, and 3.5% of residues, respectively, from the hydropathy plot of the *N. crassa* H^+-ATPase. This morphological model, together with knowledge that the nucleotide-reactive sites lie in the large cytoplasmic segment, support the view of the enzyme as a bifunctional molecule: one in which energization (phosphorylation) is remote from transport itself (passage of an ion through the membrane-dielectric barrier). The point is reinforced by the fact that some active transport systems which do not use nucleotides for energy — e.g. the H^+-coupled lactose carrier ("lac permease") in *Escherichia coli* — consist almost entirely of transmembrane helices, with only small segments of the polypeptide looping into the cytoplasm or into the extracellular space (Kaback 1986). The helices have generally been *presumed* to form a transmembrane channel which is somehow gated; in the case of the PM ATPases, gating would follow upon phosphorylation at the remote site.

6.2 A Kinetic Model

A different sort of model is suggested by Fig. 2, or rather by the pump behavior shown in the upper curve of that figure. Such an I–V curve is analogous to the familiar velocity-vs-concentration plots for ordinary enzymes, in being a steady-state kinetic diagram. It therefore evokes kinetic reaction schemes, of which the simplest has two states: M_1, the pre-transport state, with the H^+-binding site facing the cell interior; and M_2, the post-transport state, with the H^+ site facing the cell exterior. One transition, say $M_1 \rightarrow M_2$, is assumed to carry charge and therefore to be voltage-dependent;

and the other, then $M_2 \rightarrow M_1$, is electroneutral. [Though it anticipates the later discussion slightly, this simple scheme can be rationalized in terms of actual chemistry by reference to Fig. 6 (top line). There, M_1 would be $H^+ \cdot E_1 \sim P$, M_2 would be $H^+ \cdot E_2 \cdot P$, and charge transport would be the single arrow between them. The electroneutral transition (or repriming) would lump together all of the other reactions, from $H^+ \cdot E_2 \cdot P$ to E_1 and from E_1 back to $H^+ \cdot E_1 \sim P$.]

Analysis of the I−V data in Fig. 2 via such a model has led to several interesting conclusions, of which the most important is that charge transport is very fast, of the order of 5000/sec, once the enzyme is primed by $\sim P$ and H^+. Since the overall cycling rate of the enzyme is much slower, approximately 200/sec, charge transport clearly is not rate-limiting, and one or more of the chemical steps lumped into repriming must be slow. Another interesting conclusion is that reversal of both steps is slow, but the ratio of rate constants − forward/reverse is much larger for charge transport than for repriming. This means that the *major release of energy* (about 7 kcal/mol $\approx$ 300 mV) − from the bond structure of the phosphorylated enzyme into the bulk electrochemical gradient for H^+ − *occurs during actual charge transit* of the membrane. A much smaller release (about 2 kcal/mol $\approx$ 100 mV) is associated with repriming, which effectively represents the apparent change in pK_a for the transported proton. To draw a mechanical analogy from such results, we could say that the proton "pump" behaves more like a snap-gate or a gun, than like a pump. An important confirmation of these results from I−V analysis is the deduction of Amory et al. (1982), from ^{18}O-exchange measurements of the S. *pombe* ATPase, that the transition $M_1 \rightarrow M_2$ occurs with a specific rate constant of nearly 10^4/sec. In this respect the fungal PM H^+-ATPases may differ substantially from the other cation ATPases, for which the corresponding rate constants are in the range 50−500/sec (McIntosh and Boyer 1983; Nakao and Gadsby 1986).

6.3 A Chemical Reaction Model

Chemical investigations of partial reactions in P-type ATPases have been carried out most extensively with the Na^+,K^+-ATPase and with the Ca^{2+}-ATPase of sarcoplasmic reticulum. Those studies have led to complex reaction models with as many as 20 discrete reaction states, required to accommodate multiple ion binding and release (Karlish et al. 1978). On the *assumption* of a fundamentally common reaction mechanism for all P-type ATPases, a much simpler model of the fungal ATPases can be drawn (Fig. 6 [top]) for transport of a single proton. In the presence of sufficient cytoplasmic free protons, ATP binds to the enzyme (E_1) so as to trap a transportable proton and transfer the γ-PO_4 to Asp^{378} thus forming the "$E_1 \sim P$" state ($H^+ \cdot E_1 \sim P$), with conservation of energy. Quickly thereafter, the trapped proton is either ejected or exposed to the *trans* side of the membrane with concomitant destabilization of the phosphate bond, forming the "$E_2 \cdot P$" state. Subsequently, H^+ and/or phosphate is lost, returning the enzyme to its initial state, in a reaction which may be catalyzed by *extracellular* protons.

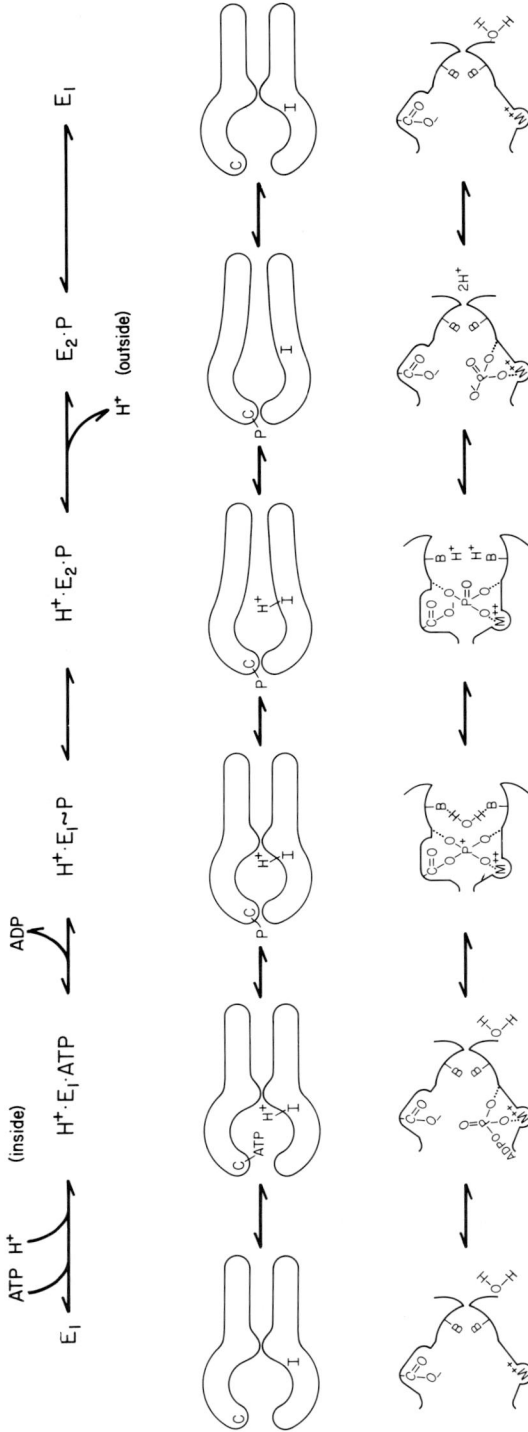

Fig. 6. Three models for the reaction mechanism of fungal plasma-membrane H$^+$-ATPase. *Top line*: Likely chemical reaction model, adapted from that for the Na$^+$, K$^+$-ATPase (Karlish et al. 1978) by deletion of steps associated specifically with multiple-ion and ion-return functions. *Middle line*: Mechanical analogue for proton transport, showing sequential closing and opening of two ion "gates" as originally suggested by Jardetsky (1966). Vertical positions designate explicit correspondence to the chemical reaction model. Modified from Slayman and Sanders (1985). *Bottom line*: A "non-transport" model of the fungal proton pump, emphasizing phosphate chemistry. Pumped protons would originate from the water of hydrolysis. Redrawn from Scarborough (1985)

6.4 A Composite Model

Although physical trapping of *protons* within the transport enzyme would be difficult to demonstrate, trapping of transported ions by the Na^+,K^+-ATPase (judged by decreased exchangeability of ions within the membrane) is firmly established (Glynn and Richards 1982; Forbush 1987). This suggests a rather simple correspondence between the morphological model of the enzyme, described above, and the chemical reaction model; this is represented in Fig. 6 (middle line; after Jardetsky 1966). With respect to the structural picture in Fig. 5, clustered transmembrane helices would form a channel adjacent to an internal cavity in the cytoplasmic domain of the ATPase. The channel and cavity would be connected, however, only *after* energization of the enzyme by ATP. And the thermodynamic requirements of active transport generally demand that opening of one gate, between the enzyme cavity and the transmembrane channel, co-ordinate with closure of another gate (i.e. between the cytoplasm and the enzyme cavity) so that at least *one gate is always closed*.

6.5 Other Points of View

The models just presented treat protons (or hydronium ions) like all other ions — Na^+, K^+, and Ca^{2+} — which the P-type ATPases transport: as if they are literally moved through the protein, from one side of the membrane to the other, via a generalized ionic channel. The essential reason for such treatment is the very fact that several different cations are indeed transported by closely related enzymes. Numerous authors, however, have invoked theoretical arguments toward quite different mechanisms of *proton* movement. One such mechanisms was proton relay via hydrogen-bond displacement and rotation: a so-called proton wire. This idea emerged from experiments on oxidative phosphorylation (Morowitz 1978; Freund 1981) and recently received interesting experimental support from site-directed mutation studies on the lac permease of *E. coli* (Carrasco et al. 1986). It has not been taken seriously, however, for the P-type ATPases.

Another idea, depicted in Fig. 6 (lower line), is that for protons *no ion* is transported *through* the protein or membrane. Rather, water — taken into the catalytic cavity of the enzyme — would be split during hydrolysis of aspartyl phosphate; the resultant hydrogen ions would be released extracellularly, and the oxide anion (combined into phosphate) would exit intracellularly. This is one concrete example of a theoretical model for active transport first discussed by Osterhout in the 1930's and later carefully analyzed via linear-coupling theory (Blumenthal et al. 1967). Its current incarnation, derived from a suggestion by Mitchell (1974), has been proposed by Scarborough (1985) to unify the chemistry of transport ATPases and ATPsynthases with that of ordinary ATPases. It would use a single hinge movement to access the catalytic cavity, and could obviate the transmembrane proton channel either by substituting static asymmetry (inner face $HPO_4^=$ permeable, outer face H^+ permeable) or by using a bridging phosphate as a gate (see above). Although the model has been proposed as the fundamental basis for phosphate-linked active transport, it cannot eliminate bona fide transmembrane pathways for any ion other than H^+, and it has not yet prompt-ed an explicit experimental test.

7 Current Directions

Because the cation ATPases in general, and the fungal plasma-membrane ATPase in particular, are now subjects of very active study, it is not possible to provide an up-to-date review. Horizons of research on the plasma-membrane ATPases are expanding rapidly by application of new technologies in molecular biology (including, especially, yeast genetics), spectroscopy, protein chemistry, immunology, electrophysiology, and crystallography.

Identified-site mutagenesis of the structural gene for fungal plasma-membrane ATPase is yielding important preliminary results in several laboratories (see, e.g. Ghislain et al. 1987; McCusker et al. 1987; Cid and Serrano 1988), but has thus far been limited by the labor required and by a lack of convenient expression systems ac-cessible to both biochemical and physiological analysis. Attempts are being made, however, to get around the latter limitation by application of patch-recording methods to yeast protoplasts (see e.g. Gustin et al. 1986), by attachment of catalytically active liposomes to bilayer membranes (Ziegler et al. 1988; G. Nagel and C.L. Slayman, un-published experiments), and by heterologous expression of cloned ATPase genes (Perona and Serrano 1988; Money et al. 1988). A tentative but particularly interesting recent finding is the presence of an ATP-dependent K^+ channel associated with one mutant of the *S. cerevisiae* ATPase (H. Lecar, cited by Haber 1988). Regrettably, no major efforts appear underway to define the partial reaction chemistry of fungal plas-ma-membrane ATPases, probably because of a general perception that the real excite-ment will be elsewhere. Several laboratories are pursuing biogenetic studies of the en-zyme, which are important in their own right, but do not directly relate to transport mechanism. Practically everyone with a computer is now 'playing' sequence and fold-ing games, in order to relate the known primary structures both to hydropathy data and to the crystal structures of soluble ATPases. Much of the activity is focussed upon the nucleotide-binding regions of the enzyme, and has led to several interesting specu-lations (Serrano 1988). A comprehensive analysis of likely folding and coiling pat-terns for all of the P-type ATPases is also being carried out (Green 1988). Unques-tionably, however, now that the primary structures of several fungal plasma-mem-brane ATPases are known, the single most important line of research is crystallization of the enzyme. This is also the most difficult and unpredictable line, which − though not much discussed − is well underway in several laboratories, and is yielding nearly purified protein for rigorous physical studies (Hennessey and Scarborough 1988).

Acknowledgements. The authors are indebted to Dr. Carolyn Slayman for much helpful discussion, to her and Dr. K. M. Hager for use of Fig. 5, to Drs. R. J. Brooker and M. R. Blatt for use of the unpublished data in Figs. 1 C and 2. Supported by Research Grant GM-15858 from the National Insti-tute of General Medical Sciences.

References

Addison R (1986) Primary structure of the *Neurospora* plasma membrane H^+-ATPase deduced from the gene sequence. J Biol Chem 261:14896–14901

Addison R, Scarborough GA (1982) Conformational changes of the *Neurospora* plasma membrane H^+ATPase during its catalytic cycle. J Biol Chem 257:10421–10426

Amory A, Goffeau A, McIntosh DB, Boyer PD (1982) Exchange of oxygen by the plasma membrane ATPase from the yeast *Schizosaccharomyces pombe*. J Biol Chem 257:12509–12516

Blatt MR, Rodriguez-Navarro A, Slayman CL (1987) Potassium-proton symport in *Neurospora*: Kinetic control by pH and membrane potential. J Membr Biol 98:169–189

Blumenthal R, Caplan SR, Kedem O (1967) The coupling of an enzymatic reaction to transmembrane flow of electric current in a synthetic "active transport" system. Biophys J 7:735–757

Bowman BJ, Mainzer SE, Allen KE, Slayman CW (1978) Effects of inhibitors on the plasm membrane and mitochondrial adenosine triphosphatases of *Neurospora crassa*. Biochim Biophys Acta 512:13–28

Bowman BJ, Blasco F, Slayman CW (1981) Purification and characterization of the plasma membrane ATPase of *Neurospora crassa*. J Biol Chem 256:12343–12349

Brandl CJ, Green NM, Korczak B, MacLennan DH (1986) Two Ca^{2+}-ATPase genes: homologies and mechanistic implications of deduced amino acid sequences. Cell 44:597–607

Carrasco N, Antes LM, Poonian MS, Kaback HR (1986) *Lac* permease of *Escherichia coli*: Histidine-322 and glutamic acid-325 may be components of a charge-relay system. Biochemistry 25:4486–4488

Chapman JB, Johnson EA, Kootsey JM (1983) Electrical and biochemical properties of an enzyme model of the sodium pump. J Membr Biol 74:139–153

Cid A, Serrano R (1988) Mutations of the yeast plasma membrane H^+-ATPase which cause thermosensitivity and altered regulation of the enzyme. J Biol Chem 263:14134–14139

Dame JB, Scarborough GA (1981) Identification of the phosphorylated intermediate of the *Neurospora* plasma membrane H^+-ATPase as a β-aspartyl phosphate. J Biol Chem 256:10724–10730

Engelman DM, Steitz TA, Goldman A (1986) Identifying non-polar transbilayer helices in amino acid sequences of membrane proteins. Annu Rev Biophys Biophys Chem 15:321–353

Forbush B III (1987) Rapid release of ^{42}K or ^{86}Rb from two distinct transport sites on the Na, K-pump in the presence of P_i or vanadate. J Biol Chem 262:11116–11127

Freund F (1981) Proton highlife and midway tunneling. TIBS 6:142–145

Ghislain M, Schlesser A, Goffeau A (1987) Mutation of a conserved glycine residue modifies the vanadate sensitivity of the plasma membrane H^+-ATPase from *Schizosaccharomyces pombe*. J Biol Chem 262:17549–17555

Glynn IM, Richards DE (1982) Occlusion of rubidium ions by the sodium-potassium pump: Its implications for the mechanism of potassium transport. J Physiol 330:17–43

Goffeau A, Slayman CW (1981) The proton-translocating ATPase of the fungal plasma membrane. Biochim Biophys Acta 639:197–223

Green NM (1988) How similar are the ion transporting ATPases? Conflicting interpretations of the transmembrane segments! Talk at International Workshop on Membrane ATPases, Osnabrück, FRG, March 1988

Gustin MC, Martinac B, Saimi Y, Culbertson MR, Kung C (1986) Ion channels in yeast. Science 233:1195–1197

Haber JE (1988) Alterations of the PM ATPase in hygromycin resistant mutants of *Saccharomyces cerevisiae*. Talk at International Workshop on Membrane ATPases, Osnabrück, FRG, March 1988

Hager KM, Mandala SM, Davenport JW, Speicher DW, Benz EJ, Slayman CW (1986) Amino acid sequence of the plasma membrane ATPase of *Neurospora crassa*: Deduction from genomic and cDNA sequences. Proc Natl Acad Sci USA 83:7693–7697

Hansen U-P, Gradmann D, Sanders D, Slayman CL (1981) Interpretation of current voltage relationships for "active" ion transport systems: I. Steady-state reaction-kinetic analysis of Class-I mechanisms. J Membr Biol 63:165–190

Hennessey JP Jr, Scarborough GA (1988) Secondary structure of the *Neurospora crassa* plasma membrane H^+-ATPase as estimated by circular dichroism. J Biol Chem 263:3123–3130

Inesi G, Scales D (1974) Tryptic cleavage of sarcoplasmic reticulum protein. Biochem 13:3298–3306

Jardetsky O (1966) Simple allosteric model for membrane pumps. Nature (London) 211:969–970

Kaback HR (1986) Active transport in *Escherichia coli*: Passage to permease. Annu Rev Biophys Chem 15:279–319

Karlish SJD, Yates DW, Glynn IM (1978) Conformational transitions between Na^+-bound and K^+-bound forms of $(Na^+ + K^+)$-ATPase, studied with formycin nucleotides. Biochim Biophys Acta 525:252–264

Leonard RT (1982) The plasma membrane ATPase of plant cells: Cation or proton pump? In: Martonosi A (ed) Membranes and transport, vol 2. Plenum, New York, 633 pp

Mandala SM, Slayman CW (1988) Identification of tryptic cleavage sites for two conformational states of the *Neurospora* plasma membrane H^+-ATPase. J Biol Chem 263:15122–15128

McCusker JH, Perlin DS, Haber JE (1987) Pleiotropic plasma membrane ATPase mutations of *Saccharomyces cerevisiae*. Mol Cell Biol 7:4082–4088

McIntosh DB, Boyer PD (1983) Adenosine 5'-triphosphate modulation of catalytic intermediates of calcium ion activated adenosinetriphosphase of sarcoplasmic reticulum subsequent to enzyme phosphorylation. Biochem 22:2867–2875

Mitchell P (1967) Proton-translocation in mitochondria, chloroplasts, and bacteria: natural fuel cells and solar cells. Fed Proc 26:1370–1379

Mitchell P (1974) A chemiosmotic molecular mechanism for proton-transporting adenosine triphosphatases. FEBS Lett 43:189–194

Money NP, Aaronson LR, Slayman CW, Slayman CL (1988) Expression of the *Neurospora* plasma membrane H^+-ATPase in *Xenopus* oocytes. Biophys J 53:137a, Item M-Pos 195

Morowitz HJ (1978) Proton semiconductors and energy transduction in biological systems. Am J Physiol 235:R99–R114

Nakaor M, Gadsby DC (1986) Voltage dependence of Na translocation by the Na/K pump. Nature (London) 323:628–630

Nicholas RA (1984) Purification of the membrane-spanning tryptic peptides of the α-polypeptide from sodium and potassium ion activated adenosinetriphosphatase labeled with 1-trithiospiro[adamantane-4,3'-diazirine]. Biochemistry 23:888–898

Perlin D, Kasamo K, Brooker RJ, Slayman CW (1984) Electrogenic H^+ translocation by the plasma membrane ATPase of *Neurospora*. J Biol Chem 259:7884–7892

Perona R, Serrano R (1988) Increased pH and tumorigenicity of fibroblasts expressing a yeast proton pump. Nature (London) 334:438–440

Sanders D, Hansen U-P, Slayman CL (1981) Role of the plasma membrane proton pump in pH regulation in non-animal cells. Proc Natl Acad Sci USA 78:5903–5907

Scarborough GA (1985) Binding energy, conformational change, and the mechanism of transmembrane solute movements. Microbiol Rev 49:214–231

Serrano R (1988) Structure and function of proton translocating ATPase in plasma membranes of plants and fungi. Biochim Biophys Acta 947:1–28

Serrano R, Kielland-Brandt MC, Fink GR (1986) Yeast plasma membrane ATPase is essential for growth and has homology with $(Na^+ + K^+)$-, K^+- and Ca^{++}-ATPases. Nature (London) 319:689–693

Slayman CL, Sanders D (1984) pH-dependence of proton pumping in *Neurospora*. In: Forte JG, Warnock DG, Rector FC (eds) Hydrogen ion transport in epithelia. Wiley, New York, pp 47–56

Slayman CL, Sanders D (1985) Steady-state kinetic analysis of an electroenzyme. Biochem Soc Symp 50:11–29

Villalobo A (1984) Energy-dependent H^+ and K^+ translocation by the reconstituted yeast plasma membrane ATPase. Can J Biochem Cell Biol 62:865–877

Warncke J, Slayman CL (1980) Metabolic modulation of stoichiometry in a proton pump. Biochim Biophys Acta 591:224–233

Ziegler W, Slayman CL, Cartwright CP (1988) Reconstitution of a plasma-membrane H^+-ATPase into bilayer lipid membranes. Proc Slov Acad Sci (in press)

Subject Index

DATE DUE

DEMCO 38-297